D1000086

ADVANCED THEORY OF SEMICONDUCTOR DEVICES

PRENTICE HALL SERIES
IN SOLID STATE PHYSICAL ELECTRONICS

Nick Holonyak, Jr., Editor

CHEO *Fiber Optics: Devices and Systems*
HAUS *Waves and Fields in Optoelectronics*
HESS *Advanced Theory of Semiconductor Devices*
NUSSBAUM *Contemporary Optics for Scientists and Engineers*
SOCLOF *Analog Integrated Circuits*
SOCLOF *Applications of Analog Integrated Circuits*
STREETMAN *Solid State Electronic Devices, 2nd Edition*
UMAN *Introduction to the Physics of Electronics*
VAN DER ZIEL *Solid State Physical Electronics, 3rd Edition*
VERDEYEN *Laser Electronics*

ADVANCED THEORY OF SEMICONDUCTOR DEVICES

KARL HESS
University of Illinois

PRENTICE HALL
Englewood Cliffs, New Jersey 07632

Library of Congress Cataloging-in-Publication Data

HESS, KARL (date)

Advanced theory of semiconductor devices.

(Prentice Hall series in solid state physical electronics)

Includes bibliographies and index.

1. Semiconductors. 2. Electronics. 3. Diodes. 4. Transistors. I. Title. II. Series.

TK7871.85.H475 1988 621.3815′2 87-25491

ISBN 0-13-011511-8

Editorial/production supervision
and interior design: *Kathryn Gollin Marshak*

Cover design: *20/20 Services*

Manufacturing buyer: *Richard Washburn*

A Division of Simon & Schuster
Englewood Cliffs, New Jersey 07632

Printed in the United States of America

10 9 8 7 6 5 4 3 2 1

ISBN 0-13-011511-8 025

PRENTICE-HALL INTERNATIONAL (UK) LIMITED, *London*
PRENTICE-HALL OF AUSTRALIA PTY. LIMITED, *Sydney*
PRENTICE-HALL CANADA INC., *Toronto*
PRENTICE-HALL HISPANOAMERICANA, S.A., *Mexico*
PRENTICE-HALL OF INDIA PRIVATE LIMITED, *New Delhi*
PRENTICE-HALL OF JAPAN, INC., *Tokyo*
SIMON & SCHUSTER ASIA PTE. LTD., *Singapore*
EDITORA PRENTICE-HALL DO BRASIL, LTDA., *Rio de Janeiro*

To the memory of my father

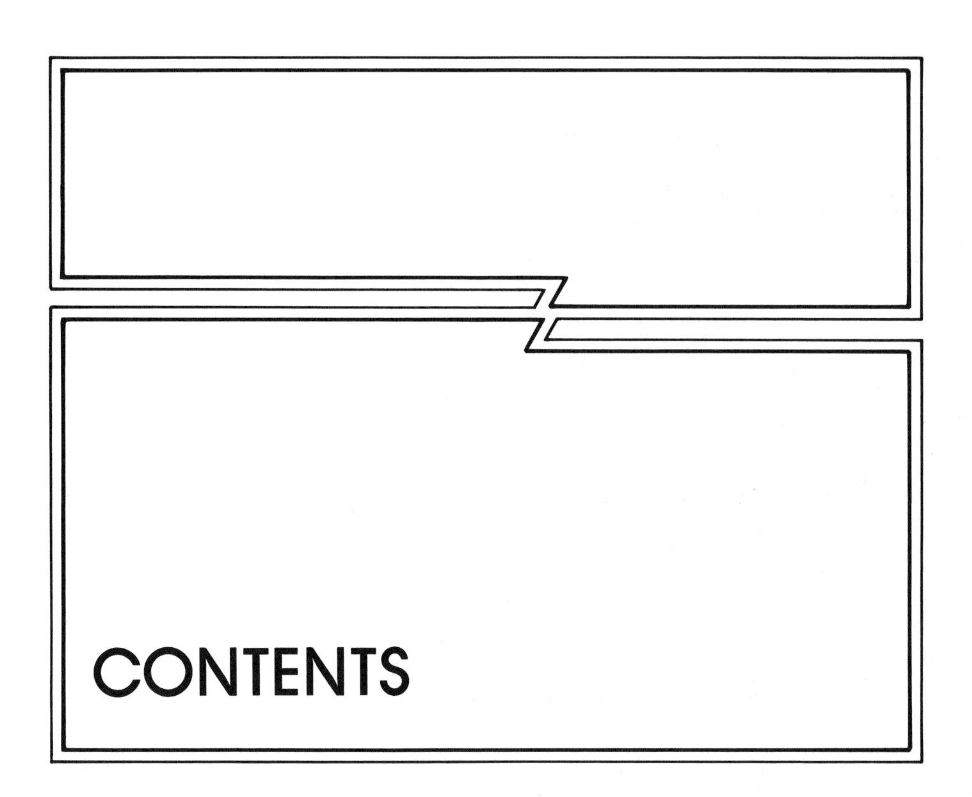

CONTENTS

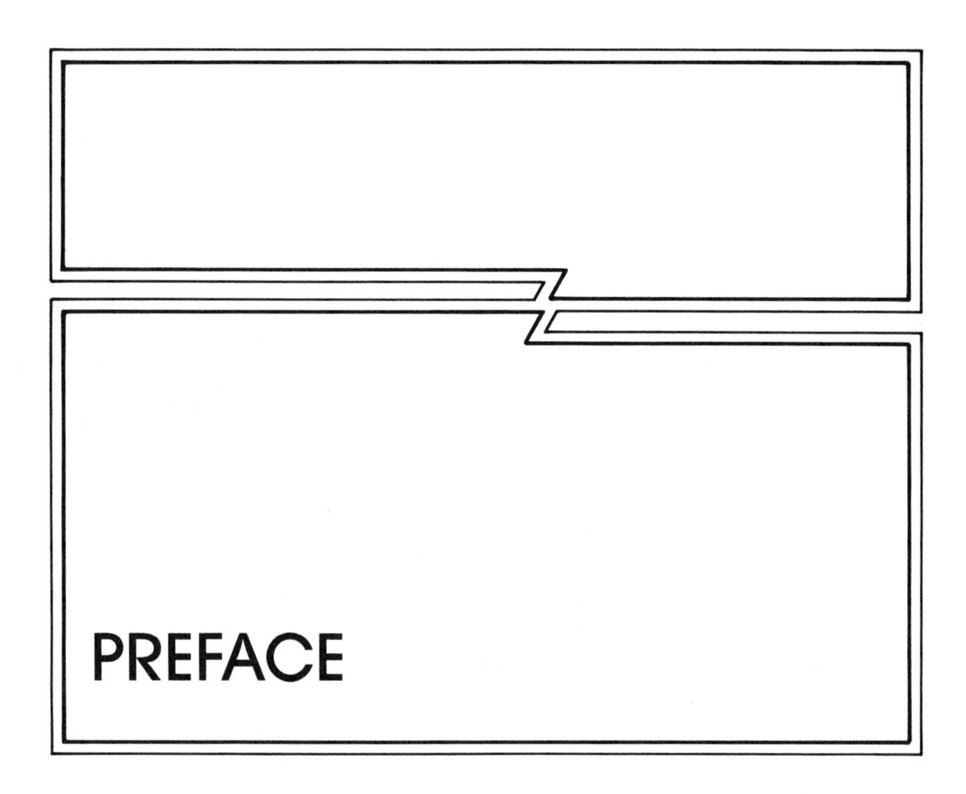

PREFACE

Since the invention of the bipolar transistor by Bardeen and Brattain in late 1947, semiconductor devices have developed at an astonishing pace. A large variety of single device components for numerous uses evolved over the first two decades of the "golden semiconductor age," while in the third decade (starting about 1970) integrated circuits revolutionized semiconductor electronics. Semiconductor memories have replaced other components and have brought us not only the video game but also the supercomputer. Current devices approach submicrometer dimensions corresponding to 10^6 elements on a chip of centimeter size. At the same time, two newer technologies of crystal growth have evolved from the vapor phase epitaxy introduced in 1960, molecular beam epitaxy (MBE) and metal organic chemical vapor deposition (MOCVD), which make it possible to grow lattice-matched semiconductor structures having a characteristic dimension of 10^{-6} cm (and less). These heterostructures (superlattices) have already shown interesting effects with high device potential caused by quantum effects.

Astonishingly enough, the theory of semiconductor devices as a branch of solid state theory has received relatively little attention. Any theory of devices must contain a careful account of (1) electron (hole) drift and diffusion, including the effect of high electric fields and high energies; (2) generation recombination (which is usually treated very cursorily in solid state texts); and (3) the self-consistent fields of variable densities of electrons, holes as well as donors and acceptors. For a proper treatment of (1)–(3) some basic knowledge of energy bands, electron impurity (lattice vibration) interaction, and basic quantum theory (tunneling, size quantization) is necessary. This background per se can be found in many solid state texts; unfortunately, however, little effort has been made to link it to electronic devices. Although much of the material and especially the depth of treatment of these texts are directly influenced by device applications, this fact is hardly ever mentioned. Often the device application is considered as something "dirty," and any remarks about it are more or less shamefully avoided. The physics student therefore usually does not understand why Gunn devices and field effect transistors are not made out of InSb or PbTe (which have higher minima and high electronic mobilities) but out of GaAs instead. I have also not found a single text on solid state that explains well why silicon is so special. On the other hand, there exist enormous amounts of information on devices with little link to the basics. In these texts, equations that describe the device operation are typically "introduced without much justification" and then are integrated to arrive at the final result. Nobody really knows just why this or that equation has been used and what has to be modified when the device operation is slightly changed or when the dimensions are shrunk, for example. Even in excellent texts on devices, principles are consistently used that are obsolete and have nothing to do with reality. For example, the electrical conductivity is always assumed to be proportional to $T^{3/2}$ (T is the absolute temperature) when ionized impurity scattering is dominant. In fact, in metal-oxide-silicon transistors this is never true and T^0 law is much better over most of the temperature range. This is only one of many examples.

To understand devices, and especially very small (submicrometer) devices, we need to establish our approach on more basic principles. We need to know about the effective mass theorem, Fermi's Golden Rule, the scattering probabilities, velocity overshoot, and the generation-recombination mechanism, to give a few examples. A book that treats all the necessary principles, including new developments and the resulting devices, would be much too long and probably unnecessary. Streetman's *Solid State Electronic Devices* provides a working knowledge of most of the important devices. It is the purpose of the present book to outline the basic unifying principles that are necessary to understand these devices in greater detail and to enable generalizations for future development. Consequently, I will discuss only certain ideal models of devices and refer to sources such as Streetman for more details on specific devices. However, much detail will be given with respect to the basic principles. Therefore, this book addresses graduate students and researchers in fields connected with semiconductor devices who wish to derive and understand the basic equations for a novel or current device, equations that are general enough to contain all the necessary basic physical effects and specific enough to make an effective solution possible. It is this quality that distinguishes a device theory from mere device modeling or from a general theory of the solid state.

In addition, I desire to show that semiconductor device theory can be a discipline by itself that contains the necessary richness and complications to attract scientific interest and provides the possibility of new developments in future decades. This richness is caused, in my opinion, by the existence and importance of generation-recombination, statistical effects (Boltzmann equation), and the possibility to control the boundary conditions on a quantum level (submicron devices, quantum well superlattice structures). These complicated effects have not only become important for small devices but can in turn be included in modern device theory and simulation, because of the enormous progress of computers. Large-scale computation will be a major tool to understand small devices, and this book is also intended to give the physical basis for large-scale computational models.

As a text for graduate courses, this book can be used in various ways. I have used it for a one-semester course for first- and second-year graduate students of physics and electrical engineering. Not all of the material can be taught in one semester. Depending on student interest, one can emphasize the solid state aspects and leave out much of the last chapters on devices, or one can emphasize the device aspects and skip much of Chaps. 3 and 9 and all of Chaps. 2 and 7. I have also used the contents of this book for a two-semester course and have supplemented the text in the second semester by a description of numerical methods in device simulations (MINIMOS, PISCES) and on another occasion by developing in class a band structure and Monte Carlo simulation. The required background for the book is an introduction to the principles of quantum mechanics and, even more so, a working knowledge of advanced calculus. The material is presented in a dense form, and, in order to really understand it, one needs to "go over it with a pencil" and in some instances consult the referenced literature.

My thinking and approach in writing this book have been much influenced by my colleagues B. G. Streetman and N. Holonyak, Jr., and by the inspiring environment at the University of Illinois.

Thanks go to E. Kesler, C. Willms, and to my wife Sylvia for their help in preparing the manuscript.

Karl Hess

ADVANCED THEORY OF SEMICONDUCTOR DEVICES

1

BRIEF REVIEW OF THE RELEVANT BASIC EQUATIONS OF PHYSICS

It is clear that from a mathematical viewpoint all equations of physics (microscopic and macroscopic) are relevant for semiconductor devices. In an absolutely strict theoretical way, we therefore would have to proceed from the fundamentals of quantum field theory and write down the $\sim 10^{23}$ coupled equations for all the atoms in the semiconductor device. Then we would have to solve these equations, including the complicated geometrical boundary conditions. However, the outcome of such an attempt is clear to everyone who has tried to solve only one of the 10^{23} equations.

Any realistic approach has to proceed differently. Based on the experience and investigations of many excellent scientists in this field, we neglect effects that would influence the results only slightly. In this way many relativistic effects become irrelevant. In my experience the spin of electrons plays a minor role in the theory of semiconductor devices and can be accounted for in a simple way (the correct inclusion of a factor of 2 in some equations).

Most effects of statistics can be understood classically, and we will need only a very limited amount of quantum statistical mechanics. This leaves us essentially with the Hamiltonian equations (classical mechanics), the Schrödinger equation (quantum effects), the Boltzmann equation (statistics), and the Maxwell equations (electromagnetics).

It is clear that the atoms that constitute a solid are coupled, and therefore the equations for the movement of atoms and electrons in a solid are coupled. This still presents a major problem. We will see, however, that there are powerful methods to decouple the equations and therefore make explicit solutions possible. My presentation attempts a delicate balance between rigor and intuitive concepts. In this way, the fundamental laws of physics are finally reduced to laws of semiconductor devices that are tractable and whose limitations are clearly stated.

Many body effects such as superconductivity are excluded from our treatment. Because of the low density of mobile particles in semiconductors, many body effects are rare, except for effects connected to screening, which will be treated in detail in Chap. 7. Effects of high magnetic fields are also excluded since they are unimportant for most device applications.

1.1 THE EQUATIONS OF CLASSICAL MECHANICS

Some time after the work of Galileo, Kepler, Copernicus, and Newton, Hamilton was able to give the laws of mechanics a very elegant and powerful form. He found that the laws of mechanics can be closely linked to the sum of kinetic and potential energy written as a function of momentumlike (p_i) and spacelike (x_i) coordinates. This function is now called the Hamiltonian function $H(p_i, x_i)$. The laws of mechanics are

$$\frac{dp_i}{dt} = -\frac{\partial H(p_i, x_i)}{\partial x_i} \tag{1.1}$$

and

$$\frac{dx_i}{dt} = \frac{\partial H(p_i, x_i)}{\partial p_i} \tag{1.2}$$

where t is time and $i = 1,2,3$. Instead of x_i, we sometimes denote the space coordinates by x, y, z.

Some simple special cases can be solved immediately. The free particle (potential energy = constant) moves according to

$$H = \sum_i p_i^2/2m$$

and we have from Eq. (1.1)

$$\frac{dp_i}{dt} = 0; \qquad p_i = \text{constant}$$

which is Newton's first law of steady motion without forces.

If we have a potential energy $V(x_1)$ that varies in the x_1 direction, we obtain from Eq. (1.1)

$$\frac{dp_1}{dt} = -\frac{\partial V(x_1)}{\partial x_1} \equiv F \tag{1.3}$$

The quantity defined as F is the force, and Eq. (1.3) is Newton's second law of mechanics.

These examples are enough for our purpose, since we will make less use of classical mechanics than we do of quantum mechanics, which is discussed in the next section.

1.2 THE EQUATIONS OF QUANTUM MECHANICS

At the beginning of the twentieth century, A. Einstein, M. Planck, L. de Broglie, E. Schrödinger, W. Heisenberg, and M. Born (to name a few) realized that nature cannot be strictly divided into waves and particles. They found that light has definite particlelike properties and cannot always be viewed as a wave, and particles such as electrons revealed definite wavelike behavior under certain circumstances. They are, for example, diffracted by gratings as if they had a wavelength

$$\lambda = h/|\mathbf{p}| \tag{1.4}$$

where $\hbar \equiv h/2\pi \simeq 6.58\ 10^{-16}$ eVs and $\mathbf{p}$ is the electron momentum.

Schrödinger showed that the mechanics of atoms can be understood as boundary value problems. In his theory, electrons are represented by a wave function $\psi(\mathbf{r})$, which can have real and imaginary parts, and follows an eigenvalue differential equation:

$$\left(-\frac{\hbar^2}{2m}\nabla^2 + V(\mathbf{r})\right)\psi(\mathbf{r}) = E\psi(\mathbf{r}) \tag{1.5}$$

The part of the left-side of Eq. (1.5) that operates on ψ is now called the Hamiltonian operator H.

Formally this operator is obtained from the classical Hamiltonian by replacing momentum with the operator $\hbar/i\ \nabla (i =$ imaginary unit), where

$$\nabla = \left(\frac{\partial}{\partial x}, \frac{\partial}{\partial y}, \frac{\partial}{\partial z}\right).$$

The meaning of the wave function $\psi(\mathbf{r})$ was not clearly understood at the time of Schrödinger; in fact, $\psi(\mathbf{r})$ was misinterpreted by Schrödinger himself. It is now agreed that $|\psi(\mathbf{r})|^2$ is the probability of finding an electron in a volume element $d\mathbf{r}$ at $\mathbf{r}$. In other words, we have to think of the electron as a point charge with a statistical interpretation of its whereabouts (the wavelike nature!) It is usually difficult to get a deeper understanding of this viewpoint of nature; even Einstein had trouble with it. It is, however, a very successful viewpoint that describes exactly all phenomena we are interested in. To obtain a better feeling for the significance of $\psi(\mathbf{r})$, we will solve Eq. (1.5) for several special cases. As in the classical case, the simplest solution is obtained for constant potential. Choosing an appropriate energy scale, we put $V(\mathbf{r}) = 0$ everywhere.

By inspection we can see that the function

$$C \exp (i\,\mathbf{k} \cdot \mathbf{r}) = C\,(\cos \mathbf{k} \cdot \mathbf{r} \; + \; i \sin \mathbf{k} \cdot \mathbf{r}) \tag{1.6}$$

is a solution of Eq. (1.5) with

$$\hbar^2 k^2 / 2m = E \tag{1.7}$$

and C a constant.

The significance of the vector $\mathbf{k}$ can be understood from analogies to well-known wave phenomena in optics and from the classical equations. Since E is the kinetic energy, $\hbar\mathbf{k}$ has to be equal to the classical momentum $\mathbf{p}$ to satisfy $E = p^2/2m$. On the other hand, in optics

$$|\mathbf{k}| = 2\pi/\lambda \tag{1.8}$$

which gives, together with Eq. (1.4),

$$\hbar\mathbf{k} = \mathbf{p}$$

which is consistent with the mechanical result.

How can the result of Eq. (1.6) be understood in terms of the statistical interpretation of $\psi(\mathbf{r})$? Apparently

$$|\psi(\mathbf{r})|^2 = |C|^2(\cos^2\, \mathbf{k} \cdot \mathbf{r} + \sin^2\, \mathbf{k} \cdot \mathbf{r}) = |C|^2$$

This means that the probability of finding the electron at any place is equal to C^2. If we know that the electron has to be in a certain volume V_{ol} (e.g., of a crystal), then the probability of finding the electron in the crystal must be one. Therefore

$$\int_{V_{ol}} |C|^2 \, d\mathbf{r} = V_{ol}|C|^2 = 1$$

and

$$|C| = 1/\sqrt{V_{ol}} \tag{1.9}$$

In other words, the probability of finding an electron with momentum $\hbar\mathbf{k}$ at a certain point $\mathbf{r}$ is the same in the whole volume and equals $1/V_{ol}$.

This is a peculiar result that can only be understood if one is either very familiar with optics (coherence conditions in interference problems) or if one understands in detail the uncertainty principle. The unfamiliar reader is referred to the literature given at the end of this section (Feynman).

By confining the electron to a volume, we have already contradicted our assumption of constant potential $V(\mathbf{r}) = 0$. (Electrons can only be confined in potential wells.) If, however, the volume is large, our mistake is insignificant for many purposes.

Let us now consider the confinement of an electron in a one-dimensional potential well (although such a thing does not exist in nature). We assume that the potential energy $V(\mathbf{r})$ is zero over the distance $(0, L)$ on the x axis and infinite at the boundaries 0 and L.

The Schrödinger equation, Eq. (1.5), reads in one dimension (x direction, $V(x) = 0$)

$$-\frac{\hbar^2}{2m}\frac{\partial^2\psi(x)}{\partial x^2} = E\psi(x) \tag{1.10}$$

Inspection shows that the function

$$\psi(x) = \sqrt{\frac{2}{L}}\sin\frac{n\pi}{L}x \qquad \text{with } n = 1,2,3,\ldots \tag{1.11}$$

satisfies Eq. (1.10) as well as the boundary conditions. The boundary conditions are, of course, that ψ vanishes outside the walls, since we assumed an infinite impenetrable potential barrier. In the case of a finite potential well, the wave function penetrates into the boundary and the solution is more complicated. If the barrier has a finite width, the electron can even leak out of the well (tunnel). This is a very important quantum phenomenon the reader should be familiar with. We will return to the tunneling effect below.

The wave function, Eq. (1.11), corresponds to energies E

$$E = \frac{n^2\pi^2\hbar^2}{2mL^2} \tag{1.12}$$

Since n is an integer, the electron can assume only certain discrete energies while other energies are not allowed. These discrete energies that can be assumed are called *quantum states* and are characterized by the *quantum number n*. The wave function and corresponding energy are therefore also denoted by ψ_n, E_n.

Think of a violin string vibrating in various modes at higher and lower tones (frequency ν), depending on the length L, and consider Einstein's law:

$$E = h\nu \tag{1.13}$$

If we compare the modes of vibration of the string with the form of the wave function for various n, then we can appreciate the title of Schrödinger's paper, "Quantization as a Boundary Value Problem."

From the above examples it is clear that the solution of the Schrödinger equation requires considerable effort and we would be in trouble if we had to find an exact solution for all kinds of potentials $V(\mathbf{r})$ that are of interest in the theory of semiconductor devices. Fortunately there is a powerful method of approximation, perturbation theory, that gives us the solution for arbitrary weak potentials. The method is very general and applies to any kind of equation.

Consider an equation of the form

$$(H_0 + \epsilon H_1)\psi = 0 \tag{1.14}$$

where H_0 and H_1 are differential operators of arbitrary complication and ϵ is a small positive number.

If we know the solution ψ_0 of the equation

$$H_0\psi_0 = 0$$

then we can assume that the solution of Eq. (1.14) has the form $\psi_0 + \epsilon\psi_1$. Inserting this form into Eq. (1.14), we obtain

$$(H_0 + \epsilon H_1)(\psi_0 + \epsilon\psi_1) = H_0\psi_0 + \epsilon H_1\psi_0 + H_0\epsilon\psi_1 + \epsilon^2 H_1\psi_1$$

We now can neglect the term proportional to ϵ^2 (because ϵ is small), and since $H_0\psi_0 = 0$, we have

$$H_1\psi_0 + H_0\psi_1 = 0 \tag{1.15}$$

This equation is now considerably simpler than Eq. (1.14) since ψ_0 is known. Therefore, ψ_1 can be determined easily if H_0 has a simple form no matter how complicated H_1 is. The form of the Schrödinger equation allows further simplification, and it is easy to show that any problem involving a small given perturbation H_1 reduces to solving a three-dimensional integral.

Assume that we know the solutions of a Schrödinger equation:

$$H_0\psi_n = E_n\psi_n \qquad n = 1,2,3,\ldots \tag{1.16}$$

and we would like to know the solutions of

$$(H_0 + H_1)\phi_m = W_m\phi_m \qquad \text{with } H_1 \ll H_0 \tag{1.17}$$

Then it is shown in elementary texts on quantum mechanics (Baym) that

$$W_m = E_m + M_{mm} + \sum_{n \neq m} \frac{|M_{nm}|^2}{E_m - E_n} \tag{1.18}$$

with

$$\phi_m = \psi_m + \sum_{n \neq m} \frac{M_{nm}}{E_m - E_n}\psi_n \tag{1.19}$$

and

$$M_{mn} = \int_{V_{ol}} \psi_n^* H_1 \psi_m \, d\mathbf{r} \tag{1.20}$$

where $d\mathbf{r}$ stands for $dxdydz$ (integration over the volume V_{ol}) and ψ_n^* is the complex conjugate of ψ_n.

We have, therefore, reduced the solution of the Schrödinger equation of the new problem to the integration of Eq. (1.20). In addition, one needs to know the solution of Eq. (1.16). Depending on the nature of the problem, this could be the solution for the free electron, the one-dimensional well, the hydrogen atom, the harmonic oscillator, or other well-known solutions.

The formalism outlined above and the examples given are independent of time, and the electrons are perpetually in appropriate (eigen) states. In many instances, however, we will be interested in the following type of problem: The electron initially is in an eigenstate of H_0, denoted, for example, by a wave vector $\mathbf{k}$ for the free electron. What is the probability that the electron will be observed in a different eigenstate characterized by the wave vector $\mathbf{k}'$, after it interacts with a potential $V(\mathbf{r}, t)$? In other words, what is the probability $S(\mathbf{k}, \mathbf{k}')$ per unit time that the interaction causes the system to make a transition from $\mathbf{k}$ to $\mathbf{k}'$?

The answer to this question is the famous Golden Rule of Fermi, which is derived in almost every text on quantum mechanics by time-dependent perturbation theory. The unfamiliar reader is urged to acquire a detailed understanding of the Golden Rule as derived, for example, in the text of Baym. Here we only illustrate its generality and discuss results for important special cases.

1. Assume that a potential $V(\mathbf{r})$ is switched on at time $t = 0$ but is time independent otherwise. One then obtains

$$S(\mathbf{k}, \mathbf{k}') = \left| \int_{V_{ol}} \psi_{\mathbf{k}'}^{*} V(\mathbf{r}) \psi_{\mathbf{k}} \, d\mathbf{r} \right|^2 \cdot \left[\frac{\sin\,(E(\mathbf{k}') - E(\mathbf{k}))t/2\hbar}{(E(\mathbf{k}') - E(\mathbf{k}))\sqrt{t}/2} \right]^2 \tag{1.21}$$

The function in brackets deserves special attention and is plotted in Fig. 1.2. Notice that as t approaches infinity, the function plotted in Fig. 1.1 becomes more and more peaked at its center ($E(\mathbf{k}') = E(\mathbf{k})$). In the limit $t \to \infty$, the so-called δ function is approached, which is defined by

$$\lim_{t \to \infty} \frac{4 \sin^2 [(E(\mathbf{k}) - E(\mathbf{k}'))\, t/(2\hbar)]}{(E(\mathbf{k}) - E(\mathbf{k}'))^2 t} = \frac{2\pi}{\hbar} \delta(E(\mathbf{k}) - E(\mathbf{k}')) \tag{1.22}$$

and can always be understood as a limit of ordinary functions.

It does have some remarkable properties, however, and the unfamiliar

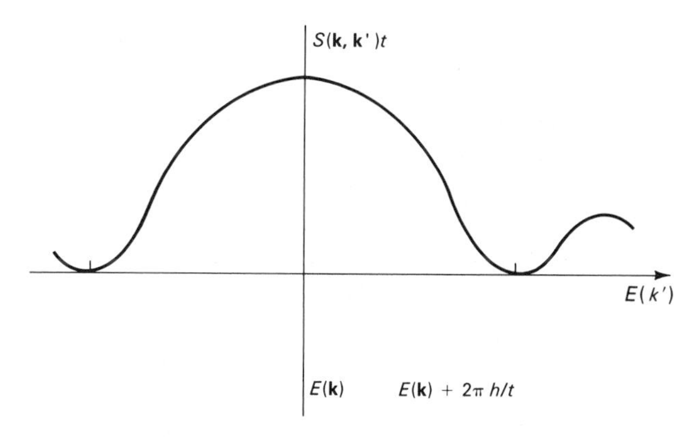

Figure 1.1 Probability of a transition from $\mathbf{k}$ to $\mathbf{k}'$ according to the Golden Rule after the potential has been on for time t.

reader should consult some of the references at the end of this section. A most important property of the δ function is the following: For any continuous function $f(E')$, we have

$$\int_{-\infty}^{+\infty} f(E')\delta(E-E')dE' = f(E) \tag{1.23}$$

2. We assume that the perturbation is harmonic, which means we have a potential of the form

$$V(\mathbf{r},t) = V(\mathbf{r})(e^{-i\omega t} + e^{i\omega t})$$

For $t \to \infty$, we obtain the transition probability

$$S(\mathbf{k},\mathbf{k}') = \frac{2\pi}{\hbar}\left|\int_{V_{ol}} \psi_{\mathbf{k}'}^{*} V(\mathbf{r})\psi_{\mathbf{k}}\, d\mathbf{r}\right|^2$$

$$\{\delta(E(\mathbf{k}) - E(\mathbf{k}') - \hbar\omega) + \delta(E(\mathbf{k}) - E(\mathbf{k}') + \hbar\omega)\} \tag{1.24}$$

It is clear that the δ function simply takes care of energy conservation. For a constant potential, we have to conserve energy as t increases. For a harmonic perturbation, the system can gain or lose energy corresponding, for example, to the absorption or emission of light.

We now turn our attention to the first term in Eq. (1.21), the matrix element, which also plays a vital role in time-independent perturbation theory.

The significance of the matrix element is best illustrated by the following special cases of well-defined potential problems (problems in which the potential is given by a certain function of coordinates):

1. $V(\mathbf{r}) =$ constant. The matrix element is then

$$\text{constant}\,\frac{1}{V_{ol}}\int_{V_{ol}} e^{-i\mathbf{k}'\cdot\mathbf{r}}e^{i\mathbf{k}\cdot\mathbf{r}}\, d\mathbf{r} \tag{1.25}$$

The integration is over the volume V_{ol} of the crystal. In many practical cases, this volume will be much larger than the de Broglie wavelength λ of the electron, which is of the order of 100 Å in typical semiconductor problems. This means that the integral of Eq. (1.25) will be very close to zero, since the cosine and sine functions to which the exponents in Eq. (1.25) are equivalent are as often positive as they are negative in the big volume. There is only one exception: In the case $\mathbf{k}' = \mathbf{k}$, the integral is equal to the volume and the matrix element is equal to constant. Therefore, we can write

$$\text{constant}\,\frac{1}{V_{ol}}\int_{V_{ol}} e^{-i\mathbf{k}'\cdot\mathbf{r}}e^{i\mathbf{k}\cdot\mathbf{r}}\, d\mathbf{r} = \text{constant}\ \delta_{\mathbf{k},\mathbf{k}'} \tag{1.26}$$

where $\delta_{\mathbf{k},\mathbf{k}'} = 1$ for $\mathbf{k} = \mathbf{k}'$ and is zero otherwise. This is known as the Kronecker delta symbol.

Consequently, the matrix element has taken care of momentum conservation; the free electron in a constant potential does not change its momentum.

2. Second, we consider an arbitrary potential having the following Fourier representation:

$$V(\mathbf{r}) = \sum_{\mathbf{q}} V_{\mathbf{q}} e^{i\mathbf{q}\cdot\mathbf{r}} \tag{1.27}$$

Then the matrix element, which we now denote by $M_{\mathbf{k},\mathbf{k}'}$ becomes

$$M_{\mathbf{k},\mathbf{k}'} = \sum_{\mathbf{q}} \frac{V_{\mathbf{q}}}{V_{\mathrm{ol}}} \int_{V_{\mathrm{ol}}} e^{i(\mathbf{k}-\mathbf{k}'+\mathbf{q})\mathbf{r}}\, d\mathbf{r}$$

$$= \sum_{\mathbf{q}} V_{\mathbf{q}} \delta_{\mathbf{k}'-\mathbf{k},\mathbf{q}} \tag{1.28}$$

How do we interpret this result? If we allow the potential of Eq. (1.27) also to have a time dependence (e.g., as $e^{i\omega t}$), then the potential can be interpreted as that of a wave (e.g., an electromagnetic wave). In this case Eq. (1.28) simply tells us that the wave vectors of all scattering agents (i.e., their momenta) are conserved, since we have

$$\mathbf{k}' - \mathbf{k} = \mathbf{q} \tag{1.29}$$

It is important to notice that Eq. (1.29) is also valid for a static, time-independent potential—that is, even a static potential "supplies" momentum according to its Fourier components—in the same way a wave does. This seems strange at first glance. To see the significance, consider the boundary of a billiards table. This boundary is an impenetrable abrupt potential step whose Fourier decomposition involves all values of $\mathbf{q}$. Indeed, the boundary can supply any momentum to the ball to make it bounce back.

The above two examples show that the Golden Rule essentially takes care of energy and momentum conservation. This is also the reason for its generality and importance. Remember, however, that this is true only for cases when time t at which we observe the scattered particle is long after the potential is switched on. For short times (in practice these are times of the order of 10^{-14} sec), the function in Eq. (1.22) cannot be approximated by a δ function, and energy need not be conserved in processes on this short time scale. This is at the heart of the energy time uncertainty relation.

To illustrate the great generality of the Golden Rule, one more example is given.

3. Consider the "tunneling problem" of Fig. 1.2. Although an electric field F is applied in the z direction, the electron in Fig. 1.2 is confined in a small well. Classically, it would stay in the well. However, since the barrier is not infinite, as assumed in Eq. (1.11), the wave function is not zero at the well boundary but penetrates the boundary. In other words, there is a finite probability of finding the electron outside the well.

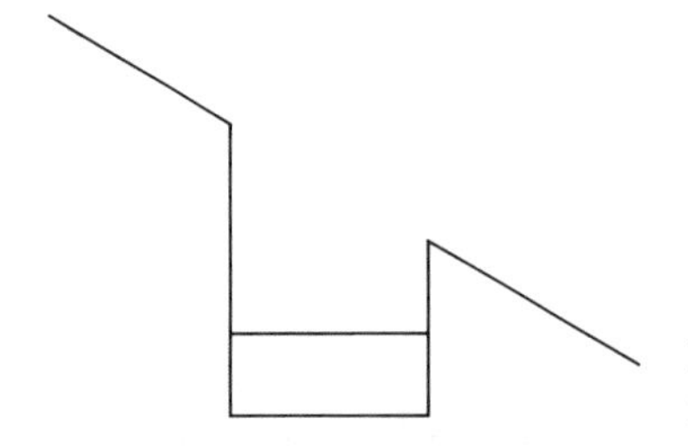

Figure 1.2 Electron in a potential well plus applied electric field F.

We can calculate the probability per unit time that the electron leaks out if we know the wave function ψ_w in the well and ψ_{ou} outside the well and regard the electric field as a perturbation. This perturbation gives a term (the potential energy) eFz in the Hamiltonian. The Golden Rule tells us that

$$S(\mathrm{w}, \mathrm{ou}) = \left| \int_{V_{ol}} \psi_{ou}^* \, eFz \, \psi_w \, d\mathbf{r} \right|^2 \cdot \frac{2\pi}{\hbar} \delta(E_{ou} - E_w) \tag{1.30}$$

In writing down this equation (which was first derived by Oppenheimer), I have swept under the rug the fact that ψ_{ou} and ψ_w are the solutions of different Hamiltonians. ψ_w is obtained from the solution of the Schrödinger equation of the quantum well and ψ_{ou} is the solution of a free electron in an electric field with

$$H = -\frac{\hbar^2 \nabla^2}{2m} - eFz$$

An exact justification of this procedure is complicated and is given in Appendix A. The matrix element is not the "normal" matrix element and we call it the transfer-matrix element, for obvious reasons (Duke).

In the following we will use a shorthand notation (introduced by Dirac) for matrix elements: For ψ_n we write $|n\rangle$ and for ψ_m^* we use $\langle m|$ to obtain

$$\langle m|H_1|n\rangle \equiv \int_{V_{ol}} \psi_m^* H_1 \psi_n \, d\mathbf{r} \tag{1.31}$$

RECOMMENDED READINGS

BAYM, G., *Lectures on Quantum Mechanics.* New York: Benjamin, 1969, p. 248.

FEYNMAN, R. P., *Lectures on Physics,* vol. III. Reading, Mass.: Addison-Wesley, 1964, pp. 1.1–4.15.

HARRISON, W. A., *Electronic Structure and Property of Solids.* San Francisco: Freeman, 1980, pp. 2–8.

DUKE, C. B., *Tunneling in Solids.* New York: Academic Press, 1969, p. 207.

PROBLEMS

1.1. Solve by perturbation theory (to first order) for y:

$$\epsilon \frac{\partial y}{\partial x}\frac{\sin(x)}{} + \frac{\partial y}{\partial x} = y$$

where ϵ is a small positive quantity.

1.2. Calculate the matrix elements for wave functions of the form $\psi = \frac{1}{\sqrt{V_{\text{ol}}}} e^{i\mathbf{k}\cdot\mathbf{r}}$ and

(a) $H_1 \propto e^{i\mathbf{q}\cdot\mathbf{r}}$

(b) $H_1 \propto \delta(\mathbf{r})$

(c) $H_1 \propto |\mathbf{r}|^{-2}$

(d) $H_1 \propto \exp\left\{\frac{-|\mathbf{r}|}{r_0}\right\} \qquad r_0 > 0$

Polar coordinates are helpful in parts **c** and **d**.

2

LATTICE VIBRATIONS

2.1 GENERAL CONSIDERATIONS

In most texts, crystal structure and the motion of electrons are treated first and the motion of the atoms, the lattice vibrations, are discussed later. However, many important concepts can be more easily understood when the methods and concepts used to describe the lattice vibrations are known. For most of the present treatment we do not need to know much about solids and crystals. It suffices for this chapter to define a crystal—and we will be mostly interested in crystalline solids—as a regular array of atoms hooked together by atomic forces. "Regular" means that the distance between the atoms is the same throughout the structure.

Since the atoms that vibrate are very heavy, many problems involving lattice vibrations can be solved by classical means (e.g., using the Hamiltonian equations). Therefore we have to derive the kinetic and potential energy. Since we would like to describe vibrations (i.e., the displacement of the atoms), we express all quantities in terms of the atomic displacements $u_i(r)$, where $i = x, y, z$ and r is the number (identification) of the atom. It is important to note that r is not equal to the continuous space coordinate r in this chapter, although it has similar significance since it labels the atoms. The displacement of an atom in a set of regularly arranged atoms is shown in Fig. 2.1.

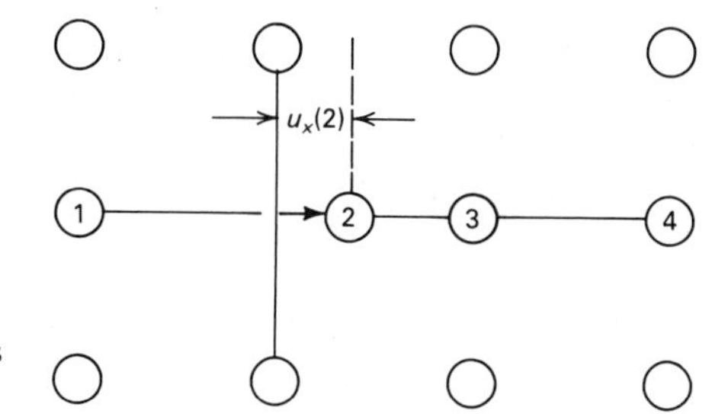

Figure 2.1 Displacement $u_i(r)$ of atoms in a crystal lattice.

The kinetic energy T is then given by

$$T = \frac{1}{2} M \sum_{ri} \dot{u}_i^2(r) \qquad \text{where} \quad \dot{u}_i(r) = \frac{\partial u_i(r)}{\partial t} \tag{2.1}$$

and M is the mass of the atoms (ions).

We assume now that the total potential energy U of the atoms can be expressed in terms of a power series in the displacements.

$$U = U_0 + \sum_{ri} B_i^r u_i(r) + \frac{1}{2} \sum_{\substack{rs \\ ij}} B_{ij}^{rs} u_i(r) u_j(s) + \cdots \tag{2.2}$$

where s also numbers the atoms as r does.

The following results rest on this series expansion and truncation, which makes a first principle derivation (involving many body effects) unnecessary.

Equation (2.2) is, of course, a Taylor expansion with

$$B_i^r = \frac{\partial U}{\partial u_i(r)} \tag{2.3}$$

Since the crystal is in equilibrium, that is, at a minimum of the potential energy U, the first derivative vanishes and

$$B_i^r = 0 \tag{2.4}$$

We further have

$$B_{ij}^{rs} = \frac{\partial^2 U}{\partial u_i(r) \partial u_j(s)} = B_{ji}^{sr} \tag{2.5}$$

We now use the fact that the crystal is translationally invariant—that is, we can shift the coordinate system by s atoms (start to count s atoms later), and the crystal is transformed into itself (at least if it is infinite). Therefore

$$B_{ij}^{rs} = B_{ij}^{(r-s)0} \tag{2.6}$$

Furthermore, a rigid displacement (all u_i equal) of the crystal does not change U and we therefore have, from Eqs. (2.2) and (2.6),

$$\sum_r B_{ij}^{0r} = 0 \tag{2.7}$$

To derive Eq. (2.6), we have assumed an infinite crystal. We also could have introduced so-called periodic boundary conditions, that is, continue the crystal by repeating it over and over. In one dimension, this means we consider only rings of atoms (see Fig. 2.2). This approach amounts to neglecting any surface effects or other effects that are sensitive to the finite extension of crystals.

We can now derive the equations of motion by using Eqs. (1.1) and (1.2) with coordinates $u_i(r)$ instead of x_i:

$$\dot{p}_i(r) = -\frac{\partial H}{\partial u_i(r)} \tag{2.8a}$$

and

$$p_i(r) = M\dot{u}_i(r) \tag{2.8b}$$

Equation (2.8a) gives

$$\dot{p}_i(r) = -\frac{1}{2}\frac{\partial}{\partial u_i(r)}\left[\sum_{\substack{mn \\ ij}} B_{ij}^{mn} u_i(m) u_j(n)\right] \tag{2.9}$$

Here, also, the indices m, n are used to number the atoms (as r, s above). Therefore

$$\dot{p}_i(r) = -\sum_{s,j} B_{ij}^{rs} u_j(s)$$

and together with Eq. (2.9) one obtains

$$M\ddot{u}_i(r) + \sum_{sj} B_{ij}^{rs} u_j(s) = 0 \tag{2.10}$$

Remember that the index s in Eq. (2.10) runs over all the large number of atoms, that is, up to about 10^{23} in a typical crystal. r can also assume any of

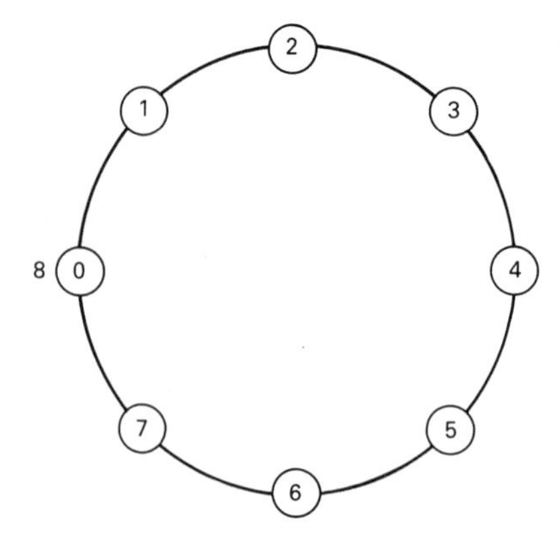

Figure 2.2 A ring of atoms representing cyclic boundary conditions.

these numbers. In other words, we have about 10^{23} coupled equations to solve. This situation is very typical for any type of solid state problem but by far not as hopeless as it may seem. Powerful methods have been developed to reduce the number of equations and the following treatment is representative.

We will assume for simplicity that the crystal is one dimensional and avoid the complicated geometrical arrangement of atoms in a real crystal. (We will learn more about this when we discuss the electrons and their motion in crystals.)

In the three-dimensional case, Eqs. (2.4)–(2.7) are very helpful since they reduce the numbers of parameters. Without going into details, I would like to mention that this reduction of parameters is generally accomplished by group theoretical arguments, and Eq. (2.6) is a direct consequence of the translational invariance (group of translations).

To proceed explicitly with our one-dimensional model, we need to make the drastic assumption that each atom interacts only with its nearest neighbor. (We can use the same method also for second, third . . . nearest neighbor interaction, if we proceed numerically and use a high speed computer.) Our assumption means

$$B^{r,s} \neq 0 \qquad \text{only for} \qquad s = r \pm 1$$

Notice that we dropped the indices i, j because this is a one-dimensional problem. Without any loss of generality, we may assume $r = 0$. Then we have

$$B^{0s} \neq 0 \qquad \text{for } s = \pm 1$$

and

$$B^{0s} = 0 \qquad \text{otherwise} \tag{2.11}$$

Furthermore

$$B^{01} = B^{-10}$$

according to Eq. (2.6) and

$$B^{-10} = B^{0(-1)}$$

according to Eq. (2.5). Therefore

$$B^{01} = B^{0(-1)}$$

From Eq. (2.7) we obtain

$$B^{00} = -2B^{01} \tag{2.12}$$

It is now customary to denote $B^{01} = B^{0(-1)}$ by $-\alpha$ (α is the constant of the "spring" forces that hold the crystal together) and therefore B^{00} by 2α. The equation of motion, Eq. (2.10), then becomes for any r

$$M\ddot{u}(r) = -2\alpha u(r) + \alpha u(r-1) + \alpha u(r+1) \tag{2.13}$$

2.2 SOLUTION FOR LINEAR CHAINS

Equation (2.13) leaves us still 10^{23} coupled differential equations. However, these equations are now in *tridiagonal form,* and such a form can be reduced to one equation by skillful substitution. The substitution can be derived from Bloch's theorem, which we will discuss later. It also can be guessed:

$$u(r) = ue^{iqra} \tag{2.14}$$

Note that the amplitude u is still a function of time. Here a is the distance between atoms (i.e., the lattice constant). Equation (2.13) becomes

$$M\ddot{u}e^{iqra} = -2\alpha ue^{iqra} + \alpha ue^{iqra}e^{-iqa} + \alpha ue^{iqra}e^{iqa} \tag{2.15}$$

which gives

$$M\ddot{u} = \alpha u(2\cos qa - 2)$$

and

$$\ddot{u} = \frac{-4\alpha u}{M}\sin^2\left(\frac{aq}{2}\right)$$

This gives

$$\ddot{u} + \nu^2 u = 0 \tag{2.16}$$

with

$$\nu = 2\left|\sin\frac{aq}{2}\right|\sqrt{\frac{\alpha}{M}} \tag{2.17}$$

This means that the atoms are oscillating in time with frequency ν, which is a function of the wave vector q. The function is shown in Fig. 2.3.

There are several important points to notice. First, at $q = -\pi/a$ and

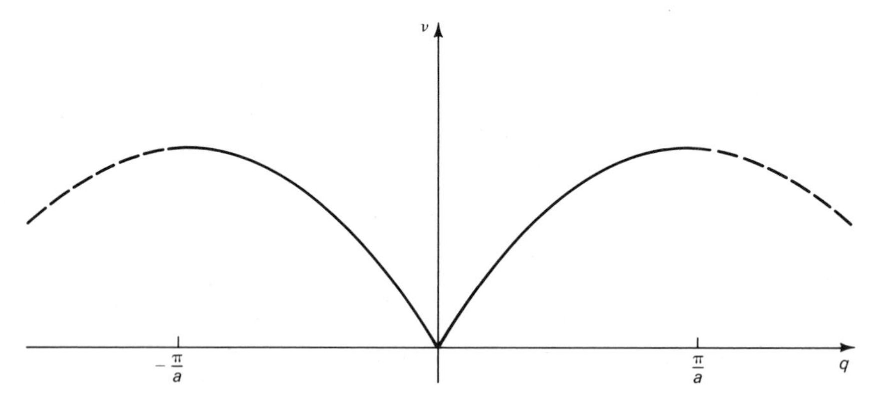

Figure 2.3 "Dispersion relation" $\nu(q)$ for lattice vibrations. (Remember, $E = h\nu$.)

$q = \pi/a$, the energy has its highest value. For these q, the wavelength $\lambda = 2\pi/q$ has the value $\lambda = 2a$. As can be seen from Fig. 2.4, this is the shortest wavelength that we really need to describe the physics of the lattice vibrations. Shorter wavelengths lead only to "wiggles" between the atoms but the displacements are actually the same. For example, if $q = 3\pi/a$ and $\lambda = 2/3a$, the atoms are displaced in exactly the same way as for $q = \pi/a$. In other words, for any q outside the zone $-\pi/a \leq q \leq \pi/a$, which is called the *Brillouin zone*, we can find a q inside the zone that describes the same displacement, energy, and so on. Notice that in a real crystal the arrangement of atoms is different in different directions. Therefore, the three-dimensional Brillouin zone is usually a complicated geometrical figure. (See Chap. 3.)

Second, q is not a continuous variable because of the boundary conditions. Consider, for example, the ring of Fig. 2.2 with eight atoms and

$$u(0) = u(8)$$

or, in general, for N atoms, we have

$$u(N) = u(0) \qquad \text{and} \qquad u(N) = u(0)e^{iqNa}$$

Therefore e^{iqNa} equals one, and we conclude that $q = 2\pi l/Na$ where l is an integer. If we restrict q to the first Brillouin zone, we have $-N/2 \leq l \leq N/2$. This means that q assumes only discrete (not continuous) values. However, because of the large number N, it can almost be regarded as continuous.

Third, without emphasizing it, we have developed a microscopic theory of sound propagation in solids. For small wave vectors q (i.e., large λ) we have

$$\sin\frac{qa}{2} \approx \frac{qa}{2}$$

and

$$\nu = \sqrt{\frac{\alpha}{M}}\,qa$$

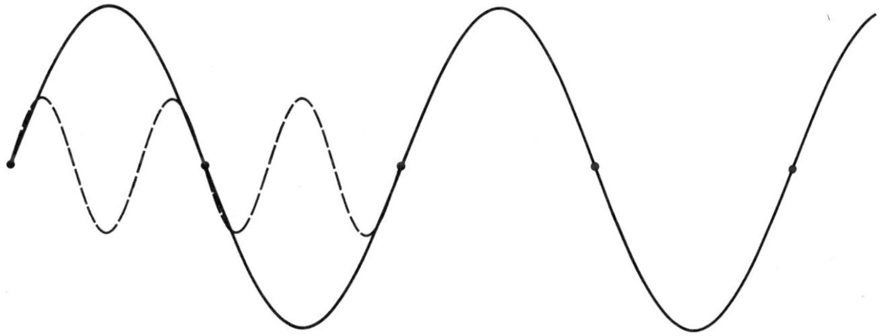

Figure 2.4 Illustration of shortest possible physical wavelength of lattice vibrations.

Using $\lambda\nu = v_s$, where v_s is the velocity of sound we obtain

$$v_s = 2\pi a\sqrt{\frac{\alpha}{M}}$$

which is a microscopic description of the sound velocity.

In real crystals additional complications arise from the fact that we can have two or even more different kinds of atoms. These atoms may oscillate as the identical atoms in the above example. There are, however, different modes of oscillation possible. If we think of a chain with two different kinds of atoms, it can happen that one kind of atom (black) oscillates against the other kind (white).

Such an oscillation can take place, and indeed does, at very high (optical) frequency, and the corresponding lattice vibrations are called the optical phonons. It is very important to note that in principle all black atoms can oscillate in phase against the white ones. This means that we can have high frequencies (energies) even if the wavelength is very large or the q vector is very small (see Fig. 2.5).

The energy versus q relation can then have two branches, the acoustic and the optic, as shown in Fig. 2.6.

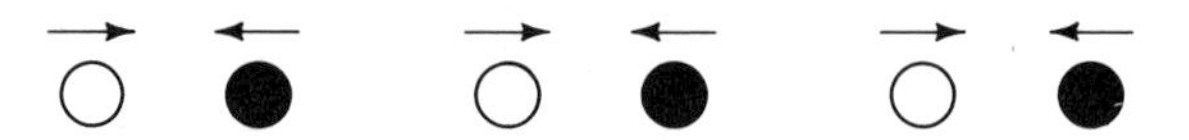

Figure 2.5 Two different kinds of atoms oscillating against each other. This represents a wave with high energy (frequency) and small wave vector.

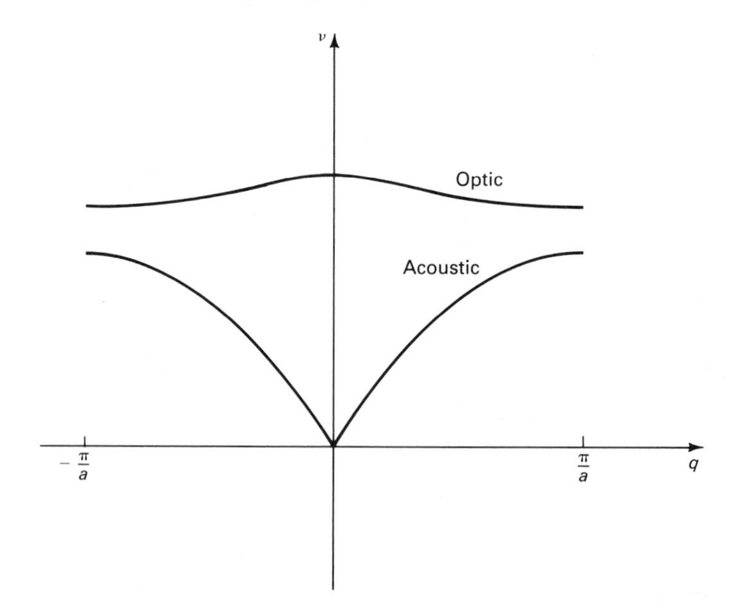

Figure 2.6 Schematic $\nu(q)$ diagram for acoustic and optic phonons in one dimension.

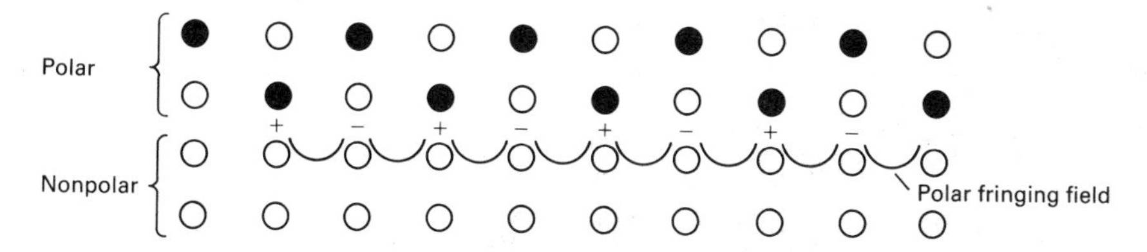

Figure 2.7 Polar-nonpolar material on top of each other. Notice that the phonons can create a long-range polar fringing field in the nonpolar material.

The presence of two different atoms can also cause long-range coulombic forces due to the different charge on the two atom types (ionic component). The long-range forces cannot be described by simple forces between neighboring atoms, and one calls the phonons *polar optical phonons* if these long-range forces are important. We will see later that the long-range forces give rise to an interaction with electrons that is different from the nonpolar electron-phonon interaction. To illustrate this point, we have shown in Fig. 2.7 a lattice of one kind of atom neighboring a lattice with two different kinds of atoms. A phonon causes long-range ± charges in the polar material. The fringing fields can even influence electrons in the nonpolar material, giving rise to "remote polar scattering."

Let us note in passing that in three dimensions transverse phonon modes are also possible, in addition to the longitudinal modes. We will see, however, that it is mainly the longitudinal modes that interact with electrons (and we are mainly interested in the electron-phonon interaction).

RECOMMENDED READINGS

LANDSBERG, P. T., *Solid State Theory*. New York: Wiley/Interscience, 1969, pp. 327–56.

The remote phonon interaction is described in K. Hess and P. Vogl, *Solid State Communications*, 30 (1979), 807.

PROBLEM

2.1. Consider a one-dimensional crystal lattice with two ions (atoms) repeated in a circular arrangement. The two ions (atoms) are identical, with mass M, but are connected by springs of alternating strength (D_1, D_2).

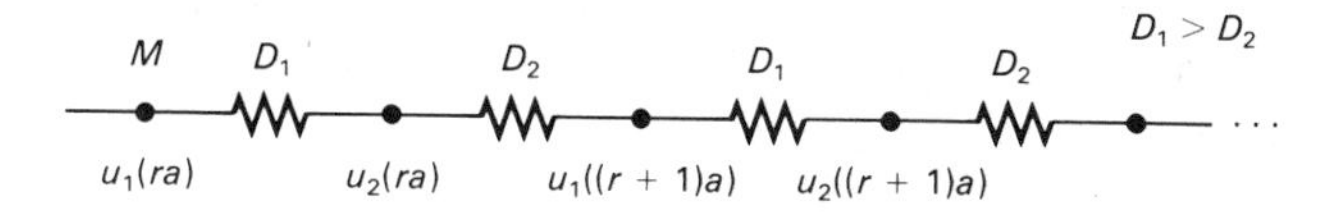

(a) Derive the equations of motion. (Consider only nearest neighbor interactions, where the force is proportional to the difference in displacements.)

(b) Find and sketch the dispersion relation of the possible vibrational modes. (Assume all displacements are traveling waves with sinusoidal time dependence, that is, $u_i(ra) = \epsilon_i e^{i(qra - \omega t)}$.)

(c) Discuss the form of the dispersion relation and the nature of the modes for $q \ll \pi/a$ and $q = \pi/a$, where q is the wave vector.

(d) Find the velocity of sound (ω/q for $q \to 0$).

(e) Show that the group velocity $\partial\omega/\partial q$ becomes zero at the Brillouin zone boundary. (This is a general result.)

3

THE SYMMETRY OF THE CRYSTAL LATTICE

3.1 CRYSTAL STRUCTURES

In Chap. 2 we discussed the lattice vibrations without defining exactly what a crystal lattice is. Here we give this definition and we see that a crystal is an object of high symmetry. This symmetry can be used to obtain general information about the properties of crystals and also to abbreviate complicated algebra. Full use of the symmetry requires knowledge of group theory. This knowledge is not required for the reader of this book. Nevertheless an attempt is made here to introduce group theoretical techniques via practical examples.

A crystal consists of a *basis* and a so-called Bravais lattice. The basis can be anything ranging from atoms to giant molecules, such as desoxyribonucleic acid (DNA). The Bravais lattice is a set of points $\{\mathbf{R}_l\}$ that is generated by three noncoplanar translations $\mathbf{a}_1$, $\mathbf{a}_2$, $\mathbf{a}_3$, which are vectors of three-dimensional space.

$$\mathbf{R}_l = l_1\mathbf{a}_1 + l_2\mathbf{a}_2 + l_3\mathbf{a}_3 \tag{3.1}$$

and the l_i are integers.

According to the properties of this lattice, under reflection, rotation, and so on, one can distinguish 14 types of Bravais lattices. For us, only the cubic types matter; see Fig. 3.1. The important semiconductors are characterized by a *tetrahedral arrangement* of the nearest neighbor atoms. Their lattice

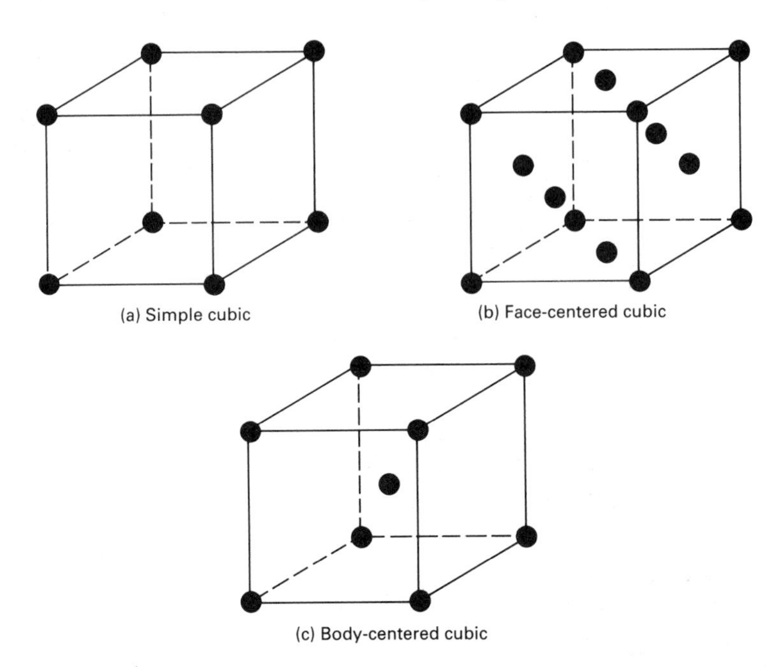

Figure 3.1 The three types of cubic Bravais lattices.

can be viewed as a face-centered cubic lattice with a basis of two atoms. For silicon and germanium, these two atoms are equal; for GaAs and the III–V compounds, the two atoms are different. This is illustrated in Fig. 3.2.

The question now arises of whether we can view the lattice of silicon under all circumstances as a face-centered cubic crystal with a basis of two atoms. The answer is, in principle, yes; the symmetry of one face-centered cubic crystal is present. For some effects, however, the existence of the two basis atoms is vital. Consider, for example, lattice vibrations. It is clear that the two basis atoms are connected by different force (spring) constants α than two atoms on the side of the cube (Fig. 3.2). Therefore, optical phonons will exist (as discussed earlier). Instead of two different kinds of atoms vibrating against each other, the two *sublattices* associated with the two basis atoms can vibrate against each other. By two sublattices, we mean that we can also view the silicon crystal as two interconnected face-centered cubic lattices (sublattices), each having one basis atom.

The three translation vectors $\mathbf{a}_1$, $\mathbf{a}_2$, $\mathbf{a}_3$ that generate a cubic face-centered lattice are shown in Fig. 3.3. It is important to note that these vectors are different from the vectors that generate the simple cubic lattice. If these vectors (along the sides of the cube) were chosen to generate the lattice points $\mathbf{R}_l$, some points could not be reached with integer values of l.

Figures 3.4a and 3.4b are photographs of a gallium-arsenide crystal

Figure 3.2 Crystal structure of silicon (or GaAs if two kinds of atoms are on the appropriate lattice sites). Notice the tetrahedral arrangement of nearest neighbor atoms and the equivalence to a face-centered cubic lattice (if a two-atom basis is assumed).

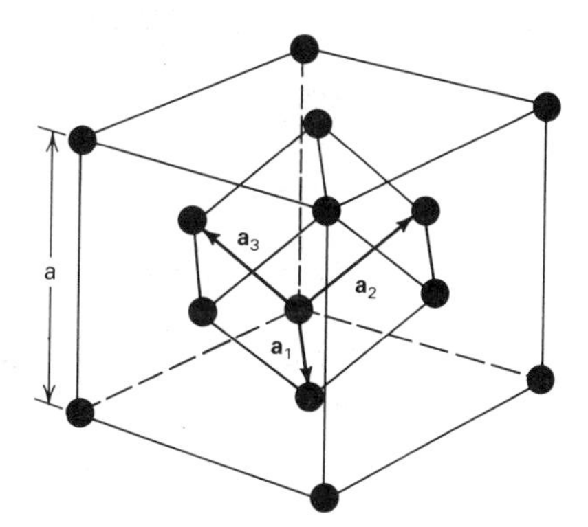

Figure 3.3 Vectors $\mathbf{a}_1$, $\mathbf{a}_2$, $\mathbf{a}_3$ generating the face-centered cubic lattice.

model in the [110] and [100] directions. These photographs illustrate the anisotropy of a crystal. In other words, lattice waves or electrons traveling in different directions find, in general, different patterns (and therefore a different Brillouin zone boundary).

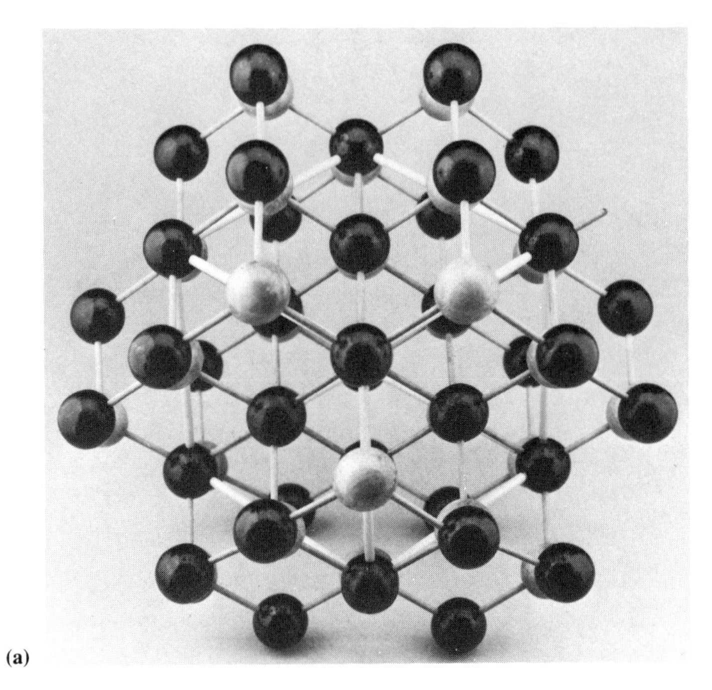

(a)

(b)

Figure 3.4 Photographs of a GaAs crystal model in (a) [100] and (b) [110] crystallographic direction.

3.2 ELEMENTS OF GROUP THEORY

3.2.1 Point Group

From a more mathematical viewpoint, it is important to note that crystal lattices—and with them all their physical properties—are transformed into themselves by certain geometrical operations (rotations, reflections, etc.). The set of all these operations is called the point group of the crystal lattice. Forty-eight such operations (translations are excluded for the moment) transform a cube into itself. The group of these 48 operations is called O_h. Twenty-four of the 48 operations that form the subgroup T_d are shown in Table 3.1. In this table the operation $Q_3(\bar{x}_1 x_2 \bar{x}_3)$ means, for example,

$$Q_3 f(x_1, x_2, x_3) = f(-x_1, x_2, -x_3)$$

for any function f of the coordinates.

Operating on these 24 transformations with the inversion $Q_0(\bar{x}_1 \bar{x}_2 \bar{x}_3)$ gives another 24 symmetry operations, which, together with the operations in Table 3.1, form the 48 operations of O_h.

If $f(x_1, x_2, x_3)$ is a physical property of the crystal, and the crystal has the symmetry O_h, then we obtain the value of f at many other points simply by applying the symmetry operations. This can save enormous amounts of computation time. (We explain the use of band-structure calculations in Chap. 4.) The following example gives also a clear illustration of the advantageous use of the symmetry operations.

Bardeen, Schrieffer, and Stern realized that electrons can form a two-dimensional gas at the interface between Si and SiO_2 in a metal-oxide-semiconductor (MOS) transistor. Figure 3.5 shows the basic geometry of a MOS transistor (which is described in detail in Chap. 14).

It is known that bulk silicon exhibits an isotropic conductivity σ. The interesting questions arose, then, of whether the conductivity of the two-dimensional electron sheet is also isotropic in the interface plane and if the conductivity depends on the crystallographic surface orientation of silicon. To settle these questions experiments were performed on (100), (110), and (111) surfaces (the reader should be familiar with the Miller indices) by fabricating many transistors on various wafers of these three surface orientations. It was

TABLE 3.1 Elements of the Point Group T_d

$Q_1(x_1 x_2 x_3)$	$Q_2(x_1 \bar{x}_2 \bar{x}_3)$	$Q_3(\bar{x}_1 x_2 \bar{x}_3)$	$Q_4(\bar{x}_1 \bar{x}_2 x_3)$	$Q_5(x_2 x_3 x_1)$	$Q_6(\bar{x}_2 x_3 \bar{x}_1)$
$Q_7(\bar{x}_2 \bar{x}_3 x_1)$	$Q_8(x_2 \bar{x}_3 \bar{x}_1)$	$Q_9(x_3 x_1 x_2)$	$Q_{10}(\bar{x}_3 \bar{x}_1 x_2)$	$Q_{11}(x_3 \bar{x}_1 \bar{x}_2)$	$Q_{12}(\bar{x}_3 x_1 \bar{x}_2)$
$Q_{13}(\bar{x}_1 x_3 \bar{x}_2)$	$Q_{14}(\bar{x}_1 \bar{x}_3 x_2)$	$Q_{15}(\bar{x}_3 \bar{x}_2 x_1)$	$Q_{16}(x_3 \bar{x}_2 \bar{x}_1)$	$Q_{17}(x_2 \bar{x}_1 \bar{x}_3)$	$Q_{18}(\bar{x}_2 x_1 \bar{x}_3)$
$Q_{19}(x_1 x_3 x_2)$	$Q_{20}(x_1 \bar{x}_3 \bar{x}_2)$	$Q_{21}(x_3 x_2 x_1)$	$Q_{22}(\bar{x}_3 x_2 \bar{x}_1)$	$Q_{23}(x_2 x_1 x_3)$	$Q_{24}(\bar{x}_2 \bar{x}_1 x_3)$

After D. J. Morgan in P. T. Landsberg, ed., *Solid State Theory: Methods and Applications*, Table C.7.1 copyright © 1969 John Wiley & Sons, Ltd., Reprinted by permission.

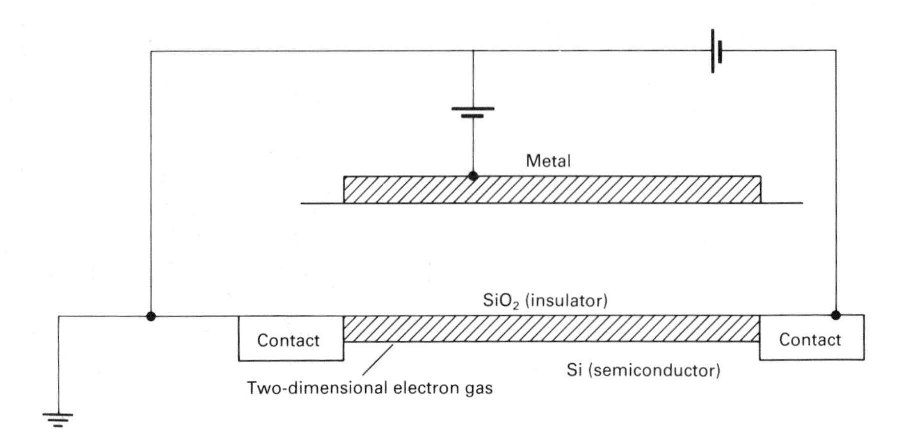

Figure 3.5 MOS transistor with a two-dimensional sheet of electrons at the Si-SiO_2 interface.

found that on (100) and (111) surfaces the conductivity is isotropic. The (110) surface, however, shows an anisotropic electrical conductivity.

Below we will see that this result can be obtained by a "back of an envelope" calculation. The current density **j** as a function of electric field **F** is given by Ohm's law

$$\mathbf{j} = \sigma \mathbf{F} \tag{3.2}$$

In isotropic materials the conductivity σ is a scalar quantity. If we allow for anisotropy, σ becomes a matrix and Eq. (3.2) assumes the form (in two dimensions x, y)

$$\begin{pmatrix} j_x \\ j_y \end{pmatrix} = \begin{pmatrix} \sigma_{xx} & \sigma_{xy} \\ \sigma_{yx} & \sigma_{yy} \end{pmatrix} \begin{pmatrix} F_x \\ F_y \end{pmatrix} \tag{3.3}$$

Denoting the conductivity matrix by $\hat{\sigma}$ and the current density and electric field again by their vectors **j** and **F**, we have

$$\mathbf{j} = \hat{\sigma} \mathbf{F} \tag{3.4}$$

We now apply to Eq. (3.4) one of the symmetry operations Q of a regular square (our system is two dimensional and the (100) surface has the symmetry of a square). The specific symmetry operation we choose is a rotation by 90°. In the notation of Table 3.1 and in three dimensions, such a rotation would be $Q_0 Q_{14}$.

Since we have given the conductivity in matrix form, we also would like to express the rotation in matrix form. From calculus, we know that a rotation by an angle ϕ can be represented by

$$Q_\phi = \begin{pmatrix} \cos\phi & \sin\phi \\ -\sin\phi & \cos\phi \end{pmatrix}$$

which gives for $\phi = 90°$

$$Q_{90} = \begin{pmatrix} 0 & 1 \\ -1 & 0 \end{pmatrix}$$

Applying the operation Q_{90} to Eq. (3.4) from the left, we obtain

$$Q_{90}\mathbf{j} = Q_{90}\hat{\sigma}Q_{90}^{-1}Q_{90}\mathbf{F} \tag{3.5}$$

Here we have inserted before the field vector the operation $Q_{90}^{-1}Q_{90}$. Q_{90}^{-1} is the inverse operation of Q_{90} and therefore $Q_{90}^{-1}Q_{90}$ is the identity matrix

$$Q_{90}^{-1}Q_{90}\mathbf{F} = \mathbf{F}$$

From a physical point of view, it is now important to note that for $\phi = 90°$, we have done nothing but turned the (100) surface into itself. Consequently, the current density has to be related to the field in the same way before and after the rotation. Therefore, the conductivity matrix must be the same and

$$Q_{90}\hat{\sigma}Q_{90}^{-1} = \hat{\sigma} \tag{3.6}$$

which reads in matrix form

$$\begin{pmatrix} 0 & 1 \\ -1 & 0 \end{pmatrix}\begin{pmatrix} \sigma_{xx} & \sigma_{xy} \\ \sigma_{yx} & \sigma_{yy} \end{pmatrix}\begin{pmatrix} 0 & -1 \\ 1 & 0 \end{pmatrix} = \begin{pmatrix} \sigma_{xx} & \sigma_{xy} \\ \sigma_{yx} & \sigma_{yy} \end{pmatrix} \tag{3.7}$$

Performing the multiplication on the left-hand side of Eq. (3.7), we have

$$\begin{pmatrix} \sigma_{yy} & -\sigma_{yx} \\ -\sigma_{xy} & \sigma_{xx} \end{pmatrix} = \begin{pmatrix} \sigma_{xx} & \sigma_{xy} \\ \sigma_{yx} & \sigma_{yy} \end{pmatrix}$$

from which it follows that

$$\sigma_{yy} = \sigma_{xx} \tag{3.8}$$

and

$$-\sigma_{yx} = \sigma_{xy} \tag{3.9}$$

Using, in addition to the 90° rotation, a reflection symmetry of the (100) surface, that is, applying $Q_0(\bar{x}_1, x_2)$ (most conveniently in matrix form), we obtain

$$\sigma_{xy} = \sigma_{yx} \tag{3.10}$$

Equation (3.10) together with Eq. (3.9) gives $\sigma_{yx} = \sigma_{xy} = 0$, which together with Eq. (3.8) proves that $\mathbf{j}$ and $\mathbf{F}$ point in the same direction on a (100) surface—that is, the surface is isotropic.

We can do the same proof for a (111) surface that is turned into itself by a rotation of $\phi = 120°$. However the (110) surface has only a $\phi = 180°$ symmetry and this is not enough to prove that this surface exhibits isotropic behavior. Indeed, experiments show that the (110) surface conductivity is anisotropic.

These results are a special case of a more general rule: Any physical property that can be represented as a matrix of rank r_a is scalar in crystallographic systems that can be transformed into themselves by a number of rotations (around all main axes) larger than r_a.

The rank of a matrix is given by the number of indices. Thus the conductivity is of rank two; a matrix with elements a_{ikem} would be of rank 4. In our example the number of rotations transforming the surface into itself was four for the (100) surface (90° rotation), three for the (111) surface (120° rotation), but only two for the (110) surface (180° rotation), which is not larger than the rank of the conductivity matrix.

3.2.2 Translational Invariance

We have not yet discussed in detail another type of symmetry: the symmetry of translations. If we apply a translation $\mathbf{R}_l$ [see Eq. (3.1)] to a crystal, the crystal is transformed into itself (except at the boundaries, which we disregard). We can now argue (as we did for the point group) that any physical property, denoted by $f(\mathbf{r})$, of the crystal has to be the same before and after this translation:

$$f(\mathbf{r} + \mathbf{R}_l) = f(\mathbf{r}) \tag{3.11}$$

where $\mathbf{r}$ is the space coordinate and $\mathbf{R}_l$ is a lattice vector. In other words $f(\mathbf{r})$ is a function that is periodic with respect to all lattice vectors $\mathbf{R}_l$. The periodicity is a hint that Fourier expansion will be a powerful mathematical tool to treat all those functions $f(\mathbf{r})$.

We therefore Fourier expand the function $f(\mathbf{r})$ as

$$f(\mathbf{r}) = \sum_{\mathbf{K}_h} A_{\mathbf{K}_h} e^{i\mathbf{K}_h \cdot \mathbf{r}} \tag{3.12}$$

with

$$A_{\mathbf{K}_h} = \frac{1}{\Omega} \int_{\Omega} f(\mathbf{r}) e^{-i\mathbf{K}_h \cdot \mathbf{r}} \, d\mathbf{r} \tag{3.13}$$

Ω is the basic volume that generates the crystal when repeated over and over by $\mathbf{R}_l$. The subscript h of $\mathbf{K}_h$ is an integer and labels the vectors $\mathbf{K}$.

The choice of the basic volume is not unique. We could, for example, choose the cell that has the three vectors $\mathbf{a}_1$, $\mathbf{a}_2$, $\mathbf{a}_3$ as boundaries. We can also choose the so-called Wigner-Seitz cell, which is obtained as follows. Connect all nearest neighbor atoms by lines ($\mathbf{a}_1$, $-\mathbf{a}_1$, $\mathbf{a}_2$, $-\mathbf{a}_2$, . . .) and cut the connections in half by planes. The geometrical figure enclosed by all these planes is the Wigner-Seitz cell. Note that this cell looks very different for face-centered cubic, body-centered cubic, and simple cubic lattices.

However we choose the cell, the volume is the same and is given by

$$\Omega = \mathbf{a}_1 \cdot (\mathbf{a}_2 \times \mathbf{a}_3) \tag{3.14}$$

Using Eq. (3.11), we have

$$f(\mathbf{r} + \mathbf{R}_l) = \sum_{\mathbf{K}_h} A_{\mathbf{K}_h} e^{i\mathbf{K}_h(\mathbf{r} + \mathbf{R}_l)} = f(\mathbf{r}) \tag{3.15}$$

and therefore

$$e^{i\mathbf{K}_h \cdot \mathbf{R}_l} = 1 \tag{3.16}$$

From Eq. (3.16), it follows that

$$\mathbf{K}_h \cdot \mathbf{R}_l = 2\pi \text{ times integer} \tag{3.17}$$

It can be shown (by inspection) that any vector

$$\mathbf{K}_h = h_1 \mathbf{b}_1 + h_2 \mathbf{b}_2 + h_3 \mathbf{b}_3 \tag{3.18}$$

with

$$\mathbf{b}_1 = \frac{2\pi}{\Omega} \mathbf{a}_2 \times \mathbf{a}_3,$$

$$\mathbf{b}_2 = \frac{2\pi}{\Omega} \mathbf{a}_3 \times \mathbf{a}_1,$$

$$\mathbf{b}_3 = \frac{2\pi}{\Omega} \mathbf{a}_1 \times \mathbf{a}_2 \text{ and integer numbers } h_1, h_2, h_3$$

satisfies Eq. (3.17).

The vectors $\mathbf{K}_h$ are called reciprocal lattice vectors (their dimension is cm^{-1}). These vectors also generate a lattice, the so-called reciprocal crystal lattice, which is complementary to the crystal lattice. Cubic lattices have reciprocal cubic lattices. However, the reciprocal lattice of a face-centered cubic crystal is body-centered cubic, and vice versa.

3.3 BRAGG REFLECTION

We learned in Chap. 2 about the importance of a reciprocal lattice vector, the vector $2\pi/a$ (in one dimension) or any multiple of it. We have seen that the wave vector q of phonons is basically restricted to a zone $-\pi/a \leq q \leq \pi/a$ and assumes the discrete values $2\pi l/Na$, where $-N/2 \leq l \leq N/2$.

For a three-dimensional crystal, the possible values that the wave vector $\mathbf{q}$ can assume are

$$\mathbf{q} = \mathbf{K}_h / N \qquad \text{with } 0 \leq |h_1|, |h_2|, |h_3| \leq N/2 \tag{3.19}$$

Here we have assumed that the crystal contains N repetitions of the basis in each of the main directions $\mathbf{a}_1$, $\mathbf{a}_2$, and $\mathbf{a}_3$.

As already mentioned, the largest physical values of $|\mathbf{q}|$ define an area called the Brillouin zone (see Fig. 3.6). This zone is the Wigner-Seitz cell of the reciprocal lattice [as can be seen from (Eq. 3.19)]. The concept of the Brillouin zone is not only important for phonons but also for electrons. The relevance and significance of the zone concept for electrons can be seen from Bloch's theorem, which we state without proof (it follows from the translation invariance):

The wave function ψ of an electron in a crystal can be described by a wave vector $\mathbf{k}$ (analogous to $\mathbf{q}$ for the phonons) and fulfills the relation

$$\psi(\mathbf{k},\mathbf{r}+\mathbf{R}_l) = e^{i\mathbf{k}\cdot\mathbf{R}_l}\,\psi(\mathbf{k},\mathbf{r}) \tag{3.20a}$$

for any lattice vector $\mathbf{R}_l$. It can be shown that Eq. (3.20a) is equivalent to [see Eq. (2.14)]

$$\psi(\mathbf{k},\mathbf{r}) = e^{i\mathbf{k}\cdot\mathbf{r}}\,u_{\mathbf{k}}(\mathbf{r}) \tag{3.20b}$$

where $u_{\mathbf{k}}(\mathbf{r})$ is a function periodic with respect to $\mathbf{R}_l$. That is

$$u_{\mathbf{k}}(\mathbf{R}_l+\mathbf{r}) = u_{\mathbf{k}}(\mathbf{r}) \tag{3.20c}$$

It is important to note that the wave function is unchanged if we replace $\mathbf{k}$ by $\mathbf{k}+\mathbf{K}_h$, where $\mathbf{K}_h$ is a reciprocal lattice vector. This can be immediately seen

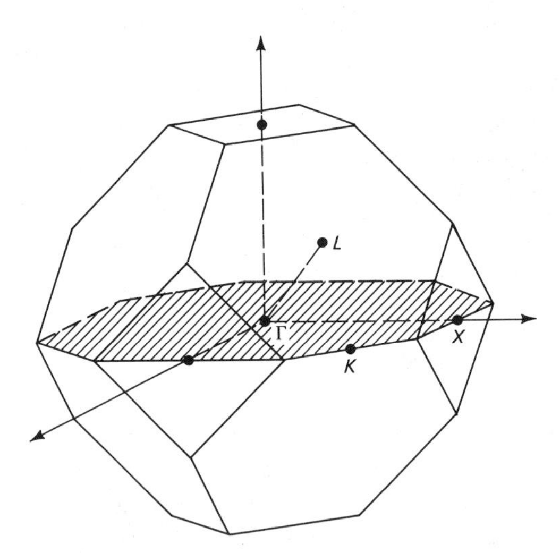

Figure 3.6 Brillouin zone of the face-centered cubic lattice (body-centered cubic reciprocal lattice). Notice that the zone extends to $2\pi/a$ in one direction (X) in contrast to the one-dimensional zone. The labels Γ, X, L, and K denote symmetry points. Γ is the zone center [0,0,0]; X, the endpoint in [1,0,0] direction; and L and K, the zone endpoints in [1,1,1] and [1,1,0] directions, respectively.

from Eqs. (3.16) and (3.20a). Therefore it is clear that the wave vector $\mathbf{k}$, which for a free electron is proportional to the momentum, must have a different meaning in the crystal.

To explore this meaning we will use perturbation theory by regarding the crystal as a perturbation of the free electron behavior. To calculate the matrix elements, we Fourier decompose the periodic crystal potential $V(\mathbf{r})$:

$$V(\mathbf{r}) = \sum_h B_{\mathbf{K}_h} e^{i\mathbf{K}_h \cdot \mathbf{r}} \tag{3.21}$$

The matrix element is then

$$\langle \mathbf{k}' | V(\mathbf{r}) | \mathbf{k} \rangle = \sum_h B_{\mathbf{K}_h} \delta_{\mathbf{k} - \mathbf{k}', -\mathbf{K}_h} \tag{3.22}$$

and vanishes except for

$$\mathbf{k} + \mathbf{K}_h = \mathbf{k}' \tag{3.23}$$

The collision of the "light" electron with the "huge" crystal lattice will be elastic (we ignore lattice vibrations), and therefore the energy before and after the collision will be the same, which is equivalent to

$$|\mathbf{k}| = |\mathbf{k}'| \tag{3.24}$$

Squaring Eq. (3.23) and using Eq. (3.24), we have

$$-2\mathbf{k} \cdot \mathbf{K}_h = |\mathbf{K}_h|^2 \tag{3.25}$$

As is well known, Eq. (3.25) represents the condition for Bragg reflection. It is also clear that Bragg reflection occurs for $\mathbf{k}$ at the Brillouin zone boundary [solve Eq. (3.25) in one dimension]. Bragg reflection simply means that the electrons are reflected by crystal planes, so Eq. (3.23) is valid (see Fig. 3.7).

We return now to the Bloch theorem [Eq. (3.20b)] and put $u_{\mathbf{k}}(\mathbf{r})$ as a constant. One then recovers the wave function of the free electron, and $\hbar\mathbf{k}$ is the momentum of the free electron. For a free electron, momentum is conserved. In a crystal, $\mathbf{k}$ is not conserved because the crystal itself can contribute vectors $\mathbf{K}_h$, as can be seen from Eq. (3.23). We now understand that there is not only one $\mathbf{k}$ attributed to the wave function, but all $\mathbf{k} + \mathbf{K}_h$—or, as stated before, all wave functions with wave vectors $\mathbf{k} + \mathbf{K}_h$—are equivalent. In other

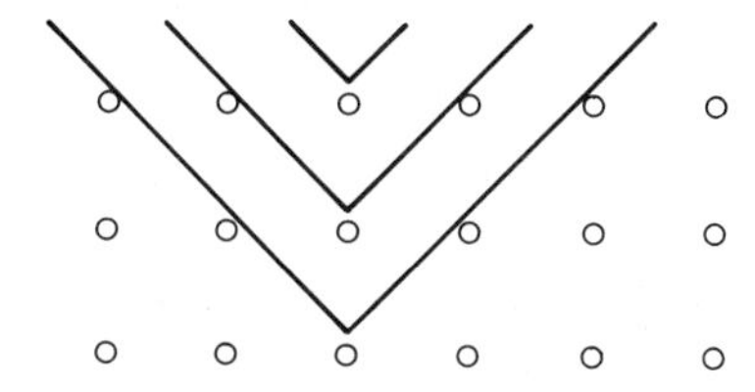

Figure 3.7 Bragg reflection of electrons (waves) by crystal planes.

words, again we can restrict ourselves to use **k** within the Brillouin zone. If **k** lies outside the zone, we deduct $\mathbf{K}_h$ until we obtain a value inside the zone.

What happens to the energy of the electrons? Is there a maximum energy at the Brillouin zone boundary as in the case of lattice vibrations? The answer is that there is not; the electron energy as a function of **k** is multiple valued and only limited within certain bands. A helpful analogy is the following.

We compare energy and wave vector with the true kinetic energy and rotational frequency of a spinning wheel in a movie. As the wheel spins faster and faster, the picture of the wheel seems to stop as soon as the frequency of the moving pictures ν_0 is the same as the spinning frequency of the wheel ν. Increasing ν leads to a picture in which the wheel seems to spin in the opposite direction until $\nu = 2\nu_0$ and then in forward direction again for $2\nu_0 \leq \nu \leq 3\nu_0$ and so on. If we see only the movie, we do not know which kinetic energy or frequency ν the wheel really has except if somebody tells us in which frequency range

$$n\nu_0 \leq \nu \leq (n+1)\nu_0 \tag{3.26}$$

the wheel is spinning.

This situation is analogous in crystals with respect to the energy. For any given **k** vector, the energy can be viewed as multiple valued and we need to specify a range, or *band,* in order to find the energy E from **k**. The theory of $E(\mathbf{k})$, band-structure theory, is treated in Chap. 4.

RECOMMENDED READING

MADELUNG, O., *Introduction to Solid State Theory*. New York: Springer-Verlag, 1978, pp. 36–55.

PROBLEMS

3.1. Show the equivalence of Eqs. (3.20a) and (3.20b)

$$\psi(\mathbf{r} + \mathbf{R}_l) = e^{i\mathbf{k}\cdot\mathbf{R}_l}\,\psi(\mathbf{r})$$

and

$$\psi = e^{i\mathbf{k}\cdot\mathbf{r}}\,u(\mathbf{r})$$

where $u(\mathbf{r})$ has the period of the lattice.

3.2. Find the surface atomic densities of (100), (110), and (111) planes in a silicon structure with a lattice constant of a. Which plane has the highest density?

3.3. **(a)** Find the reciprocal lattice of the three-dimensional face-centered cubic lattice. Use as lattice vectors ($\hat{\mathbf{x}}$, $\hat{\mathbf{y}}$, $\hat{\mathbf{z}}$ are the unit vectors in the respective direction).

$$\mathbf{a}_1 = \frac{a}{2}(\hat{\mathbf{x}} + \hat{\mathbf{y}})$$

$$\mathbf{a}_2 = \frac{a}{2}(\hat{\mathbf{y}} + \hat{\mathbf{z}})$$

$$\mathbf{a}_3 = \frac{a}{2}(\hat{\mathbf{z}} + \hat{\mathbf{x}})$$

(b) Sketch the first Brillouin zone.
(c) Find the volume of the first Brillouin zone.
(d) Repeat parts **a–c** for the body-centered cubic lattice with primitive lattice vectors.

$$\mathbf{a}_1 = \frac{a}{2}(\hat{\mathbf{x}} + \hat{\mathbf{y}} - \hat{\mathbf{z}})$$

$$\mathbf{a}_2 = \frac{a}{2}(-\hat{\mathbf{x}} + \hat{\mathbf{y}} + \hat{\mathbf{z}})$$

$$\mathbf{a}_3 = \frac{a}{2}(\hat{\mathbf{x}} - \hat{\mathbf{y}} + \hat{\mathbf{z}})$$

4

THE THEORY OF ENERGY BANDS IN CRYSTALS

4.1 COUPLING ATOMS

In Chap. 3 we hinted at a band structure for the $E(\mathbf{k})$ relation from rather formal arguments. We now introduce the bands from phenomenological considerations.

Consider a series of quantum wells, as shown in Figs. 4.1a–4.3c. The wells in Fig. 4.1a are separated and essentially independent. Each well has, therefore, a series of discrete levels. In Fig. 4.1b, the wells are closer together and coupled by the possibility of tunneling. This coupling causes a splitting of the energy levels into N closely spaced levels if we have N coupled quantum wells. The effect is much the same as the phenomenon associated with coupled oscillators (the frequency of the oscillators then splits into a series of frequency maxima) or coupled pendulums in mechanics. We can see that the single levels are, therefore, replaced by *bands* of energies.

If the wells are coupled very closely together, a new phenomenon can happen: It is possible that the bands spread more and overlap. Such an overlap is typical for metals. There is, however, an effect that can split up the bands even if we put the wells closer and closer together. This happens, in fact, in semiconductors such as diamond and silicon. The effect is known as bonding-antibonding splitting, which is schematically explained by Figs. 4.2a and 4.2b. The wave functions plotted in the figures give approximately the same proba-

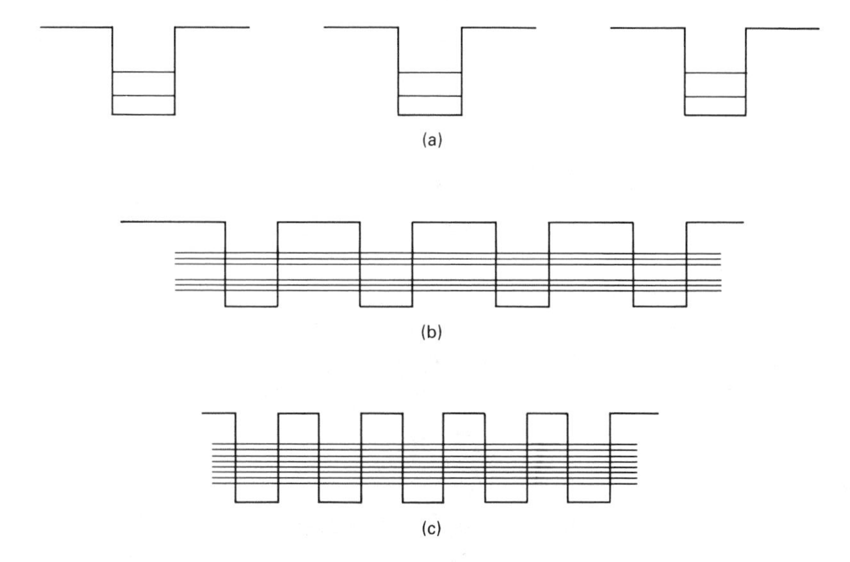

Figure 4.1 Quantum wells coupled together with increasing coupling strength.

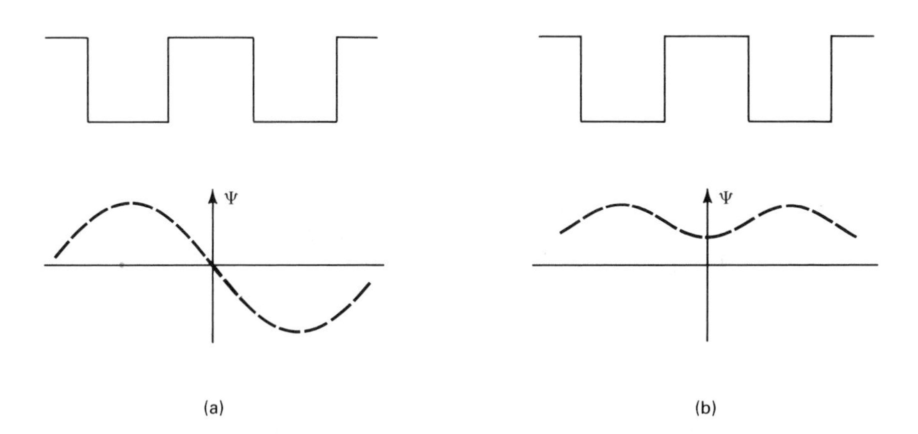

Figure 4.2 Possible forms of wave function for two coupled wells.

bility for finding an electron in either well. However, the probability of finding an electron between the wells vanishes in Fig. 4.2a and is finite in Fig. 4.2b. The situation is known from molecules where the electron holds together the positive nuclei of the ions by being in between them. Therefore, a state resembling that of Fig. 4.2b is called a bonding state, whereas Fig. 4.2a (with no probability of finding an electron in between) represents an antibonding state.

Under certain circumstances, this separation into bonding and antibonding states can lead to an additional splitting of the bands and therefore to the appearance of additional "energy gaps"—regions without states for the electrons. (In the case of diamond and silicon, the splitting follows the formation of so-called sp^3 hybrids, which is discussed in Harrison.)

4.2 THE ALMOST FREE ELECTRON

While this discussion establishes the bands by moving the wells (atoms) closer to each other, we can also go about this another way. We can start with a free electron and introduce the crystal just by using our knowledge about Brillouin zones and restricting the energy function to this zone. We then have to plot the parabola of the free electron $E(\mathbf{k})$ relation, $E = \hbar^2\mathbf{k}^2/2m$, as shown in Fig. 4.3.

At first glance, it seems that we have done nothing more than replot a parabola in a very complicated way. However, we will see in the following that the $E(\mathbf{k})$ relation in a crystal is similar to Fig. 4.3 except that at *the intersections* and *zone boundaries* the function splits, as indicated by the dashed lines, which leads to the formation of bands and energy gaps. A rather rigorous

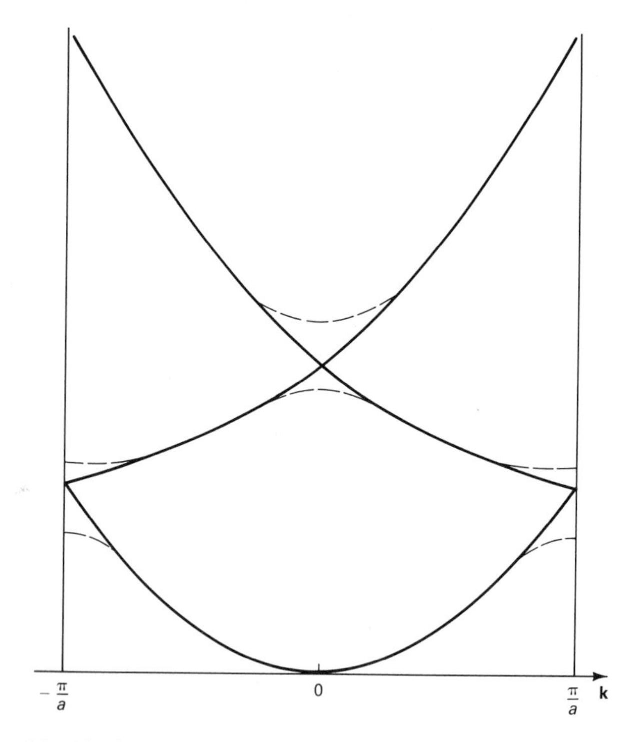

Figure 4.3 The free electron parabola plotted in the Brillouin zone. Notice that the energy now becomes a multiple valued function of **k** and we have to label the different branches (numbers 1, 2, 3 . . .) to distinguish among them.

theory of the $E(\mathbf{k})$ relation that contains most of these features automatically is described in the following discussion.

This theory is based on direct Fourier analysis of the Schrödinger equation. Bloch's theorem tells us that the wave function can be written in the form

$$\psi(\mathbf{k},\mathbf{r}) = e^{i\mathbf{k}\cdot\mathbf{r}}\, u_{\mathbf{k}}(\mathbf{r}) \tag{4.1}$$

where $u_k(\mathbf{r})$ is periodic. Therefore, we can Fourier expand $u_k(\mathbf{r})$ in terms of reciprocal lattice vectors as discussed in Eq. (3.12) to obtain

$$\psi(\mathbf{k},\mathbf{r}) = e^{i\mathbf{k}\cdot\mathbf{r}} \sum_h A_{\mathbf{K}_h} e^{i\mathbf{K}_h\cdot\mathbf{r}} \tag{4.2}$$

Inserting Eq. (4.2) into the Schrödinger equation, Eq. (1.5), with potential $V(\mathbf{r})$, we obtain

$$\frac{\hbar^2}{2m} \sum_h |\mathbf{k}+\mathbf{K}_h|^2 A_{\mathbf{K}_h} e^{i(\mathbf{k}+\mathbf{K}_h)\cdot\mathbf{r}} + V(\mathbf{r}) \sum_h A_{\mathbf{K}_h} e^{i(\mathbf{k}+\mathbf{K}_h)\cdot\mathbf{r}} = E(\mathbf{k}) \sum_h A_{\mathbf{K}_h} e^{i(\mathbf{k}+\mathbf{K}_h)\cdot\mathbf{r}} \tag{4.3}$$

The method of solving differential equations by Fourier transformation proceeds now by multiplying Eq. (4.3) by a term $e^{-i(\mathbf{k}+\mathbf{K}_l)\cdot\mathbf{r}}$ and integrating over the volume V_{ol} of the crystal. Using Eqs. (1.26) and (1.31), we then obtain

$$(E_l^0 - E(\mathbf{k}))A_{\mathbf{K}_l} + \sum_h A_{\mathbf{K}_h} \langle \mathbf{K}_l | V(\mathbf{r}) | \mathbf{K}_h \rangle = 0 \tag{4.4}$$

where we have used the definition

$$\frac{\hbar^2}{2m} |\mathbf{k}+\mathbf{K}_l|^2 \equiv E_l^0 \tag{4.5}$$

Since Eq. (4.4) is homogeneous, a nonzero solution for the $A_{\mathbf{K}_h}$ exists only if the determinant of the coefficients vanishes; that is

$$\det \left| \sum_h \left((E_l^0 - E(\mathbf{k}))\delta_{l,h} + \langle \mathbf{K}_l | V(\mathbf{r}) | \mathbf{K}_h \rangle \right) \right| = 0 \tag{4.6}$$

where l and h are integers ($l, h = 0, \pm 1, \pm 2, \ldots$). Equation (4.6) is called the secular equation and gives us the possible values of the energy $E(\mathbf{k})$ as a function of wave vector $\mathbf{k}$.

Equation (4.6) can only be solved numerically since usually a large number of Fourier components ($\mathbf{K}_h$) is necessary to give a reasonable description of energy bands in semiconductors. However, we can see some significant features of the solution if we restrict ourselves to a one-dimensional model and to two reciprocal lattice vectors $0 (h = 0)$ and $-2\pi/a$ ($h = -1$), which corresponds to one incident and one reflected wave. Then the only matrix element that matters is $\langle 0|V(\mathbf{r})|-1\rangle$, which we denote by M. The secular equation then reads (putting $\langle 0|V(\mathbf{r})|0\rangle = \langle -1|V(\mathbf{r})|-1\rangle = 0$ by proper choice of the energy scale):

$$\begin{vmatrix} E_0^0 - E(\mathbf{k}) & M \\ M^* & E_{-1}^0 - E(\mathbf{k}) \end{vmatrix} = 0 \tag{4.7}$$

which gives

$$E(\mathbf{k}) = \frac{1}{2}(E_0^0 + E_{-1}^0) \pm \frac{1}{2}\sqrt{(E_0^0 - E_{-1}^0)^2 + 4|M|^2} \tag{4.8}$$

with M^* being the complex conjugate of M. Plotting this $E(\mathbf{k})$ relation gives exactly the first two bands (0 and 1) of Fig. 4.3 with the splitting at the Brillouin zone boundary of magnitude $2|M|$. This splitting is called the energy gap E_G.

The calculation, as we have presented it, looks rather simple, and the question arises why the band structures of a broad class of semiconductors have been accurately calculated only after 1965. One of the reasons for the problems in band-structure calculations is the fact that it is difficult to calculate the potential $V(\mathbf{r})$. It is clear that $V(\mathbf{r})$, the potential an electron "feels," is generated not only by the atomic nuclei but also by all the other electrons in the crystal. Therefore, it is not even true that the potential can be written as a locally defined function $V(\mathbf{r})$. However, experience has shown that this is not too bad an approximation for most semiconductors. But how do we determine $V(\mathbf{r})$?

From Eq. (4.6) we can see that it is not even necesary to know $V(\mathbf{r})$. All one needs are the matrix elements—that is, its Fourier components with respect to reciprocal lattice vectors. Cohen and Bergstresser have demonstrated that only a few Fourier components are necessary to obtain relatively accurate $E(\mathbf{k})$ relations if they are chosen wisely. Below we discuss how this choice should be made. We start from the fact that the potential $V(\mathbf{r})$ must be a sum of all the contributions of the atoms that constitute the crystal.

If the atoms are located at $\mathbf{R}_l'$ we can write by introducing a suitable function ω

$$V(\mathbf{r}) = \sum_l \omega(\mathbf{r} - \mathbf{R}_l') \tag{4.9}$$

$\mathbf{R}_l'$ is not necessarily a lattice vector. In the case of silicon, we have two atoms in the Wigner-Seitz cell and $\mathbf{R}_l'$ has to reach both. The matrix element can be written as a Fourier coefficient obtained from

$$M_{mn} \equiv \langle \mathbf{K}_n|V(\mathbf{r})|\mathbf{K}_m\rangle = \frac{1}{V_{\text{ol}}}\int_{V_{\text{ol}}} \sum_l \omega(\mathbf{r} - \mathbf{R}_l')\, e^{-i\mathbf{K}\cdot\mathbf{r}}\, d\mathbf{r} \tag{4.10}$$

where we have properly normalized the wave function to the crystal volume, the integration extends over this volume, and $\mathbf{K} = \mathbf{K}_m - \mathbf{K}_n$ (m and n are integer indices as l and h above).

We now rewrite Eq. (4.10) as

$$M_{mn} = \frac{1}{V_{ol}} \sum_l e^{-i\mathbf{K}\cdot\mathbf{R}'_l} \int_{V_{ol}} e^{-i\mathbf{K}\cdot(\mathbf{r}-\mathbf{R}'_l)} \omega(\mathbf{r}-\mathbf{R}'_l)\, d(\mathbf{r}-\mathbf{R}'_l) \tag{4.11}$$

Notice that we can replace $\mathbf{r} - \mathbf{R}'_l$ by $\mathbf{r}'$ in the integral, which then does not depend on the index l. Remember also that we have two atoms (ions) in the Wigner-Seitz cell. To account for this, we label the position of the second atom by $\mathbf{R}_l + \boldsymbol{\tau}$ and position the first atom at $\mathbf{R}_l$. $\mathbf{R}_l$ is now a lattice vector. (It is also common to position $\mathbf{R}_l$ in between the two atoms and to add and subtract a vector $\boldsymbol{\tau}/2$ to reach the two atomic positions.) Therefore, using Eq. (3.17), we can rewrite Eq. (4.11) as

$$M_{mn} = \sum_l (1 + e^{-i\mathbf{K}\cdot\boldsymbol{\tau}}) \frac{1}{V_{ol}} \int_{V_{ol}} e^{-i\mathbf{K}\cdot\mathbf{r}'} \omega(\mathbf{r}')\, d\mathbf{r}' \tag{4.12}$$

where the label l now runs over all Wigner-Seitz cells (not atoms). The function $\omega(\mathbf{r}')$ is the same in each Wigner-Seitz cell.

We assume also that $\omega(\mathbf{r}')$ vanishes rapidly if we go outside the cell and therefore

$$\int_{V_{ol}} e^{-i\mathbf{K}\cdot\mathbf{r}'} \omega(\mathbf{r}') d\mathbf{r}' = \int_{\Omega} e^{-i\mathbf{K}\cdot\mathbf{r}'} \omega(\mathbf{r}') d\mathbf{r}' \tag{4.13}$$

where Ω is the volume of the Wigner-Seitz cell given by Eq. (3.14). Since the volume of the crystal $V_{ol} = N\Omega$, we finally obtain from Eq. (4.12)

$$M_{mn} = (1 + e^{-i\mathbf{K}\cdot\boldsymbol{\tau}}) \frac{1}{\Omega} \int_{\Omega} e^{-i\mathbf{K}\cdot\mathbf{r}'} \omega(\mathbf{r}') d\mathbf{r}' \tag{4.14}$$

The first term in Eq. (4.14) is called the structure factor $S(\mathbf{K})$

$$S(\mathbf{K}) = (1 + e^{-i\mathbf{K}\cdot\boldsymbol{\tau}}) \tag{4.15}$$

while the second term is the form factor. The structure factor can easily be calculated from the reciprocal lattice vectors $\mathbf{K}_m - \mathbf{K}_n = \mathbf{K}$.

It remains to determine the form factors. Cohen and Bergstresser assumed that only reciprocal lattice vectors $\mathbf{K}$ of squared magnitude 0, 3, 4, 8, and 11 (in units of $2\pi/a$) contribute and all the other Fourier components vanish. Therefore, the form factor is replaced by five unknowns. These five unknowns are adjusted to obtain an $E(\mathbf{k})$ relation that fits best the existing experiments (optical absorption, etc.). Band-structure calculations are then reduced to the solution of the system of Eqs. (4.4) and (4.6), which can be achieved with standard numerical routines.

In this way, Cohen and Bergstresser found the $E(\mathbf{k})$ relation of many semiconductors. Their results for Si, GaAs, and other materials are shown in Fig. 4.4. The $\mathbf{k}$ vectors are plotted along some of the major directions in the Brillouin zone (see Fig. 3.6)—from Γ to X, Γ to L, and so on. Not all bands are shown in the figure but only the most important ones that contribute to

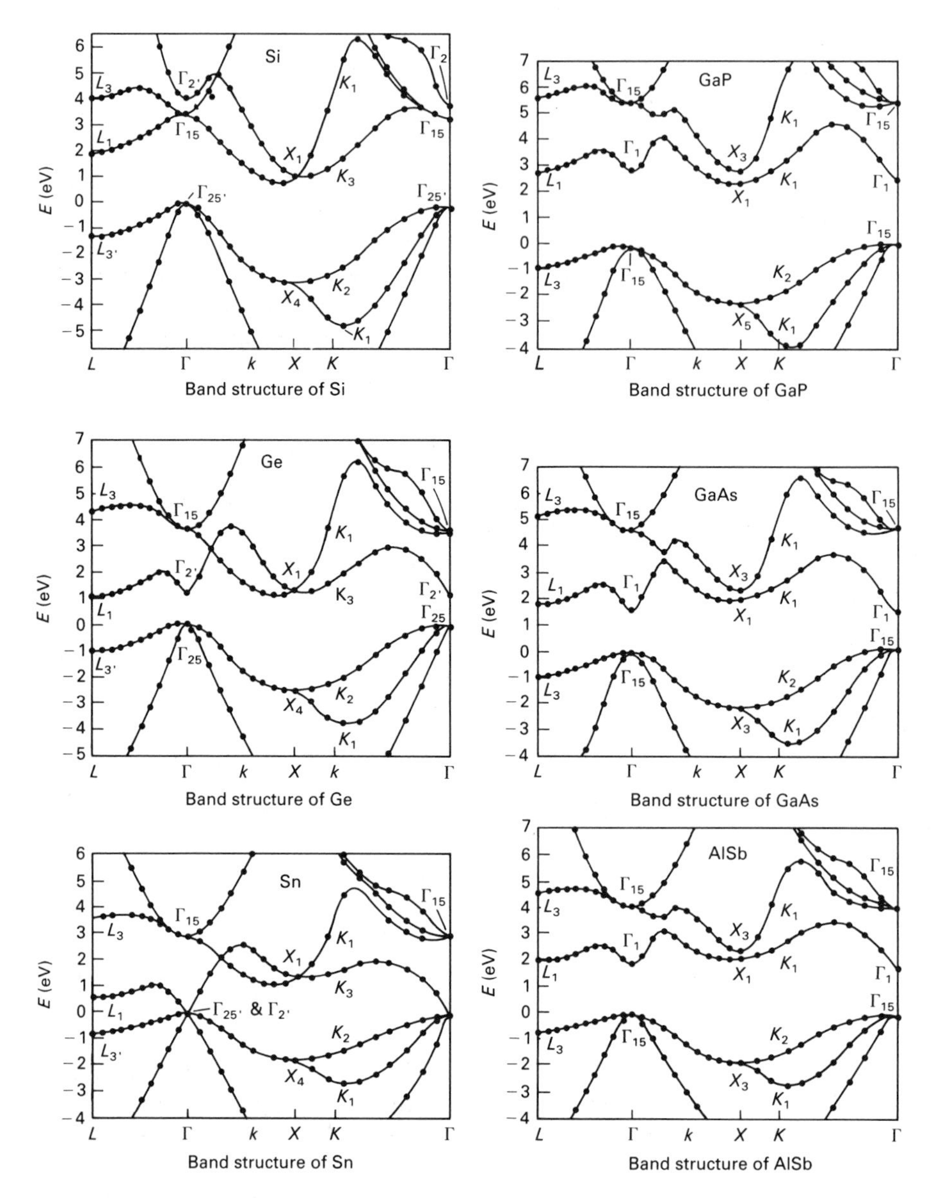

Figure 4.4 Band structure of important semiconductors. (After Cohen and Bergstresser, Fig. 1-6.)

electronic conduction. These are the highest bands that are still filled with electrons, the valance band, and the next higher band separated by an energy gap E_G, the conduction band.

The indices of Γ, *L*, and *X* in Fig. 4.4 denote the symmetry of the wave functions and their behavior with respect to transformations of the coordinates *x*, *y*, *z*. Γ_1 means, for example, that the wave function at this point has *s* symmetry (i.e., is spherically symmetric); Γ_{15} means that the wave function has p symmetry (i.e., is cylindrically symmetric around the x, y, z axes); and so on. The method described above is called the empirical pseudo-potential method.

To obtain a more complete view of the band structure, it is customary to plot lines or surfaces of constant energy in **k** space. Such a plot is shown in Fig. 4.5. The label on the curves is the energy of the particular curve in electron volts. It is important to note that for free electrons the lines of equal energy would be circles (disregarding relativistic effects). The Bragg reflection from crystal planes gives rise to the very complicated pattern of Fig. 4.5. Only at very low energies in certain bands do the lines approach circles (for silicon, even this is not true).

In actual calculations, it is costly to compute the $E(\mathbf{k})$ relation for all points of the Brillouin zone. However, we know from Chap. 3 that 48 operations transform the cube into itself and the same 48 operations transform the Brillouin zone of a cubic lattice into itself. We, therefore, need to calculate the $E(\mathbf{k})$ relation only in 1/48 of the Brillouin zone and then apply the symmetry operations of Table 3.1 (replacing $x_1 x_2 x_3$ by $\mathbf{k}_x\ \mathbf{k}_y\ \mathbf{k}_z$) to obtain $E(\mathbf{k})$ everywhere, since

$$QE(\mathbf{k}) = E(\mathbf{k}') = E(\mathbf{k}) \tag{4.16a}$$

The 1/48 part of the zone that can be turned into the full Brillouin zone by the operations Q is given by the conditions

$$0 \leq \mathbf{k}_z^* \leq \mathbf{k}_x^* \leq \mathbf{k}_y^* \leq 1 \tag{4.16b}$$

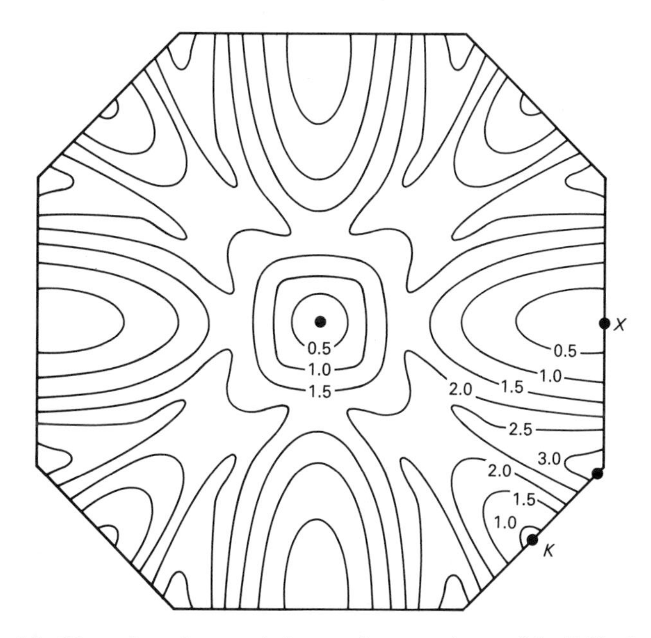

Figure 4.5 Lines of equal energy in **k** space for a certain cut of the Brillouin zone and the conduction band of GaAs. (After Shichijo and Hess, Fig. 4.)

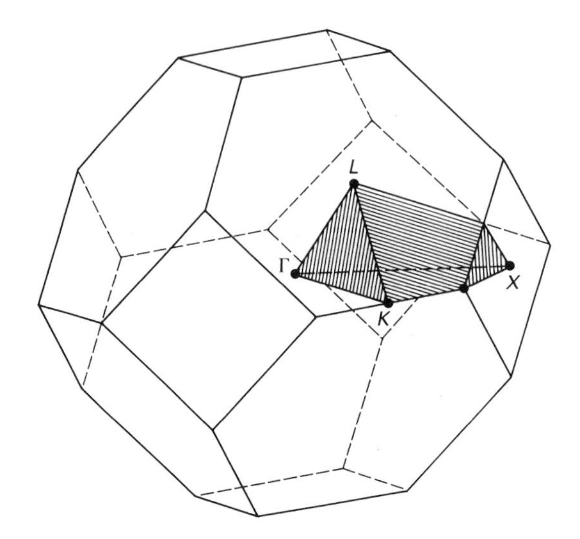

Figure 4.6 Sampling region for the calculation of the band structure. The region is a 1/48 part of the Brillouin zone.

and

$$\mathbf{k}_x^* + \mathbf{k}_y^* + \mathbf{k}_z^* \leq 3/2 \tag{4.16c}$$

Here the starred **k** components are normalized by $2\pi/a$, for example, $\mathbf{k}_z^* = \mathbf{k}_z/(2\pi/a)$. The volume defined by Eqs. (4.17a) and (4.17b) is called the irreducible wedge, and is shown in Fig. 4.6.

It should be noted that the translational invariance leads also to an important law for $E(\mathbf{k})$:

$$E(\mathbf{k} + \mathbf{K}_h) = E(\mathbf{k}) \tag{4.16d}$$

for the very same reasons which lead to Eq. (4.16a).

4.3 THE EQUATIONS OF MOTION IN A CRYSTAL

How does an electron move in such a complicated energy band? The equations of motion are simpler than one would expect. We sketch only the derivation since the full derivation takes considerable space. It is shown in Appendix B that if we restrict ourselves to the consideration of one particular band (one band approximation), the Schrödinger equation can be written as

$$(E(-i\nabla) + eV_{\text{ext}})\psi = E\psi \tag{4.17a}$$

if a weak external potential V_{ext} is applied to the crystal.

Here $E(-i\nabla)$ simply means that we take the function $E(\mathbf{k})$ and Taylor expand it. Then we replace all $\mathbf{k}$ by $-i\nabla$. One always defines functions of operators in this way. To give a one-dimensional example, we have

$$E\left(-i\frac{\partial}{\partial x}\right) = E(0) - \left.\frac{\partial E}{\partial k}\right|_{k=0} i\frac{\partial}{\partial x} - \frac{1}{2}\left.\frac{\partial^2 E}{\partial k^2}\right|_{k=0}\frac{\partial^2}{\partial x^2}\cdots \tag{4.17b}$$

This equation, viewed as a differential equation, is, of course, more difficult to solve than the original Schrödinger equation. The given form, however, is useful for the general considerations that are discussed below.

To proceed, we need to know one more major theorem of mechanics. This is the correspondence principle, which tells us that the center motion of a wave (actually a wave packet) can be obtained from the classical Hamiltonian equations that are obtained by replacing $-i\hbar\nabla$ in the Hamiltonian operator by the momentum $\mathbf{p} = \hbar\mathbf{k}$. In other words, the Hamiltonian function becomes (notice that the crystal potential has disappeared from our equations in the one band approximation)

$$H(\mathbf{p}, \mathbf{r}) = E(\mathbf{p}/\hbar) + eV_{\text{ext}} \tag{4.17c}$$

and from Eq. (1.2) we obtain for the center motion of the wave packet

$$\mathbf{v} = \dot{\mathbf{r}} = \frac{1}{\hbar}\nabla_{\mathbf{k}}E(\mathbf{k}) \tag{4.18}$$

where

$$\nabla_{\mathbf{k}} = \left(\frac{\partial}{\partial k_x}, \frac{\partial}{\partial k_y}, \frac{\partial}{\partial k_z}\right)$$

The momentum changes only in time if we apply an external electric (or magnetic field), and from Eq. (1.1) and $\mathbf{F} = -\nabla V_{\text{ext}}$ we obtain

$$\hbar\dot{\mathbf{k}} = -e\mathbf{F} \tag{4.19}$$

where $\mathbf{F}$ is the electric field.

Equation (4.19) is identical to the equation for a free electron and is true only because we restrict ourselves to one band. Equation (4.18) tells us that the velocity $\mathbf{v}$ of the electron points always perpendicular to the curves of constant energy in $\mathbf{k}$ space [since it is proportional to the gradient of $E(\mathbf{k})$]. Therefore, the motion of an electron in a crystal can be much more complicated than the motion of a free electron.

The question arises, then: Why can we describe the conductivity of a metal and semiconductor usually in simple terms?

After understanding the complicated relation between energy and momentum as a consequence of Bragg reflection in the crystal, it is extremely surprising that simple models that regard the electrons as free (and assume a quadratic relation between energy and momentum) have worked so well in the past. In fact, elementary introductions to semiconductor physics and electronics can be given without much knowledge of the band structure at all. The reason is that electrons reside, for many practical cases, close to the minima (maxima) of the $E(\mathbf{k})$ relation.

In this case we can expand the $E(\mathbf{k})$ relation in a Taylor series (in one dimension):

$$E(k') = E(0) + \left.\frac{\partial E}{\partial k'}\right|_{k'=k_0}(k'-k_0) + \frac{1}{2}\left.\frac{\partial^2 E}{\partial k'^2}\right|_{k'=k_0}(k'-k_0)^2 + \cdots \quad (4.20)$$

where k_0 designates the location of the minimum (maximum) of the $E(k)$ relation. We now can choose the energy scale so that $E(0) = 0$. Furthermore, since the first derivative vanishes at an extremum, we may write by renaming $(k' - k_0)$ by k and the number

$$\left.\frac{\partial^2 E}{\partial k'^2}\right|_{k=k_0} \quad \text{by} \quad \frac{\hbar^2}{m^*}:$$

$$E(k) = \hbar^2 k^2/2m^*$$

which is the equation for the free electron with the mass replaced by an "effective mass" m^*.

If the surfaces of equal energy in $\mathbf{k}$ spaces are ellipsoidal (as they are in silicon and germanium), one obtains by the same reasoning, using the coordinate system of the main ellipsoidal axes

$$E(\mathbf{k}) = \sum_j \frac{\hbar^2}{2m_j^*} k_j^2 \quad (4.21)$$

In silicon, the minima of the conduction band do have an ellipsoidal shape with the two masses $m_l^* = 0.91\ m_0$ and $m_t^* = 0.19\ m_0$ (l stands for the longitudinal and t for the transverse axis of the ellipsoid of revolution). This seems puzzling because we know from group theory that the conductivity of silicon is isotropic. The solution of the puzzle is that silicon has six equivalent ellipsoids (minima) in the [100], [−100], [0 1 0][0 −1 0] and [0 0 1][0 0 −1] directions, and the average over all ellipsoids gives an isotropic conductivity mass m_σ of

$$\frac{1}{m_\sigma^*} = \frac{1}{3}\left(\frac{1}{m_l^*} + \frac{2}{m_t^*}\right) \quad (4.22)$$

This effective mass treatment close to a minimum can be put in a very general form, because not only functions of numbers but also functions of operators can be written as a Taylor series [Eq. (4.17b)].

This general form is the *effective mass theorem*. The effective mass theorem is derived by expanding the one band approximation formula, Eq. (4.17a), into the Taylor series of Eq. (4.17b) and truncating after the third term (the term proportional to $\partial^2/\partial x^2$). Weak external potentials V_{ext} can be simply added to the differential operator on the left-hand side of Eq. (4.17b). Of course, they must not be strong enough to perturb the band structure itself, which is the case if all their Fourier components (with respect to the reciprocal lattice vectors) are smaller than the Fourier components of the crystal potential energy $V(\mathbf{r})$.

The effective mass theorem can then be stated in the following form.

Assume we have a periodic crystal potential energy $V(\mathbf{r})$ and additional (*weaker in all Fourier components*) external potentials V_{ext}. If one is interested only in the properties near an extremum $E(\mathbf{k}_0)$ of the $E(\mathbf{k})$ relation, then the Schrödinger equation

$$\left(-\frac{\hbar^2}{2m}\nabla^2 + V(\mathbf{r}) + eV_{ext}\right)\psi = E\psi \tag{4.23}$$

can be replaced by

$$\left(-\sum_j \frac{\hbar^2}{2m_j^*}\frac{\partial^2}{\partial x_j^2} + eV_{ext}\right)\psi = (E - E(\mathbf{k}_0))\psi \tag{4.24}$$

That is, the periodic crystal potential has disappeared and the mass has been replaced by an effective one. Notice, also, that the energy is counted from $E(\mathbf{k}_0)$ (the first term of the Fourier expansion).

Examples of the usefulness of the effective mass theorem are numerous, three of which are given below.

1. Assume V_{ext} is the potential of a single charged impurity in a crystal:

$$V_{ext} \approx -\frac{e}{4\pi\epsilon_0\epsilon r}$$

where ϵ is the dielectric constant, ϵ_0 is the dielectric constant of vacuum, and e is the magnitude of the electronic charge. Then, for the isotropic case, we have a Schrödinger equation

$$\left(-\frac{\hbar^2}{2m^*}\nabla^2 + \frac{e^2}{4\pi\epsilon_0\epsilon r}\right)\psi = (E - E_c)\psi \tag{4.25}$$

where $E(\mathbf{k}_0) = E_c$ is the energy of the band edge of the conduction band. This equation is identical to the equation of the hydrogen atom except for the appearance of the dielectric constant $\epsilon_0\epsilon$ and the effective mass m^*. It follows that such an impurity behaves like a hydrogen atom with the energy measured from E_c.

2. Let V_{ext} be a square well; then close to the band edge the electron will have energy levels in this square well exactly as in free space but with an effective mass instead of the real mass.

3. Generally, any solved potential problem is solved for crystal electrons close to the band minimum as long as the Fourier components of V_{ext} are smaller than the components of $V(\mathbf{r})$. Assume, for example, that we have a regular lattice of impurities embedded in the host lattice of the semiconductor,

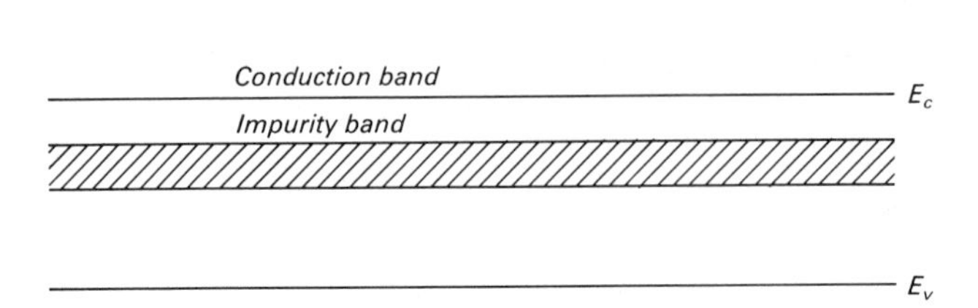

Figure 4.7 Impurity band below conduction band edge in a semiconductor.

that is, V_{ext} is periodic, denoted by $V_{ext}^{per}(\mathbf{r})$. Then the Schrödinger equation for isotropic effective mass reads

$$\left(-\frac{\hbar^2}{2m^*}\nabla^2 + V_{ext}^{per}(\mathbf{r})\right)\psi = (E - E_c)\psi \tag{4.26}$$

which is the equation for a *band structure*.

Indeed, for large numbers of impurities, one observes an "impurity band" below the conduction band of a semiconductor, as shown in Fig. 4.7.

Let's check if we really can use the effective mass theorem for an impurity band. One would assume that the Fourier components of the impurity potentials (N_I impurities per cubic centimeter) are small as long as the spacing d between the impurities is much larger than the crystal lattice constant; that is, we have to postulate

$$d = N_I^{-1/3} \gg a \tag{4.27}$$

For typical lattice constants $a \approx 5 \times 10^{-8}$ cm, this means that the impurity concentration is $N_I \lesssim 10^{19}$ cm^{-3}.

For higher densities there is another powerful theorem that gives us again a "method" to "transform away" complications introduced by the crystal structure. This is the virtual crystal approximation, which is discussed later.

4.4 MAXIMA OF ENERGY BANDS—HOLES

Up to now we have mainly looked at the minima of energy bands. Let us look at maxima in the valence band. The new feature is that the effective mass is now *negative*. This brings us to the concept of *holes in semiconductors*.

Assume a valence band as found in the zincblende (GaAs) and diamond type (Si) and shown in Fig. 4.8. First note that there are two $E(\mathbf{k})$ relations present in this band. This means we can have two types of electrons close to the top of the valence band (there is a deeper split-off band caused by spin-orbit interaction, which is not discussed here). We can treat these two types of electrons separately (in Chap. 6 we discuss how to determine their respective numbers). For reasons that will be clarified below, the upper curve is called heavy hole curve, while the lower curve is the $E(\mathbf{k})$ relation of the light holes.

We label below all relevant quantities (such as the wave vector) by the subscript el for electrons since we are now introducing holes as additional "quasiparticles."

Each parabola of Fig. 4.8 can be described by

$$E_{\text{el}} = -\frac{\hbar^2 \mathbf{k}_{\text{el}}^2}{2|m^*|} \tag{4.28}$$

Remember that the effective mass at the maximum is negative (second derivative). The electric current $\mathbf{I}$ is obtained by summing over all velocities of filled states $\mathbf{v}_{\mathbf{k}}^{\text{el}}$ multiplied by the elementary charge (which is negative for electrons)

$$\mathbf{I} = -e \sum_{\substack{\mathbf{k}_{\text{el}} \\ \text{(full)}}} \mathbf{v}_{\mathbf{k}}^{\text{el}} \tag{4.29}$$

Now we use the following fact.

The total wave vector of the electrons in a filled band is zero: $\Sigma \mathbf{k} = 0$. This result follows from the geometrical symmetry of the Brillouin zone: Every fundamental lattice type has symmetry under the inversion operation about any lattice point, and thus it follows that the Brillouin zone of the lattice also has inversion symmetry. If the band is filled, all pairs of orbitals $\mathbf{k}$ and $-\mathbf{k}$ are filled, and the total wave vector is zero. Therefore

$$\sum_{\substack{\mathbf{k}_{\text{el}} \\ \text{(full)}}} \mathbf{v}_{\mathbf{k}}^{\text{el}} + \sum_{\substack{\mathbf{k}_{\text{el}} \\ \text{(empty)}}} \mathbf{v}_{\mathbf{k}}^{\text{el}} = 0 \tag{4.30}$$

Equations (4.29) and (4.30) give

$$\mathbf{I} = +e \sum_{\substack{\mathbf{k}_{\text{el}} \\ \text{(empty)}}} \mathbf{v}_{\mathbf{k}}^{\text{el}} \tag{4.31}$$

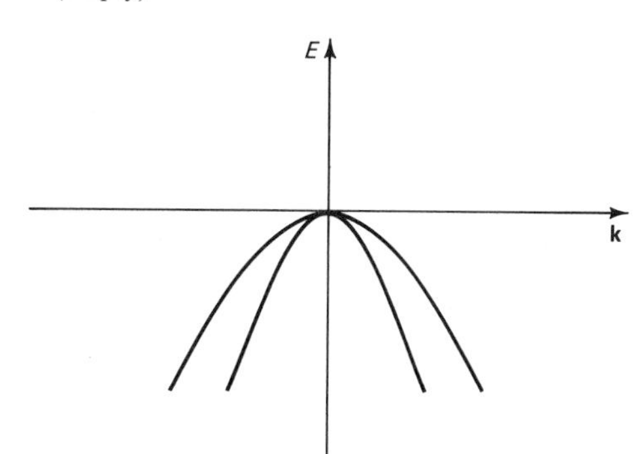

Figure 4.8 Typical form of $E(\mathbf{k})$ relation at the top of the valence band. The deeper split-off band due to the spin-orbit interaction is not shown.

In most practical cases Eq. (4.31) is much more convenient than Eq. (4.29), since one has fewer empty states than full states. The question arises of whether we can describe the current by a fictitious particle with positive charge. Also, we would prefer particles with positive mass (as we are used to) and we would like to attribute the empty states to these quasi- (fictitious) particles because we prefer to sum over the smaller number of states. Does such a quasiparticle exist? Before we can answer that question, we have to ask ourselves how the empty states evolve in time, because our quasiparticles have to do exactly the same. We note the following.

The unoccupied levels in a band evolve in time under the influence of applied fields precisely as they would if they were occupied by real electrons. This is so because, given the values of **k** and **r** at $t = 0$, the semiclassical Eqs. (4.18) and (4.19), being six first-order equations in six variables, uniquely determine **k** and **r** at all subsequent (and all prior) times, just as in ordinary classical mechanics.

The position and momentum of a particle at any instant determine the entire orbit in the presence of specified external fields. In Fig. 4.9 we indicate schematically the orbits determined by the semiclassical equations, as lines in a seven-dimensional **rk**t space. Because any point on an orbit uniquely specifies the entire orbit, two distinct orbits can have no points in common. We can, therefore, separate the orbits into occupied and unoccupied orbits (see Fig. 4.9) according to whether they contain occupied or unoccupied points at $t = 0$. At any time after $t = 0$, the unoccupied levels will lie on unoccupied orbits, and the occupied levels on occupied orbits. Thus the evolution of both oc-

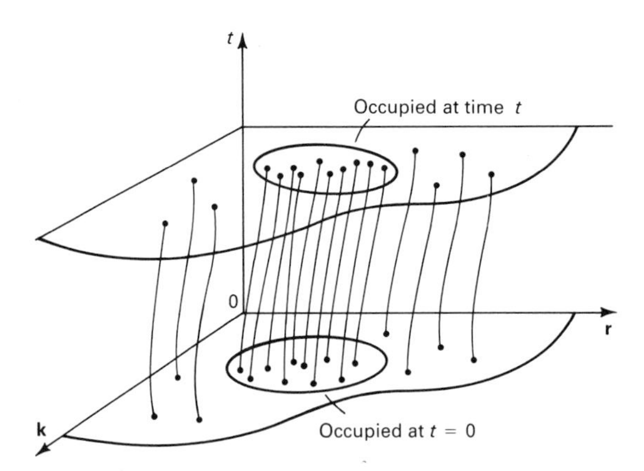

Figure 4.9 Schematic illustration of the time evolution of orbits in semiclassical phase space (here **r** and **k** are each indicated by a single coordinate). The occupied region at time t is determined by the orbits that lie in the occupied region at time $t = 0$. (From *Solid State Physics* by Neil W. Ashcroft and N. David Mermin. Copyright © 1976 by Holt, Rinehart & Winston. Reprinted by permission.)

cupied and unoccupied levels is completely determined by the structure of the orbits.

This structure depends only on the form of the semiclassical equations and not on whether an electron happens actually to be following a particular orbit. This means the empty states evolve in the seven-dimensional **rk***t* space according to Eqs. (4.18) and (4.19)

$$\frac{d\hbar\mathbf{k}_{\mathrm{el}}}{dt} = -e\mathbf{F} \tag{4.32}$$

(**F** is the electric field), as well as Eq. (4.18)

$$\mathbf{v}_{\mathbf{k}}^{\mathrm{el}} = \frac{1}{\hbar}\nabla_{\mathbf{k}_{\mathrm{el}}} E_{\mathrm{el}}(\mathbf{k}_{\mathrm{el}}) \tag{4.33}$$

Now the question is: Can a quasiparticle, which we will call a "hole," have positive charge and positive mass and fulfill Eqs. (4.31), (4.32), and (4.33), and can it be identified with the missing electron(s)? In other words, are the following equations equivalent to Eqs. (4.31), (4.32), and (4.33)?

$$\mathbf{I} = e\sum_{\mathbf{k}_{\mathrm{h}}} \mathbf{v}_{\mathbf{k}}^{\mathrm{h}} \tag{4.34}$$

$$\frac{d\hbar\mathbf{k}_{\mathrm{h}}}{dt} = e\mathbf{F} \tag{4.35}$$

$$\mathbf{v}_{\mathbf{k}}^{\mathrm{h}} = \frac{1}{\hbar}\nabla_{\mathbf{k}_{\mathrm{h}}} E_{\mathrm{h}}(\mathbf{k}_{\mathrm{h}}) \tag{4.36}$$

$$E_{\mathrm{h}} = \frac{\hbar^2\mathbf{k}_{\mathrm{h}}^2}{2|m^*|} \tag{4.37}$$

(h stands for hole).

The answer is they are, if we have

$$\mathbf{k}_{\mathrm{h}} = -\mathbf{k}_{\mathrm{el}} \tag{4.38}$$

The proof is trivial since then

$$\nabla_{\mathbf{k}_{\mathrm{h}}} = -\nabla_{\mathbf{k}_{\mathrm{el}}} \tag{4.39}$$

and therefore

$$\mathbf{v}_{\mathbf{k}}^{\mathrm{h}} = \mathbf{v}_{\mathbf{k}}^{\mathrm{el}} \tag{4.40}$$

and Eq. (4.31) is clearly fulfilled and looks like an equation for positive charge. Because the time variable t was not transformed and because of Eq. (4.40), our holes evolve in the **r***t* space like electrons. In **k** space, however, this is not so; there our fictitious particles evolve on the negative side (Eq. (4.38)). That, however, does not bother us—but we remember it.

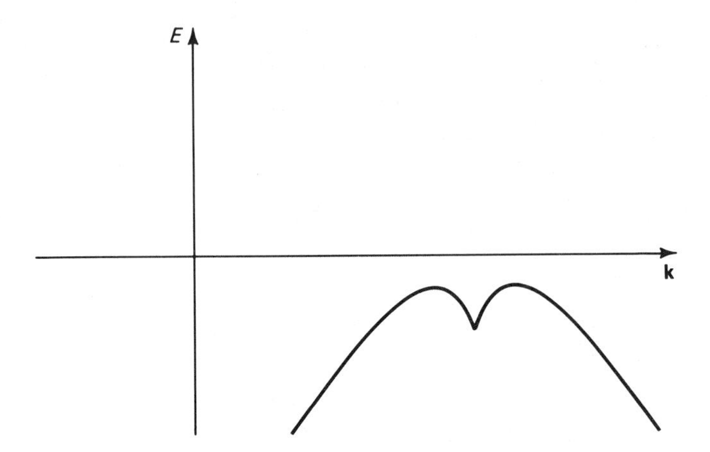

Figure 4.10 Schematic shape of the top of the valence band in tellurium.

The effective mass at the top of the valence bands of semiconductors is not always negative. In tellurium, for example, the $E(\mathbf{k})$ relation has a shape as shown in Fig. 4.10. It is then not entirely appropriate to identify missing electrons by holes. Considering the shape of Fig. 4.10, one has to name these quasiparticles differently. The effective masses can, in principle, be calculated by the pseudopotential method and Eq. (4.21). They also can be measured by cyclotron resonance.

4.5 SUMMARY OF IMPORTANT BAND-STRUCTURE PARAMETERS

It is interesting to note the generally complex anisotropic shape of the $E(\mathbf{k})$ relation in the valence band. The effective mass is very anisotropic, and any experiment will measure complicated averages corresponding to the specifics of the experiment. It is, therefore, difficult to give a single effective mass for either the heavy or the light hole band. Nevertheless, we have compiled some approximate values in Table 4.1 (the density of states masses, which we will discuss in a later chapter). Also shown in Table 4.1 are other important properties of semiconductors, such as the energy gap and the dielectric constant.

Figure 4.11 summarizes graphically the main features of conduction and valence bands in the important semiconductors.

This finishes our discussion of bulk semiconductor band structures. Before we proceed to the discussion of imperfections, however, we discuss the energy bands of alloys and of junctions of different semiconductors called heterojunctions.

TABLE 4.1 Approximate Values of Material Parameters

	IV		III–V					
	Si	Ge	AlAs	AlSb	GaAs	InP	InAs	InSb
E_G (300 K)	1.11	0.65	2.17	1.62	1.439	1.35	0.356	0.180
E_G (4.2 K)	1.21	0.74	2.22	1.70	1.52	1.42	0.409	0.235
a (Å) lattice const.	5.431	5.658	5.661	6.138	5.642	5.868	6.058	6.479
ρ(g/m^3)	2.329	5.323	3.73	4.26	5.36	4.787	5.667	5.775
$m_e^* - m_l^*/m_t^*$	0.19/0.916	0.082/1.6	0.35	0.39	0.067	0.078	0.023	0.013
m_{h1}^*	0.15	0.042	0.15	0.11	0.087		0.025	0.16
m_{h2}^*	0.52	0.34	0.76	0.5	0.462	0.8	0.41	0.43
Phonon Lo/To (meV)	51.0/57.4	28.1/33.3	50.1/44.9	42.1/39.5	36.2/33.3	42.8/37.7	30.2/27.1	24.2/22.6
Index of Refraction	3.5	4.1	3.1	3.4	3.4	3.1	3.5	3.9
$\epsilon(\omega \rightarrow \infty)$			8.16	10.24	10.92	9.52	11.8	15.7
$\epsilon(\omega \rightarrow 0)$	11.7	16.0	10.06	14.4	12.9	12.35	14.55	17.72

m_e^*, m_{h1}^*, and m_{h2}^* denote the effective density of states masses for electrons, heavy holes, and light holes, respectively [see Eq. (6.12)]. The values are given in terms of the free electron mass. $\epsilon(\omega \rightarrow \infty)$ is the dielectric constant for high frequencies ω, and $\epsilon(\omega \rightarrow 0)$ is the static dielectric constant.

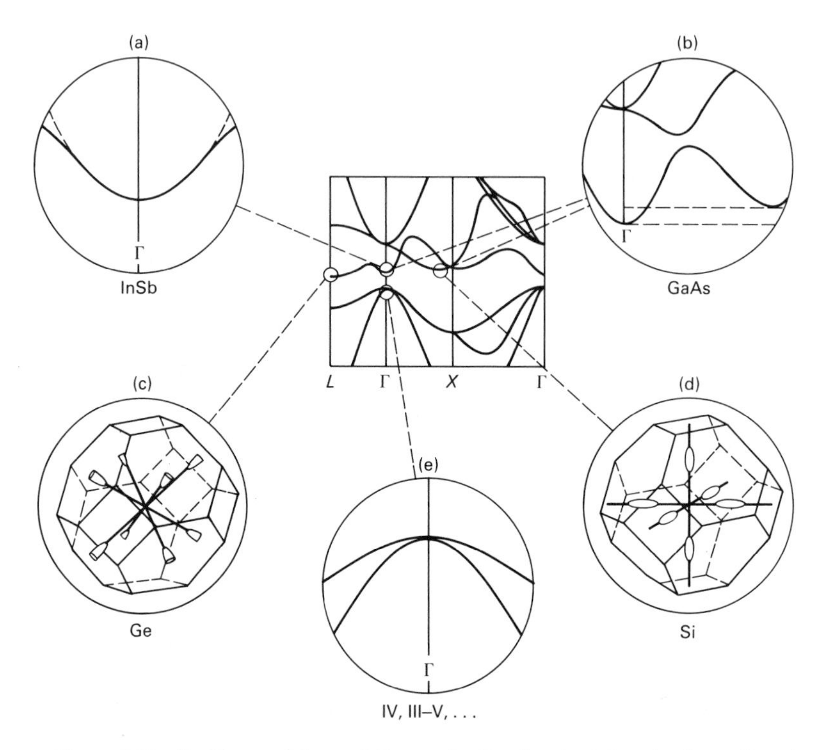

Figure 4.11 Position and form of the conduction band minima (*a, b, c, d*) and valence band maxima (*e*) of important semiconductors. Notice the ellipsoidal form for silicon (X) and germanium (L) minima. The minima at Γ are spherical. (After Madelung, Fig. 17.)

4.6 BAND STRUCTURE OF ALLOYS

In the discussion of alloys, we are mainly interested in ternary alloys, such as $Al_xGa_{1-x}As$, where we have added a certain mole fraction x of Al to GaAs. We still have a perfect crystal, but some Ga atoms are randomly replaced by Al atoms (compositional disorder). A schematic sketch of the potential of such an alloy is shown in Fig. 4.12. How do we calculate the band structure in this case?

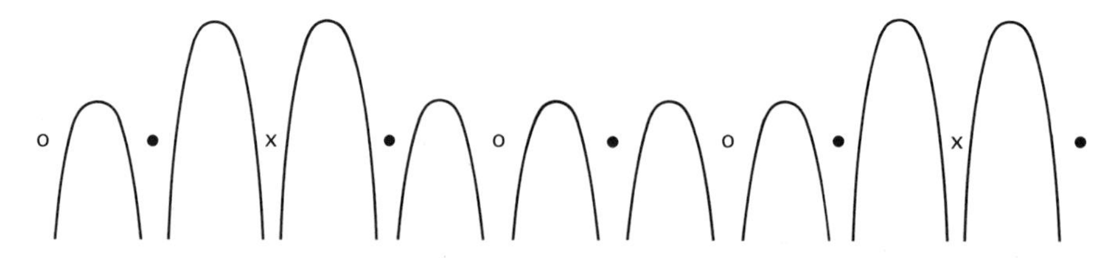

Figure 4.12 Sketch of the atomic potentials of the ternary alloy $Al_xGa_{1-x}As$.

We have seen that we cannot stretch the effective mass theorem too far to account for an "impurity crystal" of similar lattice constant comparable to the host crystal. Such a potential *cannot* be divided into $V(\mathbf{r}) + V_{ext}$ with the Fourier components of $V_{ext} \ll V(\mathbf{r})$. There is still a method, however, that allows a simple treatment of such a system. This method is called the virtual crystal approximation.

The virtual crystal approximation goes back to Nordheim, who suggested that for a crystal such as $Al_xGa_{1-x}As$, we can replace the actual potential by an average potential

$$V_{av} = (1 - x)V_{GaAs} + xV_{AlAs} \tag{4.41}$$

where V_{GaAs} and V_{AlAs} are the crystal potentials of GaAs and AlAs, respectively.

In the empirical pseudopotential method we can easily include Eq. (4.41) to calculate the necessary form factors if they are known for each compound.

In many cases a linear interpolation approximates well complex physical quantities such as the effective mass, the dielectric constant ϵ, and so on. Thus we have

$$\epsilon_{Al_xGa_{1-x}As} \approx (1 - x)\epsilon_{GaAs} + x\epsilon_{AlAs} \tag{4.42}$$

A similar equation can be written for the energy gap between the top of the valence band and the Γ, X, and L minima. The typical dependences for this type of alloy are shown in Fig. 4.13 and do not deviate too much from the linear approximation. As shown in the figure, a "disorder bowing" of the

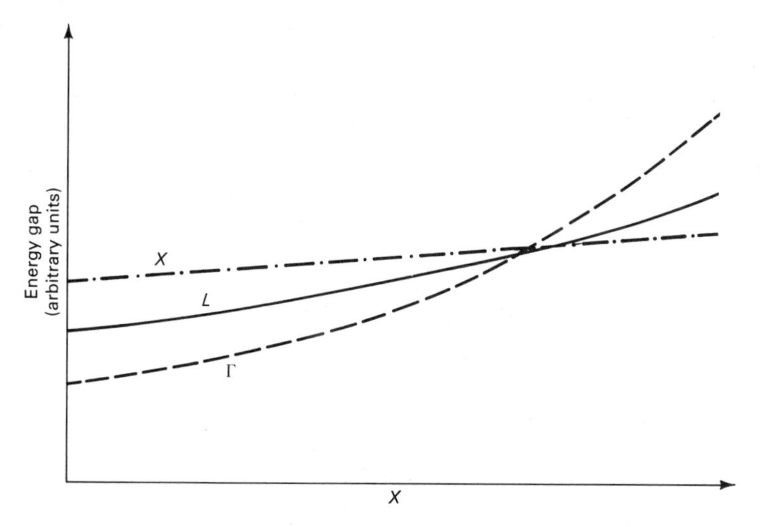

Figure 4.13 Γ, X, L conduction band minima of a typical compound $A_xB_{1-x}C$ as a function of mole fraction x.

energy gap is measured. Methods have been developed to calculate this bowing (Jones and March) but go beyond the scope of this text.

Alloys are finding increasing use in semiconductor electronics. (See Casey and Panish for an overview of many useful material systems.) Of special interest are lattice-matched layers (heterolayers) of semiconductors. Lattice-matched means that the interatomic distance is the same in two neighboring semiconductor layers, which gives the basic *possibility* of generating *ideal* interfaces without defects and distortions. Notice, however, that lattice match is only a necessary condition for perfect interfaces, not a sufficient one.

Figure 4.14 shows schematically the atomic potentials near the junction of two materials. The energies $e\chi_1$ and $e\chi_2$ are necessary to remove an electron from the respective conduction band edge to an infinite point, the *vacuum level.*

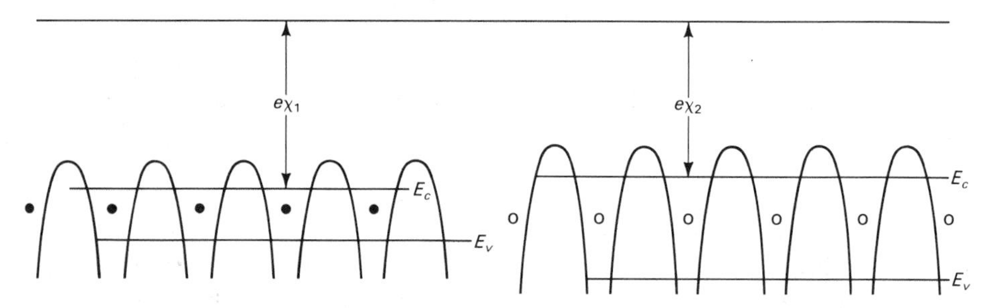

Figure 4.14 Schematic plot of the potential of atoms in two neighboring lattice-matched semiconductors. E_c and E_v denote the conduction and valence band edges, respectively.

Notice that χ_1 and χ_2 are only loosely connected to quantities that are easy to measure, such as the *photothreshold.* The reason is that image force and surface reconstruction (see end of chapter) contribute significantly to the photothreshold. Therefore, the absolute values $e\chi_1$ and $e\chi_2$ relative to the vacuum level are not very useful. The relative values, the variation from material to material $e\chi_1 - e\chi_2$, can be expected to be more meaningful, although a calculation is still difficult. Even if there were no surface and image force effects, the quantities $e\chi_1$ and $e\chi_2$ are large and their difference is small, which will give enormous requirements of accuracy to our band-structure calculation to obtain the correct difference (Harrison, p. 252). The difference between the conduction (valence) band edges in the two materials is called the band edge discontinuity ΔE_c (ΔE_v).

If a ternary alloy is the "atomic" neighbor of a binary compound, we have a natural transition region at the interface (see Fig. 4.15). It is important to realize that in this case the transition between the two materials can go over a single or many atomic layers.

It is also important to realize that a heterojunction is not the same as a

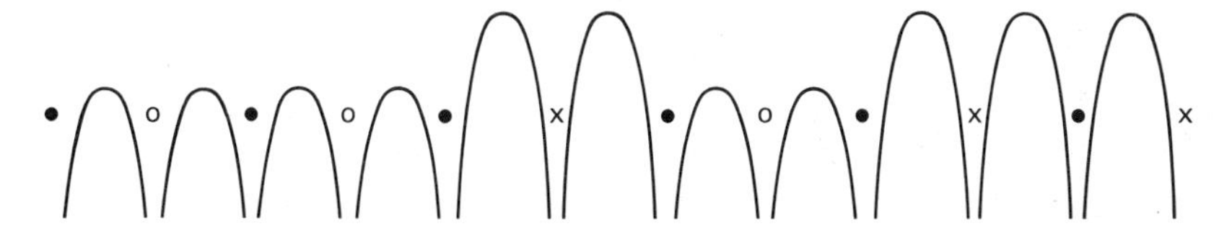

Figure 4.15 Heterointerface between binary and ternary alloys.

potential step, since the effective mass can be different on the two junction sides. If one uses effective mass theory, one has, therefore, a problem in matching wave functions at the interface since only the true (Bloch) wave functions and their derivatives are continuous, and not the effective mass (quasifree) wave functions.

One still can assume, however, that the quasi-free electron wave function (envelope function) in the two different materials away from the junction fulfill a relation of the form:

$$\begin{pmatrix} \psi_{\text{eff mass}}^{\text{GaAs}} \\ \dfrac{\partial \psi_{\text{eff mass}}^{\text{GaAs}}}{\partial z} \end{pmatrix} = \begin{pmatrix} T_{11} & T_{12} \\ T_{21} & T_{22} \end{pmatrix} \begin{pmatrix} \psi_{\text{eff mass}}^{\text{AlGaAs}} \\ \dfrac{\partial \psi_{\text{eff mass}}^{\text{AlGaAs}}}{\partial z} \end{pmatrix} \tag{4.43}$$

T_{ij} is called the transition matrix; for the $Al_xGa_{1-x}As$ and $x \lesssim 0.3$, it is close to the unit matrix, with $T_{22} \approx (m^*_{\text{GaAs}}/m^*_{\text{AlGaAs}})$ and $T_{11} \approx 1$, while $T_{12} = T_{21} \approx 0$. Of course, to obtain the exact transition matrix, the band structure of the heterojunction needs to be calculated, by the pseudopotential method, for example (Frensley and Kroemer).

RECOMMENDED READINGS

ASHCROFT, N. W., and N. D. MERMIN, *Solid State Physics.* Philadelphia: Saunders College, 1976, p. 227.

CASEY, H. C., Jr., and M. B. PANISH, *Heterostructure Lasers.* Part B. *Materials and Operating Characteristics.* New York: Academic Press, 1978, pp. 15–48.

COHEN, M. L., and T. K. BERGSTRESSER, *Physical Review,* 141, 1966, 789.

FRENSLEY, W. R., and H. KROEMER, *Journal of Vacuum Science and Technology,* 13 (810), 1976.

HARRISON, W. A., *Electronic Structure and the Properties of Solids.* San Francisco: Freeman, 1980, pp. 61–80.

JONES, W., and N. MARCH, *Theoretical Solid State Physics,* vol. 2. New York: Wiley/Interscience, 1973, p. 1093.

MADELUNG, O., *Introduction to Solid State Theory.* New York: Springer-Verlag, 1978, pp. 62–64.

———, *Grundlagen der Halbleiterphysik.* New York: Springer-Verlag, 1970, p. 38.

SHICHIJO, H., and K. HESS, *Physical Review,* B 23, 1981, 4197–4207.

PROBLEMS

4.1. Calculate the wave functions corresponding to the two solutions of Eq. (4.8) and show that their form corresponds to the bonding-antibonding case.

4.2. The spacing between successive (100) planes in NaCl is 2.82 Å; suppose that a given X ray is found to give rise to first-order Bragg reflection at a grazing angle of 11°57′. Find the wavelength of the X ray and the angle at which the second-order Bragg reflection would occur.

4.3. From the Schrödinger equation

$$\left(\frac{-\hbar^2}{2m}\nabla^2 + V\right)\psi_{\mathbf{k}} = E_{\mathbf{k}}\,\psi_{\mathbf{k}}$$

(a) Derive

$$\frac{2m}{\hbar^2}\psi_{\mathbf{k}}\nabla_{\mathbf{k}}E_{\mathbf{k}} = -2i\nabla\psi_{\mathbf{k}} - \left[\nabla^2 + \frac{2m}{\hbar^2}(E_{\mathbf{k}} - V)\right]e^{i\mathbf{k}\cdot\mathbf{r}}\nabla_{\mathbf{k}}u_{\mathbf{k}}$$

for the Bloch function $\psi_k = e^{i\mathbf{k}\cdot\mathbf{r}}u_{\mathbf{k}}(\mathbf{r})$ (*Hint:* Differentiate with respect to **k**.)

(b) Using the result of part a, show that

$$\mathbf{v} = \frac{1}{\hbar}\nabla_{\mathbf{k}}E_{\mathbf{k}}$$

where

$$\mathbf{v} = \frac{\hbar}{im}\int\psi_{\mathbf{k}}^*\nabla\psi_{\mathbf{k}}\,d\mathbf{r}$$

4.4. Find the energy gap corresponding to the potential $V_0 \cos\left(\frac{2\pi r}{a}\right)$ for a one-dimensional crystal with lattice constant a.

5

IMPERFECTIONS

With the exception of Chap. 2, we have discussed up to now only ideal crystals. Real crystals are not perfect with respect to translations and so on. Unavoidable imperfections are the phonons. An electron traveling in a crystal is scattered after a typical distance of the order of 100 Å by phonon emission or absorption. This means that the electron does not really "see" the crystal as a whole but instead only a volume of 10^{-18} cm^3, which contains about 3000 atoms. In addition to the phonons, there are a number of unavoidable imperfections. We distinguish the following:

Point defects (impurities, vacancies)
Line defects (dislocations)
Areal defects (surfaces, interfaces)
Total defects (garbage, glass, amorphous solids)

Adequate control of these imperfections is necessary if a material system is to be used to fabricate electron devices ("device grade material"). As will become clear in the following chapters, point defects need to be controlled on a sub–part per million level. Ideally, one wants to control the density of point defects on a level of 10^{14} cm^{-3}, which means having only one imperfection in about 10^8 atoms. High grade semiconductor material should be virtually free of line defects (dislocations), and indeed, silicon and gallium arsenide can be grown, if necessary, dislocation free.

As a rule of thumb, one can say that no high speed semiconductor device should include surfaces in its structure, because of the detrimental effects of surface states. Also, it is difficult to see how amorphous solids could be useful for device fabrication with the exception of devices that have to be cheap, can be slow, and have simple functions, such as solar cells.

Many early investigations concentrated on point defects that can be classified into interstitial and substitutional impurities and vacancies; see in Figs. 5.1 and 5.2.

The most important imperfections in semiconductors (in a positive sense) are the substitutional impurities that can be used as "dopants." The meaning of the word "dopant" becomes clear from the following. Imagine that we replace in a perfect silicon lattice one silicon atom by arsenic, as shown in Fig. 5.2a. Since arsenic has five valence electrons and silicon only has four, one electron is redundant and—if it is "donated" to the conduction band—can move freely in the crystal and contribute to the conductivity of silicon. The conditions that determine that the electron resides in the conduction band are discussed in Chap. 6. Let us now quantitatively evaluate what the arsenic atom introduces in the spectrum of energy levels.

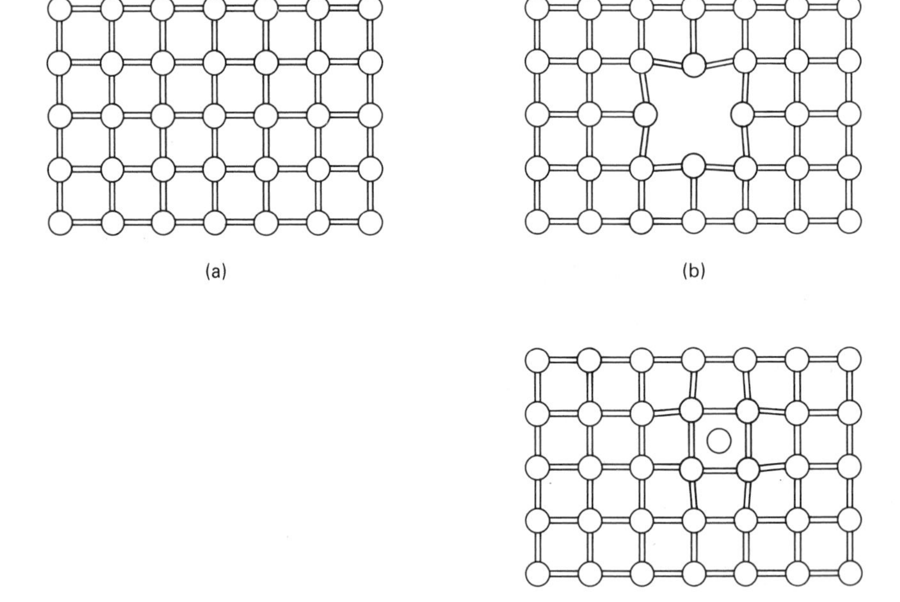

Figure 5.1 (a) Perfect crystal; (b) vacancy; (c) interstitial atom.

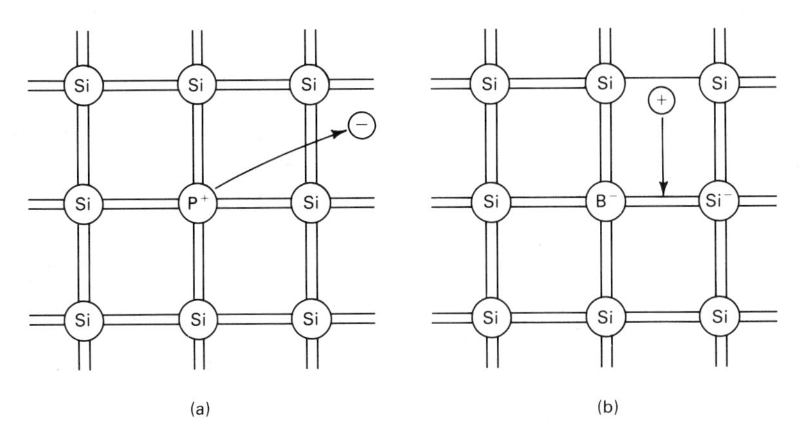

Figure 5.2 (a) Column V substitutional donor impurity; and (b) column III substitutional acceptor impurity in silicon.

If one electron has propagated away, there remains a charged arsenic ion having a potential energy

$$V_{\mathrm{As}} = -\frac{e^2}{4\pi\epsilon\epsilon_0 r} \times (\text{screening factors}) + V_2\,(\text{rearrangement of crystal atoms}) \tag{5.1}$$

The screening factors are typically of the form $e^{-r/L}$ where L is a characteristic length determined by the presence of other charges (see Chap. 7). The potential V_2 is a complicated contribution that depends on the character of the impurity atom. Its size compared to that of the host atoms will determine the rearrangement of the host atoms and therefore result in various contributions V_2. For impurities like arsenic we have, roughly, $V_2 \approx 0$ and $L \approx \infty$, so that

$$V_{\mathrm{As}} = -\frac{e^2}{4\pi\epsilon\epsilon_0 r} \tag{5.2}$$

We know, then, from the effective mass theorem that the arsenic behaves like a hydrogen atom, which gives the energy levels (see any text on quantum mechanics)

$$(E_n - E_c) = -\frac{m^* e^4}{32\pi^2 n^2 \hbar^2 \epsilon^2 \epsilon_0^2} \tag{5.3}$$

and a Bohr radius of

$$a_n = \frac{4\pi n^2 \hbar^2 \epsilon\epsilon_0}{m^* e^2} \tag{5.4}$$

It is important to note that $E_n - E_c$ is reduced (shallow impurity) compared to the free space value (~13 eV) by the factor $m^*/\epsilon^2 \sim 10^{-3}\, m$, while the Bohr radius is increased by a factor $\epsilon/m^* \sim 100/m$, where m is the mass of the free electron.

The increase in the Bohr radius justifies a posteriori our choice of the potential given in Eq. (5.2), which uses a dielectric constant and therefore assumes that the electron "sees" the crystal medium and has a large orbit around the positive charge. Indeed, a typical Bohr radius in a III–V compound is of the order of 100 Å.

It is customary to represent impurities by various graphs in real space and **k** space. These are shown in Figs. 5.3a–5.3c.

The different lengths of the ground state and the excited states in real space and **k** space correspond to the uncertainty principle: In the ground state the electron is most localized in real space (short line), which means that the uncertainty of **k** is large (long line in **k** space graph).

Similar arguments as have been advanced for As and generally elements that have one (or more) electrons too many and donate it (donors) can be given for elements that have not enough electrons. These accept (acceptors) electrons and create holes in the valence band. The theory is entirely symmetric to the donor case.

Complications of the simple model outlined above arise, for example, from the anisotropy of the silicon conduction band valleys, which is expressed

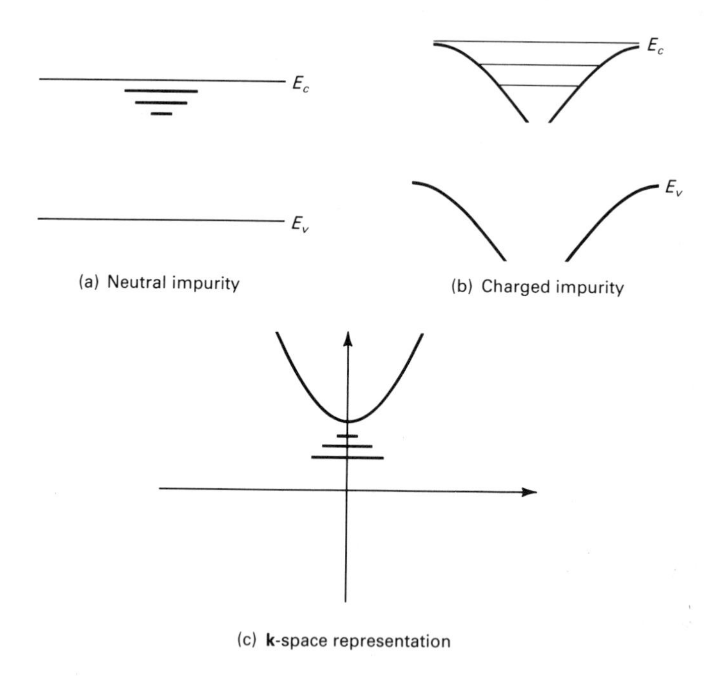

Figure 5.3 Graphs characterizing a donor impurity in a crystal. Parts a and b are graphs in real space; part c is a **k** space representation.

by the term $-\frac{\hbar^2}{2m_j^*}\frac{\partial^2}{\partial x_j^2}$ in the Hamiltonian [Eq. (4.24)]. This anisotropy causes a splitting of the excited states. The situation gets even more complicated if more than one impurity is present within several Bohr radii. Then we can have impurity molecules and aggregates of impurities. The consequences can easily be understood qualitatively using the effective mass theorem (discussed earlier).

Electrons can also be bound closer to the nucleus (e.g., if they have multiple charge or if the effective mass is large). Then the above approach is not justified and the energy levels of the impurities can be quite different from Eq. (5.3). The energies $E_n - E_c$ are larger and one speaks about deep impurity levels. The theory of deep impurity levels is involved, since it is the potential close to the nucleus and the core electrons that matters most and that binds the electrons strongly. The relevant energy scale is then electron volts (instead of millielectron volts), and small errors in a calculation will shift the energy level from the gap to the conduction band, or vice versa. Any impurity can, in principle, produce both deep and shallow levels—and deep does not necessarily mean that the level lies in the energy gap. It only means that it is connected to the more central part of the impurity potential. A theory of chemical trends has been given by Dow, and the interested reader is referred to this treatment.

One-dimensional imperfections (dislocations) have only negative aspects with respect to applications (devices) and one tries to avoid them by elaborate crystal-growth techniques. In Si and GaAs, the art has advanced to such a degree that the crystals can be produced dislocation free. Two-dimensional defects, surfaces, and interfaces, are important in semiconductor electronics and are discussed to some extent below.

A simpleminded chemist's picture of a surface is shown in Fig. 5.4. Notice that each silicon atom in the bulk of the semiconductor is surrounded by eight valence electrons and one electron is missing at the surface. As a consequence, a "surface state" is created in analogy to an impurity state. Its

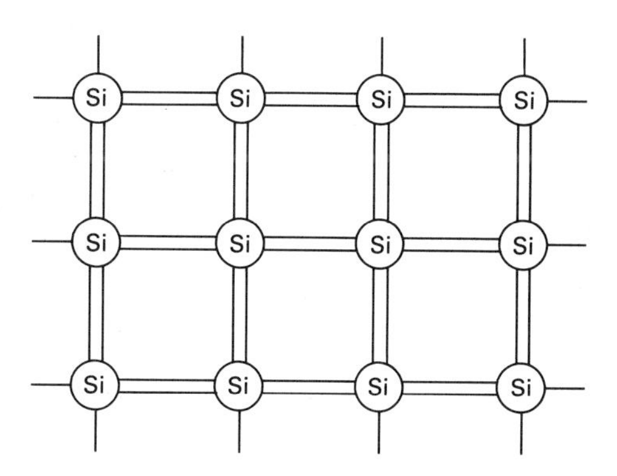

Figure 5.4 Simpleminded explanation of surface states.

energy is typically somewhere in the middle of the energy gap. These states can capture electrons and therefore are highly perturbing in the operation of semiconductor devices. *Surfaces are, therefore, avoided in semiconductor electronics and replaced by (more or less ideal) interfaces.*

Actually, this picture of surface states is artificial, and the only reason why it is still discussed in this way in almost all texts is its simplicity. Surfaces do not exist in the simple form as implied by the figure. The surface atoms attempt to relax to a state of lower energy and pair with each other. One says the surface reconstructs and patterns are formed, as demonstrated in Fig. 5.5, that allow for additional bonding of the surface atoms. When electrons are added at the surface, the reconstruction pattern may change and the electron is trapped at the surface, as implied also by the simple model. A more detailed discussion is given by Harrison.

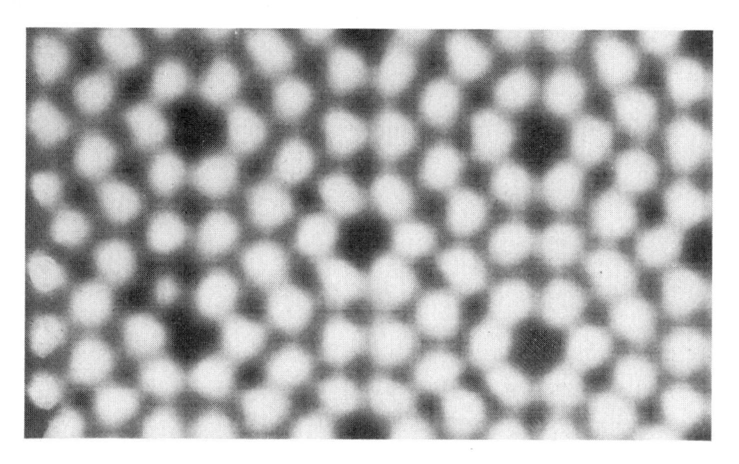

Figure 5.5 Reconstruction pattern of (111) silicon surface. (After Tromp, Hamers, and Demuth, Fig. 2f.)

To avoid trapping electrons in device applications, surfaces are replaced by interfaces. In fact, one of the biggest advantages of silicon is that it forms an almost perfect interface with its dioxide. The topology of the interface Si-SiO_2 is still not exactly known; however, it is believed that amorphous SiO_2 grows on top of Si with only few "dangling bonds" left and with very little bond angle distortions at the interface. This is shown in Fig. 5.6. The leftover dangling bonds can be saturated, for example, with hydrogen, which is made available in the oxidation process. State of the art material has only about $\lesssim 10^{10}$ cm^{-2} interface traps left (out of $\sim 10^{15}$ cm^{-2}), which is good enough for metal-oxide-silicon transistor applications.

Of course, from the viewpoint of the interface, the lattice-matched GaAs-AlAs system is still "infinitely" better.

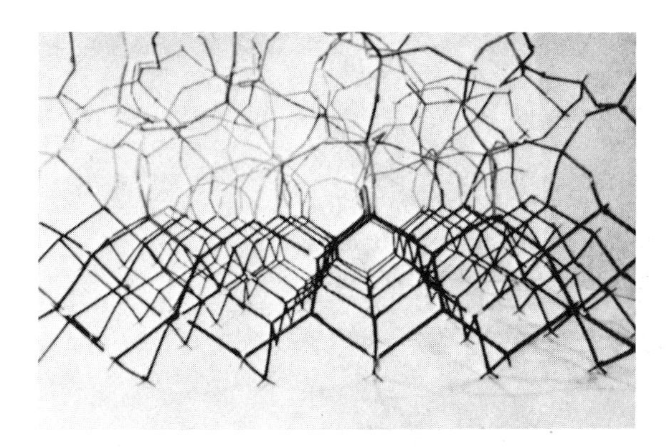

Figure 5.6 Si-SiO_2 interface model. Notice that SiO_2 growing on silicon is not a regular crystal but rather amorphous. However, on short range the SiO_2 atomic arrangement is still almost tetrahedrical and the bonding angles are distorted on average only by $\sim\pm15°$. (After S. T. Pantelides and M. Long, in *The Physics of SiO_2 and its Interfaces*, S. T. Pantelides, ed., Fig. 1. Copyright © 1978 by Pergamon Books Ltd. Reprinted with permission.)

Let me repeat in passing that lattice match is not the only criterion necessary to achieve ideal interfaces; there are other criteria. For example, if GaAs is grown on top of germanium, the Ga and As atoms have to pair with the correct atom (of the two possible) in the Ge unit cell. They do not always do this because they are usually not "smart." One, therefore, talks about the "site allocation problem" which makes the GaAs-Ge interface nonideal even though the materials are lattice-matched. There are other problems of this kind, which are discussed extensively by Kroemer.

RECOMMENDED READINGS

DOW, JOHN D., "Localized Perturbations in Semiconductors," in *Highlights of Condensed Matter Theory,* LXXXIX Corso Soc. Italiana di Fisica, Bologna, Italy, 1985.

HARRISON, W. A., *Electronic Structure and the Property of Solids.* San Francisco: Freeman, 1980, p. 229.

KROEMER, H., *Applied Physics Letters,* 36 (May 1980), 763–65.

PANTELIDES, S. T., and M. LONG, "Continuous-Random-Network Models for the Si-SiO_2 Interface," in *The Physics of SiO_2 and Its Interfaces,* ed. S. T. Pantelides. New York: Pergamon Press, 1978, pp. 339–43.

RIDLEY, B. K., *Quantum Processes in Semiconductors.* Oxford: Clarendon Press, 1982, pp. 71–81.

TROMP, R. M., R. J. HAMERS, and J. E. DEMUTH, *Physical Review,* B34 (1986), 1388.

PROBLEM

5.1. Indium arsenide has $E_G = 0.40$ eV; dielectric constant $\epsilon = 15$; electron effective mass $m^* = 0.023\ m$. Calculate

(a) the donor ionization energy

(b) the radius of the ground state orbit

(c) the minimum donor concentration at which appreciable overlap effects between the orbits of adjacent impurity atoms will occur

6

EQUILIBRIUM STATISTICS FOR ELECTRONS AND HOLES

Although we discussed the energy band structure (the electronic states) of a semiconductor in detail in the previous chapters, we did not include in our discussion whether these states are actually filled with electrons (two with opposite spin in each state are possible) or not.

In Chap. 5 we have shown, however, that full bands do not contribute to the electronic current. In a defect-free semiconductor at $T = 0$, all bands up to the so-called conduction band, are filled. The last filled band is the valence band. As the temperature increases, electrons from the valence band will be excited to the conduction band, and the electrons and holes generated in this way will be able to conduct a current (not as large as in a metal where valence and conduction bands overlap). We also can introduce electrons and holes by doping. To compute the conductivity, we need to know which states are occupied and which are empty. This knowledge is usually acquired by calculating the probability that a state is occupied and by calculating the density of states. The actual carrier occupation is then proportional to the product of these two quantities, which are treated separately below.

6.1 THE DENSITY OF STATES

Consider a crystal with periodic boundary conditions and N atoms along each of the main coordinate axes. Then, according to Eq. (3.19), the allowed wave vectors are given by

$$\mathbf{k} = \frac{\mathbf{K}_h}{N} = \frac{h_1}{N}\mathbf{b}_1 + \frac{h_2}{N}\mathbf{b}_2 + \frac{h_3}{N}\mathbf{b}_3 \tag{6.1}$$

with

$$0 \leq |h_i| \leq N/2$$

In a simple cubic crystal, $\mathbf{b}_1$ would be of the form $2\pi/a$ times unit vector (and similarly $\mathbf{b}_2$ and $\mathbf{b}_3$), and therefore, the typical form of $\mathbf{k}$ is

$$\mathbf{k} = \frac{2\pi h_1}{Na} \times \text{unit vector} + \ \ldots \tag{6.2}$$

The components of the allowed $\mathbf{k}$ values form the "lattice" shown in Fig. 6.1, where $Na = L$ is the length of the crystal in each direction.

This treatment involving the crystal length and the periodic boundary conditions, may seem somewhat artificial. However, since the end results will not depend on the quantities (N, L, etc.) connected to these artificial conditions, we need not be concerned about it. It is the advantage of periodic boundary conditions that their proper application always gives correct results for bulk properties. Of course, they are invalid when surface properties become important.

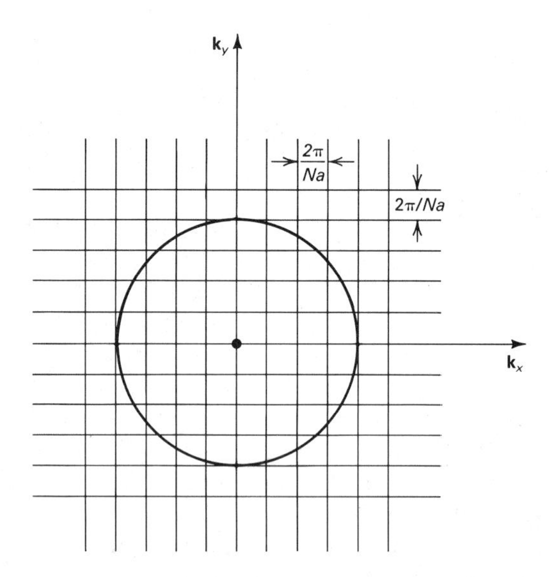

Figure 6.1 Allowed values for the **k** vector in two dimensions.

We are interested in the number of states at the energy E in the interval $[E, E + dE]$. To calculate this number, we first assume the simple case of *spherical* constant energy surfaces in **k** space. The number of allowed **k** values in the sphere in Fig. 6.1 is then equal to the number of cubes of side length $2\pi/L$ in the sphere. The volume $V_{\mathbf{k}}$ of the sphere is

$$V_{\mathbf{k}} = \frac{4\pi}{3} k^3 = \frac{4\pi}{3} \left[\frac{2m^*(E - E_c)}{\hbar^2} \right]^{3/2} \tag{6.3}$$

The number of states $\overline{N}(E)$ in this volume is

$$\overline{N}(E) = V_{\mathbf{k}} \Big/ \left(\frac{2\pi}{L} \right)^3 \tag{6.4}$$

For systems other than simple cubic, one replaces the cubes of reciprocal volume $(2\pi/L)^3$ by other zones. The outcome is the same.

Since each state can be occupied with two electrons of opposite spin, the number $N(E)$, which finally tells us how many electrons can be accommodated, is twice as large as $\overline{N}(E)$

$$N(E) = 2\overline{N}(E) \tag{6.5}$$

As can be seen from the following algebra, the quantity that is most useful is the number of states per unit sample volume at energy E in the interval $[E, E + dE]$. The quantity is called the density of states $g(E)$, which is given by

$$g(E) = \frac{1}{L^3} \frac{dN(E)}{dE} = \frac{1}{2\pi^2} \left(\frac{2m^*}{\hbar^2} \right)^{3/2} \sqrt{E - E_c} \tag{6.6}$$

To understand the importance of $g(E)$, let us assume for the moment that we know the probability $f(E)$ that an electron occupies a state with energy E. Then we can obtain the density n_c of electrons in the conduction band from

$$n_c = \sum_{\mathbf{k}} f(E) = \int_{E_c}^{\infty} f(E) g(E)\, dE \tag{6.7}$$

Equation (6.6) can be derived also in a different way: Since the allowed **k** values are separated from each other only by very small distances (L is very large), the summation Σ can be replaced by an integration $\int d\mathbf{k}$. Since the number of **k** values in the volume $d\mathbf{k} = dk_x\, dk_y\, dk_z$ is equal to $d\mathbf{k}/(2\pi/L)^3$, we obtain the number per unit volume (including a factor of 2 for the spin) as

$$2\left(\frac{1}{2\pi} \right)^3 d\mathbf{k}$$

and we have

$$\frac{1}{V_{\text{ol}}} \sum_{\mathbf{k}} \to 2 \int \left(\frac{1}{2\pi} \right)^3 d\mathbf{k} \tag{6.8}$$

Using spherical coordinates, we obtain

$$dk_x\, dk_y\, dk_z = k^2 dk \sin \vartheta d\vartheta d\phi \tag{6.9}$$

It is then easy to show that

$$\frac{2}{(2\pi)^3} \int d\mathbf{k} \rightarrow \int g(E)\, dE \tag{6.10}$$

For complicated $E(\mathbf{k})$ relations, $g(E)$ takes a form more complicated than Eq. (6.6). For example, for one of the lowest conduction band minima in silicon (one of the six ellipsoids), we have (transforming the ellipsoid to a sphere by a suitable coordinate transformation in **k** space)

$$g(E) = \frac{\sqrt{m_l^* m_t^{*2}}}{2\pi^2} \left(\frac{2}{\hbar^2}\right)^{3/2} \sqrt{E - E_c} \tag{6.11}$$

The term $\sqrt{m_l^* m_t^{*2}}$ replaces $m^{*3/2}$ of Eq. (6.6). One therefore defines a density of states mass m_d

$$m_d^* = (m_l^* m_t^{*2})^{1/3} \tag{6.12}$$

which differs from the conductivity mass of Eq. (4.22).

Higher in the conduction band, the proportionality of $g(E)$ to $\sqrt{E - E_c}$ ceases to be true because of the complicated $E(\mathbf{k})$ relation for the conduction band of semiconductors such as silicon (see Fig. 6.2).

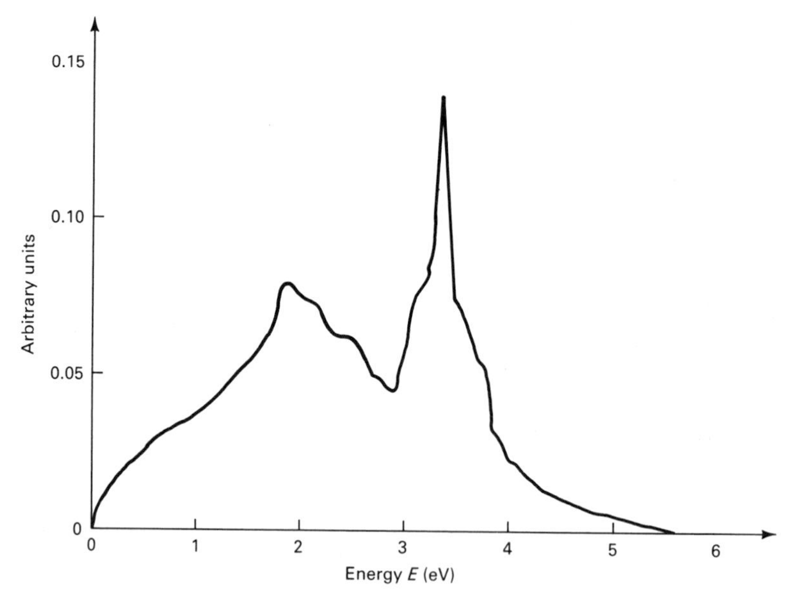

Figure 6.2 The density of states for silicon as calculated from the empirical pseudopotential model (conduction band).

Also at low energies modifications of the proportionality of $g(E)$ to $\sqrt{E - E_c}$ can be important. These modifications are usually a consequence of the doping. Figure 6.3 shows the impurity-related density of states for various degrees of doping. For light doping (Fig. 6.3a), the impurity levels are the single "hydrogenlike" levels. For higher impurity densities (Fig. 6.3b), an impurity band develops that merges at very high concentrations with the conduction band (Fig. 6.3c). In the latter case, the semiconductor behaves as a metal, that is, it is highly conducting down to the lowest temperatures.

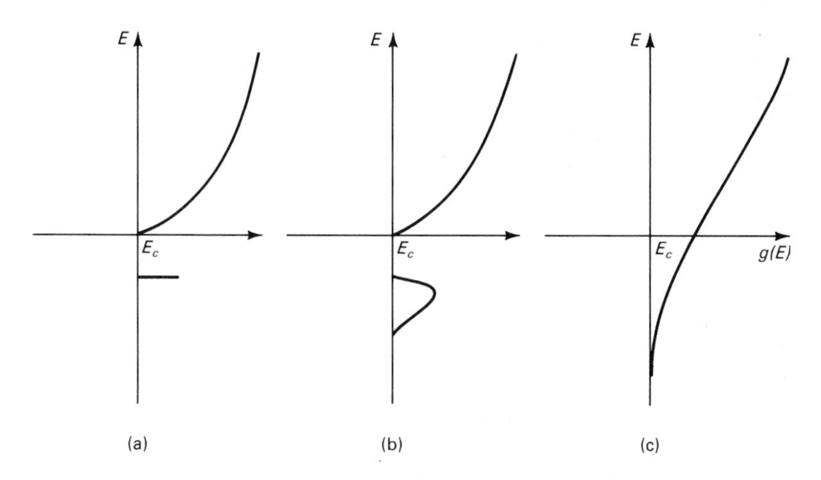

Figure 6.3 Density of states in a semiconductor including impurity levels.

A simple semiquantitative treatment of the impurity "band tail" in Fig. 6.3c has been proposed by Kane. His treatment rests on the assumption of the local conservation of the density of states. He visualizes the impurity potential as a smoothly varying potential $V_I(\mathbf{r})$, as shown in Fig. 6.4.

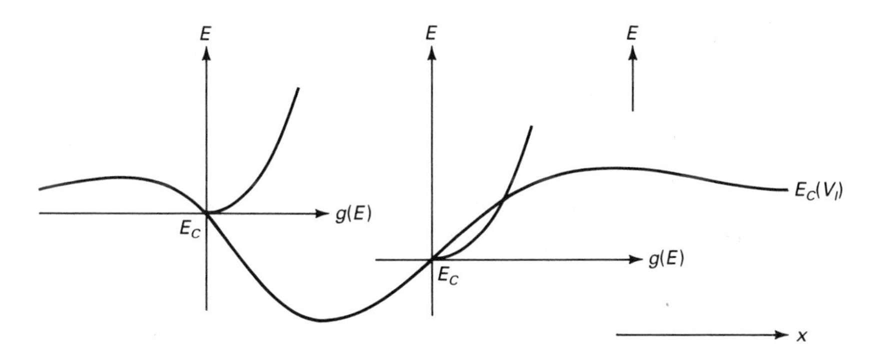

Figure 6.4 The locally conserved density of states in a slowly varying potential of impurities.

The density of states at each point **r** is the density of states of the unperturbed crystal as shown in Fig. 6.4. In other words, the conduction band edge is locally shifted and the density of states starts to increase from these shifted energy values. The fact that the conduction band edge is shifted exactly as the potential $V_I(\mathbf{r})$ follows, of course, from the effective mass theorem, which is the basis of Kane's theory.

Kane further suggests using the potential $V_I(\mathbf{r})$ as a stochastic variable obeying a Gaussian distribution F

$$F(V_I) = \frac{1}{\sqrt{\pi}\eta} \exp\left(-e^2 V_I^2/\eta^2\right) \tag{6.13}$$

The quantity η has been estimated by Kane. For our purposes, it is sufficient to regard it as parameter (a typical value for 10^{19} cm^{-3} impurities would be 50 meV). If the unperturbed density of states is denoted by $g(E)$ and the density of states including impurities by $g_I(E)$, we obtain

$$g_I(E) = \int_{-\infty}^{E} g(E_c + eV_I) F(eV_I)\, d(eV_I) \tag{6.14}$$

The equation is general enough to hold also for two-dimensional systems. Remember, however, that Eq. (6.14) is based on the effective mass theorem and will fail for very close spacing of impurities.

A very different treatment, valid for the "deeper" band tail, has been given by Halperin and Lax.

6.2 THE DISTRIBUTION FUNCTION

In equilibrium, the probability of finding an electron in a state of energy E is given by the Fermi distribution f, as derived in texts of statistical mechanics.

$$f(E) = \frac{1}{e^{(E - E_F)/kT} + 1} \tag{6.15}$$

The parameter E_F is the Fermi energy and is, as we will see later, determined by the total number of charge carriers.

Figure 6.5 shows $f(E)$ for $T = 0$ and $T \neq 0$. For zero temperature, $f(E) = 1$ below E_F and $f(E) = 0$ above E_F. This means that all the states below E_F will be filled and all above it will be empty if $T = 0$. For higher temperatures, the distribution function exhibits an exponential (Maxwell-Boltzmann) tail.

There are two important approximations to $f(E)$. For low temperatures, $f(E)$ can be approximated by the step function $H(E - E_F)$, as evident from Fig. 6.5.

$$f(E)_{T\sim 0} \approx H(E - E_F) \tag{6.16}$$

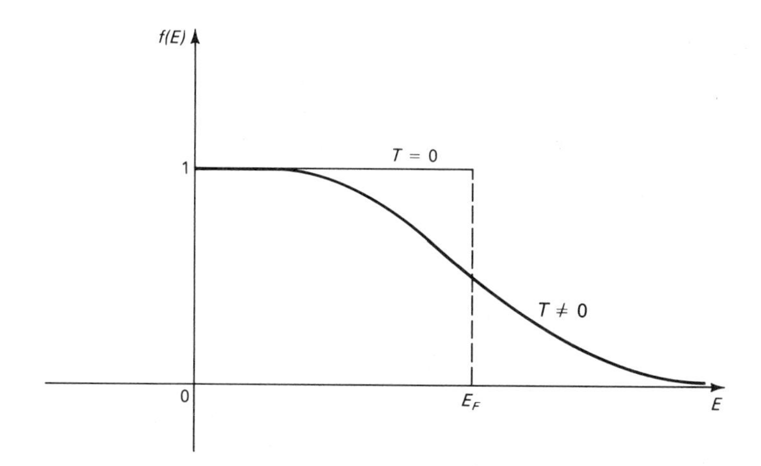

Figure 6.5 Fermi distribution as function of energy.

In this case, the derivative of f is

$$\frac{\partial f(E)}{\partial E} = -\delta(E - E_F) \tag{6.17}$$

A general useful relation is

$$\frac{\partial f(E)}{\partial E} = -f(E)(1 - f(E))/kT \tag{6.18}$$

If we are only interested in the higher energy tail of $f(E)$, then we can neglect the one compared to the exponent in Eq. (6.15), which gives

$$f(E) \approx e^{(E_F - E)/kT} \tag{6.19}$$

Notice that at finite temperatures E_F becomes a parameter which characterizes the density of electrons [see Eq. (6.20)] and represents the chemical potential, which is temperature dependent. We will still continue to call it Fermi energy. The function of Eq. (6.19) is called the Maxwell-Boltzmann distribution function.

Holes like electrons obey Fermi statistics, but as if their energy were measured "downwards." This can easily been seen from the fact that the hole distribution f_h equals $1 - f$, where f is the electron distribution.

6.3 THE ELECTRON DENSITY IN THE CONDUCTION BAND

From Eq. (6.7), by using Eq. (6.19) for the distribution function (which will be justified below), we obtain for the density n in the conduction band

$$n_c = e^{E_F/kT} \frac{1}{2\pi^2} \left(\frac{2m^*}{\hbar^2}\right)^{3/2} \int_0^\infty e^{-E/kT} \sqrt{E}\, dE \tag{6.20}$$

Here we have assumed that the conduction band extends toward infinity (the exponent decays so rapidly that this does not really matter), and we have chosen the conduction band edge as the zero of the energy. Note, also, that an effective mass of a single minimum has been used. Substituting $\bar{x}$ for E/kT and noting that

$$\int_0^\infty e^{-\bar{x}}\,\bar{x}^s\,d\bar{x} = \Gamma(s+1) \tag{6.21}$$

we obtain

$$n = \frac{1}{4}\left(\frac{2m^*kT}{\pi\hbar^2}\right)^{3/2} e^{E_F/kT} \tag{6.22}$$

Here we have used the following expression for the Γ function:

$$\Gamma(½) = \sqrt{\pi} \tag{6.23}$$

and

$$\Gamma(s+1) = s\,\Gamma(s) \tag{6.24}$$

which is valid for s being an integer multiple of ½. For s being integer, we simply have

$$\Gamma(s+1) = s! \tag{6.25}$$

For the density p of holes in the valence band, we obtain in a similar manner

$$p = \frac{1}{4}\left(\frac{2m_h^*kT}{\pi\hbar^2}\right)^{3/2} e^{(-E_G-E_F)/kT} \tag{6.26}$$

One usually defines an "intrinsic concentration" n_i by the product $n \cdot p$. This product is given (putting $E_c = 0$, E_G being the band gap) by

$$n_i^2 = n \cdot p = \frac{1}{2}\left(\frac{kT}{\pi\hbar^2}\right)^3 (m^*m_h^*)^{3/2}\, e^{-E_G/kT} \tag{6.27}$$

As can be seen, this product depends only on semiconductor material parameters and not on the Fermi energy. Equation (6.27), therefore, holds also when the semiconductor is doped. The Fermi energy E_F can be calculated from the charge neutrality condition. For the pure semiconductor this condition means

$$p = n \tag{6.28}$$

In the presence of charged donors, of density N_D^+, and charged acceptors, of density N_A^-, charge neutrality must be written as

$$N_D^+ + p = N_A^- + n \tag{6.29}$$

To solve Eq. (6.29) for the Fermi energy, we need to express N_D^+ and N_A^- as a function of E_F. The total density N_D of donors (or of acceptors N_A) can be assumed as given. The energy E_D of the donor level below the conduction band edge can be calculated from Eq. (5.3). One then would assume that

$$N_D - N_D^+ = N_D \frac{1}{e^{(E_D - E_F)/kT} + 1} \tag{6.30}$$

There are, however, subtleties such as the existence of excited donor states and the fact that a band state can contain two electrons while the charged donor is neutral after taking one electron and will therefore not accept a second electron. These details lead to a factor before the exponent in Eq. (6.30). This factor has been described in detail by Landsberg and is often ignored. The equation for the acceptors is totally analogous to Eq. (6.30).

To finish this chapter, we consider the following two examples:

1. The pure semiconductor with $m^* = m_h^*$. In this case, the equation of charge neutrality, Eq. (6.28), together with Eq. (6.22) and Eq. (6.26), gives

$$E_F = -E_G/2 \tag{6.31}$$

   That is, the Fermi energy is in the middle of the energy gap (if $m_h^* \neq m^*$, it will be slightly shifted). This justifies a posteriori the use of the Maxwellian distribution for the calculation of n_c since only the tail of the distribution function is within the conduction band. The intrinsic concentration of several semiconductors is shown in Fig. 6.6.

2. A semiconductor is doped with N_D donors and $N_D \gg n_i$. Furthermore, we assume that at high temperatures $N_D \approx N_D^+$, while at low temperatures $N_D^+ \approx 0$ (the electrons "freeze out" back to their parent donors). Then

$$n = \frac{n_i^2}{n} + N_D \approx N_D \tag{6.32}$$

   for high temperatures, while $n \approx 0$ for low temperatures. Notice, however, that if the temperature becomes very high, the term n_i^2/n in Eq. (6.32) cannot be neglected as the intrinsic concentration rises. This gives the graph n vs. T, as shown in Fig. 6.7.

It is easy to include more complicated band structures and the Fermi distribution in our calculations. For example, if we consider the silicon conduction band, we have to replace m_d^* by $(m_l^* m_t^{*2})^{1/3}$ and multiply the result by 6 because we have six degenerate ellipsoids located inside the Brillouin zone. Germanium has eight ellipsoids whose center is at the L point of the Brillouin zone. Within the zone there is only one-half of each ellipsoid and we have to multiply the result by a factor of 4 instead of by 6, as in the case of silicon.

The inclusion of the Fermi distribution instead of the Maxwell-

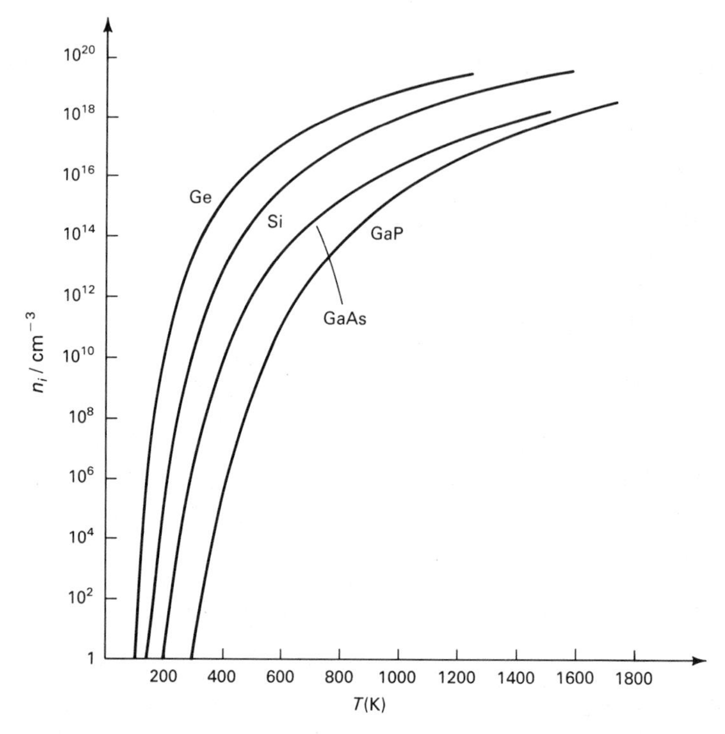

Figure 6.6 The intrinsic carrier concentration, n_i in cm^{-3}, as a function of T for Ge, Si, GaAs, and GaP. (After Thurmond, Fig. 14. Reprinted by permission of the publisher, The Electrochemical Society, Inc.)

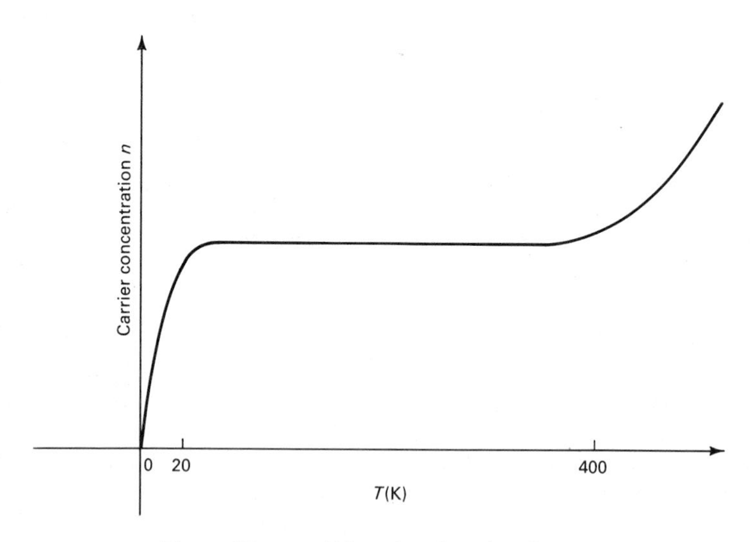

Figure 6.7 n vs. T for a doped semiconductor.

Boltzmann distribution makes numerical integrations necessary in some cases. These still can be managed, however, with a pocket calculator.

Another complication in the calculation of carrier densities is that the energy gap E_G actually depends on temperature. There are different reasons for this temperature dependence, which are described in Appendix C. The band-structure calculation presented in Chap. 4 cannot account for the temperature dependence, since in this calculation the atoms have been assumed to be "frozen" on their exact lattice sites.

RECOMMENDED READINGS

HALPERIN, B., and M. LAX, *Physical Review,* 148 (1966), 722.

KANE, E. O., *Physical Review,* 131 (1963), 79.

LANDSBERG, P. T., ed., *Solid State Theory.* New York: Wiley/Interscience, 1969, pp. 274–78.

THURMOND, C. D., *Journal of the Electrochemical Society,* 122 (1975), 1133.

PROBLEMS

6.1. Find the density of states as a function of energy for $E(\mathbf{k}) = C\left[1 - \cos\left(|\mathbf{k}|\frac{a}{2}\right)\right]$ in one, two, and three dimensions. Discuss the limiting cases $|\mathbf{k}| \to 0, \frac{\pi}{a}, \frac{\pi}{2a}$.

6.2. Calculate the density of states including a donor concentration of 10^{18} cm^{-3} using Eqs. (6.13) and (6.14).

6.3. Derive the equilibrium electron density N_i for any subband i for a single quantum well assuming two-dimensional considerations apply. Let E_i represent the lowest energy of each subband. Your answer should be valid for all temperatures (see Sec. 12.5).

7

DIELECTRIC PROPERTIES OF SEMICONDUCTORS

So far, we have treated only well-defined potential problems, that is, problems in which the potential is given and we calculate the corresponding states or transition probabilities as in Eq. (1.24). For small potentials we could always find a solution by perturbation theory [Eq. (1.28)].

We now consider a different type of problem. Assume that we put an additional charge on an atom in a crystal. What is the potential of this electronic charge? Some may find an easy answer and, in fact, we have given it in Eq. (5.1). In this equation, however, we introduced without justification a dielectric constant. In this chapter we will derive the dielectric constant and will see that it actually is not a constant but in general is a rather complex function.

The essence of our goal is to solve a many-particle problem. We introduce a charge into a solid and we would like to know its potential as a function of position. The complication is that this charge exerts forces on all the other charges in the solid and rearranges them. Our applied coulombic potential V_{ap}, therefore, causes an additional potential V_{ad} due to the rearrangement. Together the two potentials determine the true potential V_{tr} in the semiconductor.

To develop the theory for this complicated situation, we introduce two new techniques. First, we treat the problem self-consistently. This means we assume that we know the *true* impurity potential and denote it by V_{tr}. Then we calculate the redistribution of charge due to this impurity potential and obtain an additional potential V_{ad}. We then use the equation

$$V_{tr} = V_{ap} + V_{ad} \tag{7.1}$$

where V_{ap} is the applied (naked) impurity potential and determine from the definition

$$V_{tr} = \frac{V_{ap}}{\epsilon(q, \omega)} \tag{7.2}$$

the dielectric constant $\epsilon(q, \omega)$, which is a function of frequency and wave vector.

The second technique we use is the random phase approximation. This means we do the calculation using Fourier transformation; we consider only one Fourier component, and we sum all the components at the end. The implications of this approximation, which is extremely useful, have been discussed (e.g., by Ziman) and are based on the assumption that all the mixed terms in the Fourier expansions cancel because of the randomness of the phases. For simplicity, we use time-independent potentials in the derivation (i.e., $\omega = 0$). The calculation proceeds as follows.

We Fourier decompose V_{tr} into

$$\sum_{\mathbf{q}} V_{tr}^{\mathbf{q}} e^{+i\mathbf{q}\cdot\mathbf{r}}$$

and use only one component $V_{tr}^{\mathbf{q}}$. The wave function of the electron is calculated by perturbation theory. Since we have only one Fourier component, only the matrix elements involving $\mathbf{q}$ are finite and we can write the perturbed wave function as [see Eq. (1.19)]

$$\phi_{\mathbf{k}}^{m} = \psi_{\mathbf{k}}^{m} + b_{\mathbf{k}+\mathbf{q}}\psi_{\mathbf{k}+\mathbf{q}}^{l} \tag{7.3}$$

with

$$b_{\mathbf{k}+\mathbf{q}} = \frac{\langle \mathbf{k}+\mathbf{q}|eV_{tr}|\mathbf{k}\rangle}{E(\mathbf{k})_l - E(\mathbf{k}+\mathbf{q})_m} = \frac{eV_{tr}^{\mathbf{q}}}{E(\mathbf{k})_l - E(\mathbf{k}+\mathbf{q})_m} \tag{7.4}$$

here m and l are the band indices (e.g., conduction-valence band) and we consider at most two bands, $l = c$; $m = v$. The matrix element $\langle \mathbf{k}+\mathbf{q}|eV_{tr}|\mathbf{k}\rangle$ can then also be between conduction and valence band states, which introduces additional difficulties. For the present purpose we can ignore these difficulties since we will concentrate mostly on one band when evaluating the dielectric constant. The perturbed wave function allows us to calculate the new charge distribution ρ_{new}, which is given by

$$\rho_{new} = e \sum_{\mathbf{k},m,l} |\phi_{\mathbf{k}}^{m}|^2 \tag{7.5}$$

Since the "old" charge distribution is the electron distribution in the unperturbed crystal, that is, $\sum_{\mathbf{k}} e/V_{ol}$, we obtain for the change in the charge $\delta\rho$

$$\delta\rho = e \sum_{\mathbf{k},m,l} \left(|\phi_{\mathbf{k}}^m|^2 - \frac{1}{V_{ol}} \right) \tag{7.6}$$

Using Eq. (7.3) and neglecting terms in $|b_{\mathbf{k}+\mathbf{q}}|^2$, we have

$$\delta\rho = e \sum_{\mathbf{k},m,l} (b_{\mathbf{k}+\mathbf{q}} e^{i\mathbf{q}\cdot\mathbf{r}} + b_{\mathbf{k}+\mathbf{q}}^* e^{-i\mathbf{q}\cdot\mathbf{r}}) \tag{7.7}$$

Actually only *full* states can contribute to a change $\delta\rho$. Therefore, we have to multiply Eq. (7.6) by the probability that a state is occupied ($f^l(\mathbf{k})$), which gives (involving a transformation $\mathbf{k} \rightarrow \mathbf{k} + \mathbf{q}$ in part of the sum)

$$\delta\rho = \frac{e}{V_{ol}} V_{tr}^{\mathbf{q}} e^{i\mathbf{q}\cdot\mathbf{r}} \sum_{\mathbf{k},l,m} \frac{f^l(\mathbf{k}) - f^m(\mathbf{k}+\mathbf{q})}{E(\mathbf{k})_l - E(\mathbf{k}+\mathbf{q})_m} + \begin{matrix}\text{complex} \\ \text{conjugate}\end{matrix} \tag{7.8}$$

The reader is encouraged to derive Eq. (7.8) in all details.

The second step in our calculation is to determine V_{ad}. This can be done from the Poisson equation:

$$\nabla^2 V_{ad} = -\frac{1}{\epsilon_0} \delta\rho \tag{7.9}$$

Here ϵ_0 is the dielectric constant of free space.

Fourier transforming Eq. (7.9), we have

$$q^2 V_{ad}^{\mathbf{q}} = -\frac{e}{\epsilon_0} V_{tr}^{\mathbf{q}} \sum_{\mathbf{k},l,m} \frac{f^l(\mathbf{k}) - f^m(\mathbf{k}+\mathbf{q})}{E(\mathbf{k})_l - E(\mathbf{k}+\mathbf{q})_m} \tag{7.10}$$

And therefore the dielectric constant becomes

$$\epsilon(\mathbf{q}, 0) = 1 + \frac{e^2}{\epsilon_0 q^2} \sum_{\mathbf{k},l,m} \frac{f^l(\mathbf{k}) - f^m(\mathbf{k}+\mathbf{q})}{E(\mathbf{k}+\mathbf{q})_l - E(\mathbf{k})_m} \tag{7.11}$$

A special case that we can easily evaluate is that of an electron gas at a conduction band minimum in the effective mass approximation. We then have $l = m$ in Eq. (7.11) and we can perform the sum over $\mathbf{k}$ (with the help of a computer even for more complicated $E(\mathbf{k})$ relations). Here we only will evaluate the expression for $\mathbf{q} \rightarrow 0$. To do this, we expand

$$E(\mathbf{k}+\mathbf{q}) = E(\mathbf{k}) + \mathbf{q} \cdot \nabla_{\mathbf{k}} E(\mathbf{k}) \tag{7.12}$$

and

$$f(E(\mathbf{k}+\mathbf{q})) = f(E(\mathbf{k})) + \frac{\partial f}{\partial E} \mathbf{q} \cdot \nabla_{\mathbf{k}} E(\mathbf{k}) \tag{7.13}$$

which gives

$$\begin{aligned}\epsilon(\mathbf{q},0) &= 1 + \frac{e^2}{q^2\epsilon_0}\frac{2}{8\pi^3}\int \frac{\mathbf{q}\cdot\nabla E(\mathbf{k})}{\mathbf{q}\cdot\nabla E(\mathbf{k})}\left(-\frac{\partial f}{\partial E}\right) d\mathbf{k} \\ &= 1 + \frac{e^2}{q^2\epsilon_0}\int\left(-\frac{\partial f}{\partial E}\right) g(E)\, dE\end{aligned} \tag{7.14}$$

If we define

$$\frac{e^2}{\epsilon_0}\int\left(-\frac{\partial f}{\partial E}\right) g(E)\, dE \equiv q_s^2 \tag{7.15}$$

we have

$$\epsilon(\mathbf{q},0) = 1 + q_s^2/q^2 \tag{7.16}$$

q_s^2 can easily be evaluated in special cases. For example, for f being a step function and consequently $\partial f/\partial E = -\delta(E - E_F)$, we have

$$q_s^2 = \frac{e^2}{\epsilon_0} g(E_F) \tag{7.17}$$

$L_s = 2\pi/q_s$ is called the Thomas-Fermi screening length. For f being a Maxwell-Boltzmann distribution, one obtains

$$q_s^2 = \frac{e^2 n}{\epsilon_0 kT} \tag{7.18}$$

and L_s is the Debye-Hueckel screening length.

q_s is termed "screening wave vector," which becomes clear from the definition of the dielectric function: Any Fourier component $V_{\text{ap}}^{\mathbf{q}}$ with $q \ll q_s$ will result in a very small "true" Fourier component $V_{\text{tr}}^{\mathbf{q}}$, that is, the spatial change of the true potential corresponding to this or smaller wave vectors will be negligible (screened out by the other charge carriers). On the other hand, for $q \gg q_s$ the medium (electron gas) will not have an appreciable influence as $\epsilon(q, 0) = 1$.

The form of the potential in real space is given by

$$V_{\text{tr}} \propto \frac{1}{r} e^{-r/L_s} \tag{7.19}$$

This is obtained by "back" Fourier transformation. The form of the potential V_{tr} is identical to the Yukawa potential.

Equation (7.16) does not, of course, describe the dielectric function of a semiconductor since we have not taken into account the electrons in the valence band. A treatment of both bands introduces the energy gap E_G in the energy denominator of Eq. (7.11). However, since the sum over all $\mathbf{k}$ is taken in (7.11), it is the energy difference at the place of the highest density of states (jointly of conduction and valence bands) that matters. The density of states is usually very high at the energy somewhat above the X valleys. Therefore, in these cases, the most important energy difference is that around X, that is, E_G^X.

A rather tedious calculation executed by Penn gives the following result for the static dielectric constant (now denoted by ϵ_{Penn}):

$$\epsilon_{\text{Penn}} \approx 1 + \left(\frac{\hbar\omega_p}{E_G^X}\right)^2 \tag{7.20}$$

with

$$\omega_p^2 = \frac{ne^2}{\epsilon_0 m}$$

where m is the free electron mass, n the total electron density in the valence band, and ω_p the plasma frequency (Ziman). Penn neglected the conduction band electrons in the derivation. If we combine (7.20) and (7.16), we obtain roughly

$$\epsilon \approx \epsilon_{\text{Penn}} \left(1 + \left(\frac{\mathbf{q}_s}{\mathbf{q}}\right)^2 \frac{1}{\epsilon_{\text{Penn}}}\right) \tag{7.21}$$

which is sufficient to describe ϵ of a semiconductor in many practical cases and can be used to calculate any true potential in a semiconductor if the applied potential is given.

Instead of ϵ_{Penn}, one also can use the dielectric constants given in Table 4.1. Notice that the dielectric constant depends in principle also on the frequency ω, which we have neglected in our treatment. In the table, limiting values for $\omega \to 0$ and $\omega \to \infty$ (static and optic dielectric constants) are given. Screening also depends on frequency (see Problems section). In later chapters, however, we will only discuss the static case, that is, $\omega = 0$.

RECOMMENDED READING

ZIMAN, J. M., *Principles of the Theory of Solids,* 2nd ed. Cambridge: Cambridge University Press, 1972, pp. 146–60.

PROBLEM

7.1. Let a free electron gas in a positive background be perturbed by a potential

$$\delta U(\mathbf{r}, t) = U_{\mathbf{q}} e^{i\mathbf{q}\cdot\mathbf{r}} e^{i\omega t} e^{\alpha t} + cc$$

Show that the charge density fluctuations $\delta\rho(\mathbf{r}, t)$ are given by

$$\delta\rho(\mathbf{r}, t) = e \sum_{\mathbf{k}} \frac{f(\mathbf{k}) - f(\mathbf{k}+\mathbf{q})}{E(\mathbf{k}) - E(\mathbf{k}+\mathbf{q}) + \hbar\omega - i\hbar\alpha} e^{i\mathbf{q}\cdot\mathbf{r}} e^{i\omega t} e^{\alpha t} U_{\mathbf{q}} + cc$$

where cc = complex conjugate.

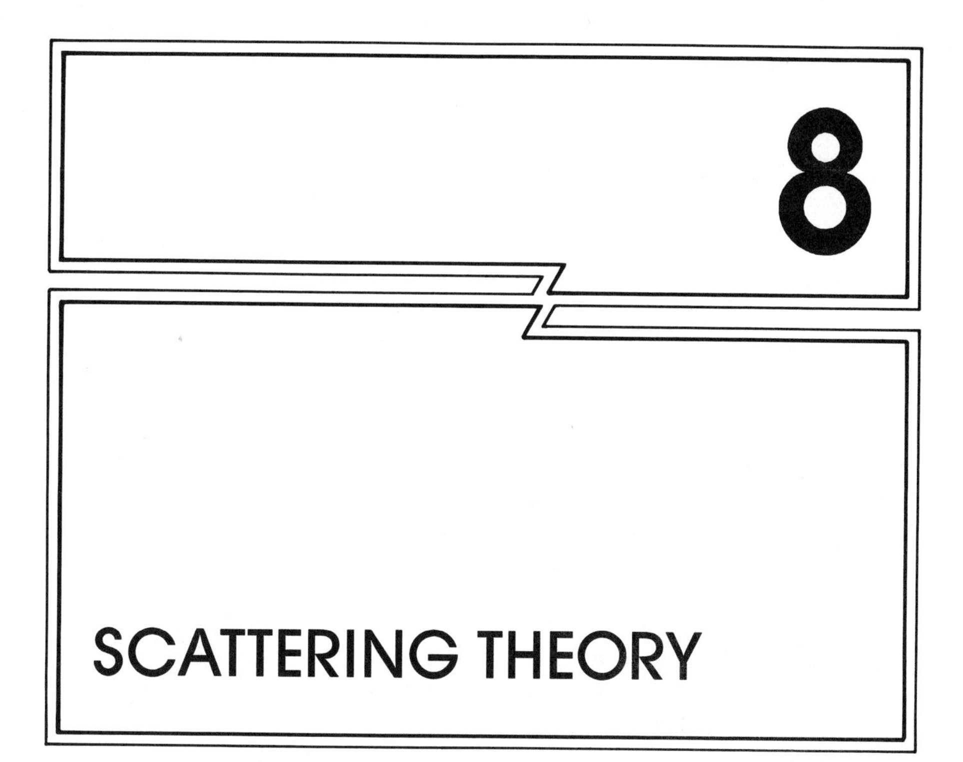

8.1 GENERAL CONSIDERATIONS—DRUDE THEORY

A precise knowledge of the motion and scattering of electrons is necessary to understand the conductivity of a solid. One would think that scattering of electrons by impurities impedes only their motion. However, this is not true. A totally perfect solid can exhibit a totally unexpected behavior, which can be seen from the following example.

The band structure of a one-dimensional (and, similarly, of a simple cubic three-dimensional) crystal is given by

$$E(k) = E_0(1 + \cos(k \cdot a + \pi))$$

Let's assume that we have one electron in the conduction band and a full valence band. The velocity of the electron is [Eq. (4.18)]

$$v = \frac{1}{\hbar}\frac{dE}{dk} = -\frac{E_0}{\hbar} a \sin(k \cdot a + \pi)$$

The k value develops as [Eq. (4.19)]

$$k = -eFt/\hbar + k_0$$

and we assume that the electron starts at $k_0 = 0$ for $t = 0$.

Therefore, we have

$$v = \frac{E_0}{\hbar} a \sin (eFat/\hbar - \pi)$$

Since the current density $j = nev$, we see that a dc electric field F gives rise to an ac current of angular frequency $\omega = eFa/\hbar$. Thus the pure semiconductor would be an excellent high frequency generator.

The angular frequency increases with lattice constant a, and, indeed, Esaki and coworkers proposed to produce very pure superlattices having rather large a. They hoped to obtain unlimited frequencies in this way. Their work stimulated the technology and we can produce very pure superlattices (lattice-matched layers of semiconductors, e.g., GaAs-AlAs). The Esaki-Tsu oscillator, however, does not yet work, for reasons not clearly understood. For most applications of semiconductors, however, a "normal" conductance is the rule, and we will see that we have a normal conductance only if scattering centers are present, that is, if the semiconductor is imperfect. By far the most important scattering agents affecting the conductivity are the lattice vibrations (phonons), since they take away energy from the electrons and do not let $\mathbf{k}$ or $E(\mathbf{k})$ grow sufficiently for the electrons to approach the Brillouin zone boundary and exhibit an oscillatory velocity.

There are other scattering mechanisms, which are elastic but still influence the electronic conduction substantially. Among these are scattering by charged impurities, neutral impurities, and surfaces. The lattice vibrations themselves are scattered by such impurities and even by the different atomic isotopes of which the crystal is composed. (These scattering processes are the reason for the slow propagation of heat, which ideally could propagate with the velocity of sound.)

Here we concentrate on the scattering of electrons. Imagine an electron propagating in a crystal and colliding with various scattering agents, as shown in Fig. 8.1. A consequence of all these collisions is a quasi-Brownian motion of the electron. As mentioned, the collisions also force the electron to stay energetically close to the conduction band edge.

Feynman introduced an idealized graphical representation of collisions that also gives additional information in the form of the $\mathbf{k}$ vector labels, as shown in Fig. 8.2. In this way also very complicated collision processes can easily be represented.

The whole theory of electric conduction would be very simple if it were not for the following complications:

1. The scattering depends on $\mathbf{k}$ and $\mathbf{q}$.
2. More than one electron is present at the same time and we have to calculate the *statistics* of electron propagation.
3. The band structure enters the equations of motion.

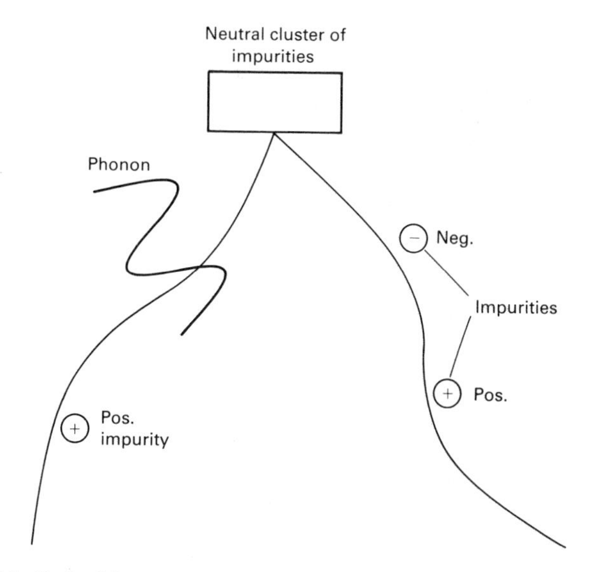

Figure 8.1 Path of the center point motion of an electron wave packet in a crystal.

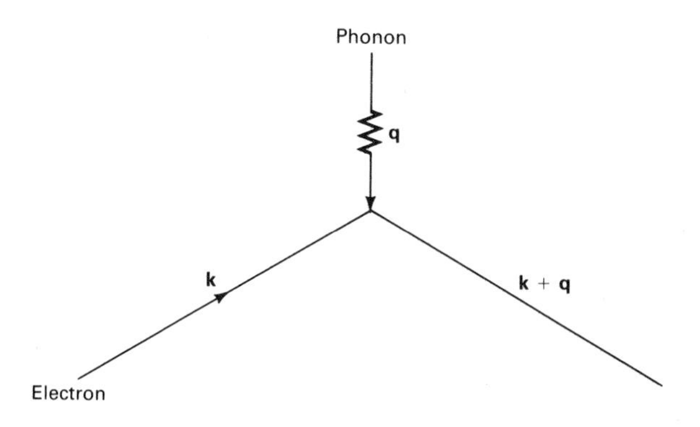

Figure 8.2 Feynman graph of collision.

We will deal with these complications later.

At this point we consider Drude's model of conduction, which circumvents these complications by simplifying assumptions. Drude assumed that all electrons move with a velocity **v** and used the equation of motion

$$\hbar \dot{\mathbf{k}} = m\dot{\mathbf{v}} = \mathbf{F}_0 \qquad (\mathbf{F}_0 = \text{force}) \tag{8.1}$$

To include the band structure one uses m^* instead of m. Since this equation leads to a continuous acceleration, Drude suggested that the scattering agents act like a friction force $\mathbf{F}_f$:

$$\mathbf{F}_f \simeq m\mathbf{v}/\tau \tag{8.2}$$

where τ is the time constant (relaxation time) of the friction force. (The meaning will be explained immediately.) The total equation of motion is, then, according to Drude

$$m\dot{\mathbf{v}} = \mathbf{F}_0 - \mathbf{F}_f = \mathbf{F}_0 - m\mathbf{v}/\tau \tag{8.3}$$

If the applied force $\mathbf{F}_0$ is set equal to zero, then

$$m\dot{\mathbf{v}} = -m\mathbf{v}/\tau \tag{8.4}$$

and

$$\mathbf{v} \propto \exp(-t/\tau) \tag{8.5}$$

Equation (8.5) shows that τ is the time constant in which $\mathbf{v}$ decays to zero. Thus the assumption of a friction force and a single velocity circumvents difficulties (1) and (2). It is only justified by its success: The Drude theory gives a qualitative explanation of almost all low field, low frequency conduction phenomena. The current density $\mathbf{j}$ is calculated by multiplying the velocity by the density of electrons:

$$\mathbf{j} = -en\mathbf{v} \tag{8.6}$$

Inserting for the applied force $\mathbf{F}_0$ the force of an electric field $\mathbf{F}$ and a magnetic field $\mathbf{B}$, one has

$$m\dot{\mathbf{v}} = -e(\mathbf{F} + \mathbf{v} \times \mathbf{B}) - m\mathbf{v}/\tau \tag{8.7}$$

This equation will now be discussed and solved for three special cases:

1. $\mathbf{F}$ is independent of time and $\mathbf{B} = 0$.

These assumptions imply a steady state solution, that is, $\dot{\mathbf{v}} = 0$ and $m\mathbf{v} = -e\mathbf{F}\tau$, which indeed gives a dc current (no Bloch oscillations):

$$\mathbf{j} = -en\mathbf{v} = \frac{e^2\tau n}{m}\mathbf{F}$$

We therefore conclude from the definition, Eq. (3.3), that the conductivity σ is scalar and given by

$$\sigma = en\mu \tag{8.8}$$

with

$$\mu = e\tau/m \tag{8.9}$$

μ is called the mobility and has the dimension cm^2/Vs. Multiplied by the electric field μ gives the negative velocity of the electrons.

2. $\mathbf{F}$ is a low-frequency electric field and $\mathbf{B} = 0$.

$$\mathbf{F} \equiv \mathbf{F}_a e^{i\omega t} + cc$$

(where cc stands for complex conjugate). To solve Eq. (8.7), we try $\mathbf{v} = \mathbf{v}_0 e^{i\omega t} + cc$, which gives

$$mi\omega\mathbf{v}_0 e^{i\omega t} = -e\mathbf{F}_a e^{i\omega t} - m\mathbf{v}_0 e^{i\omega t}/\tau$$

and a similar equation for the complex conjugate. It follows that

$$-\mathbf{v}_0 = \frac{e\mathbf{F}_a}{im\omega + \dfrac{m}{\tau}} = \frac{\mu\mathbf{F}_a}{1 + i\omega\tau} \tag{8.10}$$

that is, the velocity is equal to the dc velocity divided by $(1 + i\omega\tau)$. This means that at frequencies comparable to $1/\tau$ (which is usually above 10 GHz), the semiconductor ceases to be a resistor only and represents also an inductive delay. The imaginary part can also be viewed as a contribution of the free electrons to the dielectric constant (Maxwell's equations!). At very high frequencies, the quantum character of the phonons becomes important and a quantum mechanical treatment must be used which typically involves second-order perturbation theory.

3. $\mathbf{F}$ is independent of time; $\mathbf{B} \neq 0$, and $\mathbf{B} = (0,0,B_z)$.

If we try a solution as in example (1), then we see that the component of the velocity in the y direction is not equal to zero even if the applied field points only in the x direction. The reason, of course, is the magnetic field, which deflects electrons in the y direction ($\mathbf{v} \times \mathbf{B}$). Since we do not have any current channels in the y direction, all that happens is that electrons will accumulate at the sample boundaries and build up an electric field $\mathbf{F}_y$, which opposes further flow of electrons, as shown in Fig. 8.3.

It follows that we must admit an $\mathbf{F}_y$ component of the field and obtain from

$$-e(\mathbf{F} + \mathbf{v} \times \mathbf{B}) = m\mathbf{v}/\tau$$

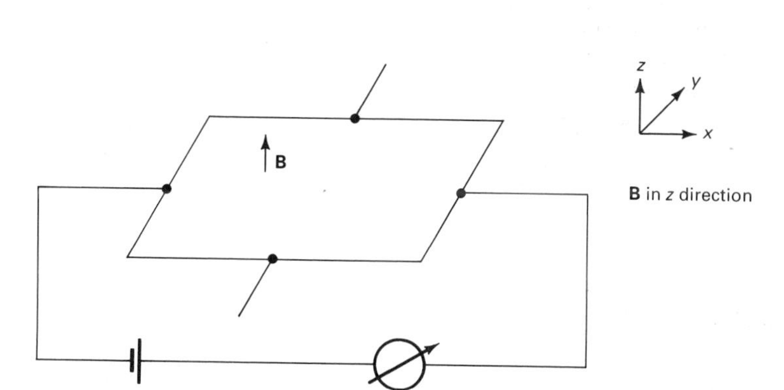

Figure 8.3 Schematic sample geometry for Hall measurements.

and

$$-e[(F_x, F_y, 0) - (0, -v_x B_z, 0)] = m(v_x, 0, 0)/\tau$$

the solution

$$-v_x = \mu F_x \tag{8.11}$$

and

$$F_y = v_x B_z \tag{8.12}$$

Since F_x, F_y, and B_z can be measured, the results allow a determination of the mobility μ and the velocity v_x. Together with the measurement of the current density $j_x = -env_x$, the electron concentration n can also be obtained. This effect (the occurrence of F_y in a magnetic field) is called the Hall effect. It is frequently used to determine μ and n.

The exact formula for the Hall field (without Drude's approximation) is almost identical to Eqs. (8.11) and (8.12) except for a statistical factor (averaging!), which is discussed in Appendix D. Some values for this factor are discussed in the review of Rode.

In the following, improvements of the Drude theory are discussed. Most of the improvements are based on a careful reassessment of the relaxation time, which is replaced by an energy-dependent function or by an integral scattering operator (see the Boltzmann equation in Chap. 9.). A central role in this theory is the scattering probability per unit time $S(\mathbf{k},\mathbf{k}')$, which can be calculated from the Golden Rule. The remainder of this chapter deals with calculations of $S(\mathbf{k},\mathbf{k}')$ and its phenomenological connection to the relaxation time and conductivity. The rigorous derivation of the conductivity from $S(\mathbf{k},\mathbf{k}')$ is discussed in Chap. 9.

8.2 SCATTERING PROBABILITY FROM THE GOLDEN RULE

We perform the calculation of $S(\mathbf{k},\mathbf{k}')$ for ionized impurity scattering and acoustic phonon scattering in detail and discuss the result for the other important mechanisms for which the derivation is totally analogous.

To obtain $S(\mathbf{k},\mathbf{k}')$ we need to calculate the matrix element $M_{\mathbf{kk}'}$ which, in the effective mass approximation, is

$$M_{\mathbf{kk}'} = \frac{1}{V_{\text{ol}}} \int_{V_{\text{ol}}} d\mathbf{r} H' e^{i(\mathbf{k}-\mathbf{k}')\cdot\mathbf{r}}$$

For impurity scattering, the perturbing part of the Hamiltonian H' is the true potential energy eV_{tr} for the impurity

$$M_{\mathbf{kk}'} = \frac{1}{V_{\text{ol}}} \int_{V_{\text{ol}}} d\mathbf{r}\, eV_{\text{tr}} e^{i(\mathbf{k}-\mathbf{k}')\cdot\mathbf{r}}$$

Using $\mathbf{k} - \mathbf{k}' \equiv \mathbf{q}$, we obtain the result that the matrix element is equal to the Fourier transformation of the true potential:

$$M_{\mathbf{kk}'} = \frac{1}{V_{ol}} eV_{tr}^{\mathbf{q}} \tag{8.13}$$

The Fourier component $V_{tr}^{\mathbf{q}}$ is easily obtained from the applied potential by Eq. (7.2) and V_{ap} can be obtained from the Poisson equation for a point charge (located at $\mathbf{r} = 0$):

$$\nabla^2 V_{ap} = -\frac{e}{\epsilon_0} \delta(\mathbf{r}) \tag{8.14}$$

Since we need to know $V_{ap}^{\mathbf{q}}$, we solve the Poisson equation by Fourier transformation, which gives

$$V_{ap}^{\mathbf{q}} = -\frac{e}{\epsilon_0 q^2} \tag{8.15}$$

Using the dielectric constant for small q as derived before, we have

$$V_{tr}^{\mathbf{q}} = \frac{e}{\epsilon_0 \epsilon (q^2 + q_s^2/\epsilon)} \tag{8.16}$$

and $M_{\mathbf{kk}'}$ becomes

$$M_{\mathbf{kk}'} = \frac{e^2}{V_{ol} \epsilon_0 \epsilon (q^2 + q_s^2/\epsilon)} \tag{8.17}$$

and therefore

$$S(\mathbf{k},\mathbf{k}') = \frac{2\pi}{\hbar} \frac{e^4}{V_{ol}^2 \epsilon^2 \epsilon_0^2 (q^2 + q_s^2/\epsilon)^2} \delta(E(\mathbf{k}) - E(\mathbf{k}')) \tag{8.18}$$

It is important to note that scattering by impurities is elastic since the impurity is so much heavier than the electron. Therefore

$$\begin{aligned} q^2 &= |\mathbf{k} - \mathbf{k}'|^2 = k^2 + k'^2 - 2kk' \cos\theta \\ &= 2k^2(1 - \cos\theta) \\ &= 4k^2 \sin^2 \frac{\theta}{2} \end{aligned}$$

where θ is the angle between k and k'. This simplifies the calculation of $S(\mathbf{k},\mathbf{k}')$ and of the total scattering rate defined by

$$\frac{1}{\tau_{tot}} \equiv \sum_{\mathbf{k}'} S(\mathbf{k},\mathbf{k}') \tag{8.19}$$

The summation over $\mathbf{k}'$ in Eq. (8.19) is conveniently transformed into an integration by choosing a polar coordinate system in which $\mathbf{k}$ points in the $\mathbf{k}_z'$

direction. This does not restrict generality and makes the angle θ between **k** and **k**$'$ equal to the polar angle. Therefore, using an isotopic effective mass, the properties of the δ function ($\delta(a(\mathbf{k}-\mathbf{k}')) \rightarrow \frac{1}{a}\,\delta(\mathbf{k}-\mathbf{k}')$) and $w = \sin\theta/2$:

$$\frac{1}{\tau^{I}_{\text{tot}}} = \frac{16\pi^2 m^* e^4 k}{\hbar^3 V_{\text{ol}} \epsilon^2 \epsilon_0^2} \int_0^1 \frac{w\,dw}{(2k^2w^2 + q_s^2/\epsilon)^2} \tag{8.20}$$

where the superscript I of τ^{I}_{tot} stands for impurity. The integration is then easily performed. (We will calculate below total rates for other scattering mechanisms).

The rate given in Eq. (8.20) is for a single impurity (also singly charged). For $\overline{N}_I$ independent impurities, the rate has to be multiplied by $\overline{N}_I$, which introduces the impurity density $\overline{N}_I/V_{\text{ol}} = N_I$ as a multiplicative factor. Much research has been devoted to impurity scattering. Approximations beyond the Golden Rule, impurity correlations, and other higher order effects have been treated. A rather complete reference list has been given by Kuchar et al. The above treatment is equivalent to that of Brooks and Herring.

The total scattering rate represents the probability per unit time that an electron with wave vector **k** is scattered into any possible state labeled by the wave vector **k**$'$.

As can be seen, $1/\tau_{\text{tot}}$ is still a function of **k**. The electrons are distributed in **k** space. The distribution, however, is not necessarily the equilibrium distribution as discussed in Chap. 6 because the application of an electric field will change this distribution function. Therefore, up to this point we are not able to calculate the average value of τ_{tot}, which we will denote by $\langle\tau_{\text{tot}}\rangle$. How this average has to be taken will be shown in Chap. 9 by solving the Boltzmann equation.

For the time being, we can argue generally that the average electron energy is given after Boltzmann by 3/2 kT, and we therefore should replace the wave vector by an average value corresponding to this energy, that is, by $k = \sqrt{3kTm^*}/\hbar$ for an isotropic effective mass m^*. This average total scattering time is closely related and approximately equal to Drude's time constant τ. The exact averaging procedure is given by Eq. (9.40).

Even more important than impurity scattering is scattering by the lattice vibrations, the phonons. Phonon scattering provides the mechanism for the energy loss. Impurity scattering is elastic, as we have seen before, and the energy of the electrons would diverge in a constant electric field. The mobility of semiconductors is also often dominated by phonon scattering, at least around room temperature.

Two new aspects are introduced by phonon scattering compared to impurity scattering. First, phonons can be absorbed and emitted, thereby changing the phonon wave function in addition to the electron wave function. The most elegant way to account for this fact is second quantization (Ziman). Here we just state the result: We can treat phonon scattering in the usual way by using Eq. (1.24) and multiplying the result by certain factors. These factors arise from the phonon wave function. It is clear that a phonon can be absorbed only if at least one phonon is present. Therefore, the result for phonon absorption as obtained by Eq. (1.24) has to be multiplied by the average number of phonons that are present, just as we multiplied the impurity scattering rate by the number (density) of impurities.

The appropriate multiplier (see Ziman) is the phonon occupation number N_q:

$$N_q = \frac{1}{e^{\hbar\omega_q/kT} - 1} \tag{8.21a}$$

In Eq. (8.21a) $\hbar\omega_q$ is the phonon energy that for acoustic phonons is given by Eq. (2.17), and is approximately independent of q for optical phonons.

The factor multiplying the phonon emission rate is $(N_q + 1)$. The emission, therefore, is composed of two contributions. One is independent of the phonon occupation number and is termed spontaneous emission (just as in the case of light). The second term is proportional to N_q and is called stimulated emission because more phonons are emitted if more "stimulating" phonons are present. Again, this is the same as in the case of light where Einstein derived the concept of stimulated emission. The similarity with light is, of course, no accident. Phonons are "bosons" (particles with integer spin) in contrast to electrons (or holes), which are fermions (half-numbered spin). Equation (8.21a) is, therefore, also called the Bose distribution.

Although Eq. (8.21a) is formally quite similar to the Fermi distribution, Eq. (6.15), the two functions behave numerically totally different. The physical reason is that only two electrons with opposite spin fit into a given energy level, while the phonons are the merrier the closer and the more of them there are.

An important approximation to Eq. (8.21a) is obtained for $\hbar\omega_q \ll kT$, which gives

$$N_q \approx \frac{1}{1 + \hbar\omega_q/kT - 1} = \frac{kT}{\hbar\omega_q} \tag{8.21b}$$

This approximation is also known as equipartition for historical reasons.

The second difference between phonon scattering and impurity scattering is in the scattering potential. Phonons distort the crystal lattice and can create in semiconductors essentially three kinds of potential energy changes—the deformation potential, the piezoelectric potential, and the polar optic potential. We will treat in detail only the deformation potential, which was introduced by Bardeen and Shockley.

The basic idea is illustrated in Fig. 8.4, which shows the conduction and valence bands as a function of the interatomic distance d. (This graph can be calculated from our pseudopotential theory by varying d.)

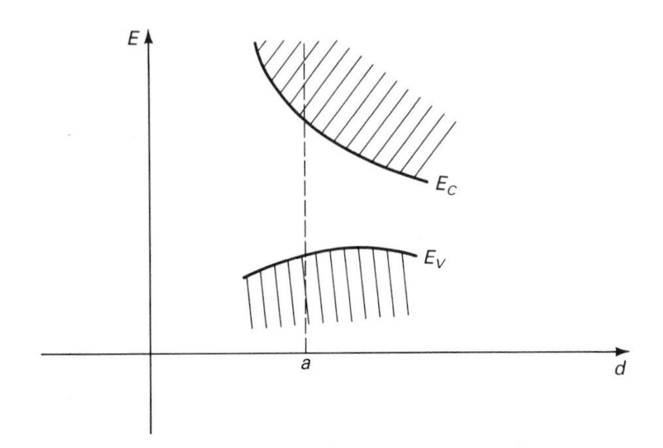

Figure 8.4 Conduction and valence bands as function of interatomic distance. The actual lattice constant of the material is denoted by a.

If the lattice is displaced by u [see Eq. (2.14)], the energy of the conduction (or valence) band will change by (a is the lattice constant)

$$\Delta E_c = E_c(a) - E_c(a + u) \tag{8.22}$$

which gives, by Taylor expansion

$$\Delta E_c = (dE_c/da) \cdot u \tag{8.23}$$

Note that we have implicitly assumed that the lattice displacement u has the same effect as expanding or compressing the whole crystal. This means our concept will actually work only if the phonon wavelength spans many lattice constants. As long as this is true, the effective mass theorem, which we are going to use, also applies.

In three dimensions the change ΔE_c is proportional to the volume change, which is given in terms of the lattice displacement $\mathbf{u}$ by

$$\frac{\Delta V_{\text{ol}}}{V_{\text{ol}}} = \nabla \cdot \mathbf{u} \tag{8.24}$$

as is known from classical mechanics. Therefore

$$\Delta E_c = V_{\text{ol}} \frac{dE}{dV_{\text{ol}}} \nabla \cdot \mathbf{u} \tag{8.25}$$

The constant $V_{\text{ol}} \dfrac{dE}{dV_{\text{ol}}}$ is usually denoted by Z_A if acoustic phonons are involved, and is called the acoustic deformation potential. Therefore, the perturbing Hamiltonian becomes

$$H_A = \Delta E_c = Z_A \nabla \cdot \mathbf{u} \tag{8.26}$$

For holes we have a change in the valence band edge ΔE_v, and the deformation potential constant is different. Rigorously, Z_A has to be replaced by a matrix because of the anisotropy of crystals (see Conwell, a). A typical value for the deformation potential is $|Z_A| \approx 8$ eV for semiconductor conduction bands and a somewhat smaller value for valence bands (Hess and Dow).

Since we have already derived **u** (Chap. 2), we can therefore derive the matrix elements. Using Eq. (2.14) in vector form and Eq. (1.28) we find

$$M_{\mathbf{k},\mathbf{k}'} = \pm Z_A i\mathbf{q} \cdot \mathbf{u}\delta_{\mathbf{k}'-\mathbf{k},\pm\mathbf{q}} V_{\mathrm{ol}} \left(N_q + \frac{1}{2} \pm \frac{1}{2}\right) \tag{8.27a}$$

Here we have chosen a + sign for the **q** vector in the case of phonon absorption and a minus sign for emission. This choice is made for convenience and has no additional meaning since **q** is a vector that can point in any direction. Notice that transversal waves (wave vector **q** perpendicular to displacement **u**) have a vanishing matrix element (the dot product is zero) and therefore do not scatter the electrons. It is only the longitudinal (**q** parallel **u**) phonons that contribute to scattering within the approximations that we have used. The factor N_q [lower sign in Eq. (8.27a)] is for absorption, while for emission a factor $(N_q + 1)$ needs to be implemented, as discussed above. We do not derive here the value of the displacement **u** and refer the interested reader to Conwell (b).

Noting that **k** is changed to $\mathbf{k} \pm \mathbf{q}$ by the scattering process and inserting the value for **u**, we can write the matrix element:

$$|\langle \mathbf{k} \pm \mathbf{q}|H_A'|\mathbf{k}\rangle|^2 = \frac{Z_A^2 \hbar\omega_q}{2V_{\mathrm{ol}}\rho v_s^2}\left(N_q + \frac{1}{2} \pm \frac{1}{2}\right) \tag{8.27b}$$

Here V_{ol} is the volume of the crystal, v_s the velocity of sound (Chap. 2), ρ the mass density (which is 2.3 g/cm^3 for silicon). The minus sign refers to phonon absorption and the plus sign to phonon emission. Notice that the approximation of Eq. (8.21b) gives a matrix element approximately independent of the phonon wave vector **q**.

For optical phonons and deformation potential interaction, one similarly obtains (now by displacing the two face-centered sublattices against each other)

$$|\langle \mathbf{k} \pm \mathbf{q}|H_0'|\mathbf{k}\rangle|^2 = \frac{Z_0^2 \hbar\omega_0}{2V_{\mathrm{ol}}\rho v_s^2}\left(N_q + \frac{1}{2} \pm \frac{1}{2}\right) \tag{8.28}$$

where the index 0 has replaced the indices q and A of Eq. (8.27). Since $\hbar\omega_0$, the optical phonon energy, is approximately independent of the wave vector **q**, N_q is constant and Eq. (8.28) does not depend on **q**.

Although the optical phonon matrix element is almost identical to the acoustic one, its origin is quite different because in the case of optical phonons the two sublattices of the crystal structure vibrate against each other. This is different from the volume change in the case of acoustic phonons. As a consequence, optical phonon scattering is sensitive to the symmetry of the particular range of the band structure in which the electron is scattered. It turns out—and this is very important to remember—that if the electron is scattered close to the Γ minimum (the same is true for X minima) and if the wave function has spherical symmetry, the matrix element vanishes, that is, optical deformation potential scattering is "forbidden" (Harrison). This is the case for the conduction band minimum of GaAs and is a reason for the high electron mobility in GaAs.

Electrons can also be scattered by polar optical phonons (Chap. 2). This mechanism is important in the Γ minimum of GaAs and also in InP because of the lack of deformation potential scattering.

The polar scattering arises from the polarities of the two different atoms in III–V compounds, as illustrated in Fig. 8.5. The displacement, illustrated in the figure, results in a macroscopic field that can scatter the electrons. Since the potential is of long range, the squared matrix element contains a coulombic factor $1/q^2$ (Madelung), which means that scattering preferably takes place at small angles ($\mathbf{k} \approx \mathbf{k}'$). The polar crystal also demonstrates very clearly that the electron is not a single entity anymore but has to be viewed together with the crystal lattice. This new entity is called (in the case of a crystal with polar component) a polaron, which is illustrated in Fig. 8.6.

This lattice distortion leads to a "renormalization" of the bare electron effective mass as calculated from the band structure of the undistorted crystal. In GaAs, InP, InSb, and so on, this renormalization is rather small (see Appendix C).

In heterostructures, more complicated forms of the polar interaction are possible. For example, at the Si-SiO_2 interface, electrons residing in the silicon can still interact with polar modes (remote polar phonon scattering), as decribed by Hess and Vogl.

The calculation of the total phonon scattering rates from the matrix elements requires still more algebra.

Optical deformation potential scattering is relatively easy to calculate. Using Eqs. (8.28) and (8.19), and considering phonon absorption only, we have

$$\frac{1}{\tau_{\text{tot}}^{\text{opd}}} = \frac{\pi Z_0^2 \hbar \omega_0 N_q}{\hbar V_{\text{ol}} v_s^2} \sum_{\mathbf{k}'} \delta(E(\mathbf{k}) - E(\mathbf{k}') + \hbar\omega_0) \tag{8.29}$$

and only the δ function argument depends on $\mathbf{k}'$.

The summation over $\mathbf{k}'$ can be converted into an energy integration

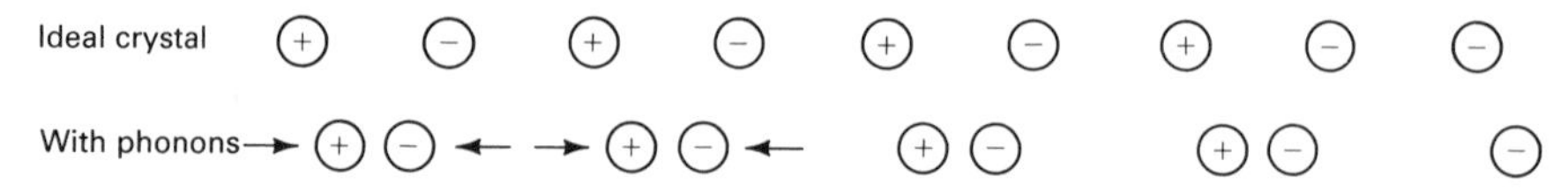

Figure 8.5 A phonon displaces the two sublattices of, for example, GaAs, against each other. For $\mathbf{q} \approx 0$, all negative atoms are displaced toward (or away from) the positive atoms.

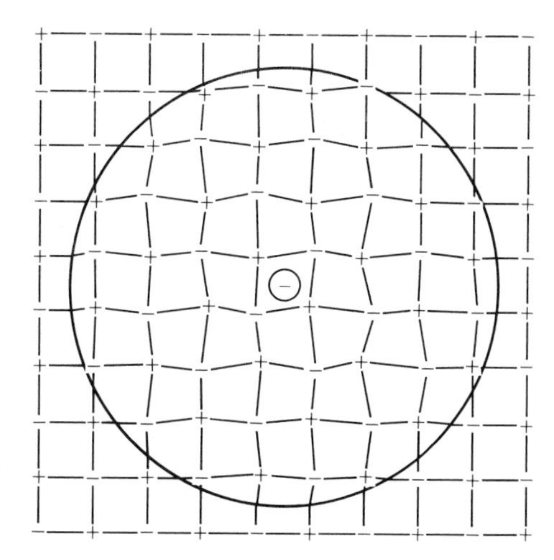

Figure 8.6 Electron plus crystal distortion forming a "polaron." (After Madelung, Fig. 59.)

according to Eq. (6.8). However, care must be taken with the spin factor. Since the electron does not change its spin when interacting with phonons, the initial spin state fixes the final one. Therefore, instead of the density of states with two possibilities for the spin, we now have only one:

$$\sum_{\mathbf{k}'}\delta(E(\mathbf{k}) - E(\mathbf{k}') + \hbar\omega_0) = \frac{V_{\text{ol}}}{2}\int \delta(E - E' + \hbar\omega_0)g(E')dE' = \frac{V_{\text{ol}}}{2}g(E + \hbar\omega_0) \tag{8.30}$$

where g is the density of states given by e.g., Eq. (6.6).

In the current literature one frequently finds a coupling constant D defined by

$$D^2 = Z_0^2\omega_0^2/v_s^2 \tag{8.31}$$

The total scattering rate for optical deformation potential scattering becomes then

$$\frac{1}{\tau_{\text{tot}}^{\text{opd}}} = \frac{D^2\sqrt{m_l^* m_t^{*2}}}{\sqrt{2}\pi\hbar^3\rho\omega_0}\left[N_q\sqrt{E + \hbar\omega_0} + (N_q + 1)\sqrt{E - \hbar\omega_0}\right] \tag{8.32}$$

This equation includes absorption and for $E \geq \hbar\omega_0$, also emission of optical phonons, for the case that the density of states is given by Eq. (6.11). For low temperatures $N_q \approx 0$, and the absorption vanishes. It is important to note, however, that spontaneous emission is still possible.

The possible scattering rate $1/\tau_{\text{tot}}^{\text{ac}}$ for acoustic phonon scattering is not as easy to compute since the phonon energy $\hbar\omega_q$ depends on $q = |\mathbf{k} - \mathbf{k}'|$. The summation (or integration) over the δ function then presents a more elaborate algebraic problem, which has been reviewed in detail by Conwell (c). The result is (in the equipartition approximation)

$$\frac{1}{\tau_{\text{tot}}^{\text{ac}}} = \frac{\sqrt{2}}{\pi}\frac{Z_A^2\sqrt{m_l^* m_t^{*2}}kT}{\hbar^4 v_s}\sqrt{E} \tag{8.33}$$

For polar optical phonon scattering, Conwell (d) gives the result

$$\frac{1}{\tau_{tot}^{po}} = \frac{2e\overline{F}}{\sqrt{2m^*E}}\left[N_q \sinh^{-1}\sqrt{\frac{E}{\hbar\omega_0}} + (N_q+1)\sinh^{-1}\sqrt{\frac{E-\hbar\omega_0}{\hbar\omega_0}}\right] \quad (8.34)$$

where $\overline{F}$ is a polar coupling field (~6000 V/cm for GaAs) and m^* is the effective mass [assumed isotropic for the calculation of Eq. (8.34)].

In contrast to deformation potential scattering, polar optical scattering becomes weaker with higher energy (i.e., the scattering rate decreases). This is because the matrix element is proportional to $1/|\mathbf{k}-\mathbf{k}'|$, which masks the proportionality to the final density of states.

As a further example, we calculate the scattering by N_δ independent potentials, which have the form of a δ function:

$$V_{ext} = V_0\delta(\mathbf{r}) \quad (8.35)$$

The matrix element is then

$$M_{\mathbf{k},\mathbf{k}'} = \frac{1}{V_{ol}}\int_{V_{ol}} e^{-i\mathbf{k}'\cdot\mathbf{r}} V_0\delta(\mathbf{r}) e^{i\mathbf{kr}} d\mathbf{r}$$

$$M_{\mathbf{k},\mathbf{k}'} = eV_0/V_{ol}$$

which means that $M_{\mathbf{k},\mathbf{k}'}$ is independent of $\mathbf{k}$ and $\mathbf{k}'$ (a δ function can provide all momenta $\mathbf{q} = \mathbf{k} - \mathbf{k}'$).

Therefore the total scattering rate becomes

$$\sum_{\mathbf{k}'} S(\mathbf{k},\mathbf{k}') = \frac{2\pi}{\hbar} e^2 \frac{V_0^2}{V_{ol}^2} \sum_{\mathbf{k}'} \delta(E(\mathbf{k}') - E(\mathbf{k})) \quad (8.36)$$

since we assume that the scattering process is elastic.

Transforming the summation into an integration over energy we get

$$\sum_{\mathbf{k}'} S(\mathbf{k},\mathbf{k}') = \frac{\pi}{\hbar} e^2 \frac{V_0^2}{V_{ol}} \int_{-\infty}^{+\infty} g(E')\delta(E'-E)dE' \quad (8.37)$$

$$= \frac{\pi}{\hbar} e^2 V_0^2 g(E)/V_{ol} \quad (8.38)$$

Equation (8.38) shows that the total scattering rate is proportional to the final density of states. This is a very important result. Of course, it is rigorously true only if the matrix element is independent of $\mathbf{k}$ and $\mathbf{k}'$. However, since this proportionality arises from the sum over all possible final states $\sum_{\mathbf{k}'}$ and since the matrix element is in many cases only weakly dependent on $\mathbf{k}$ and $\mathbf{k}'$, the proportionality to the density of final states often holds approximately.

For phonon emission we would obtain a proportionality to $g(E - \hbar\omega)$ and for absorption $g(E + \hbar\omega)$.

Notice again that Eq. (8.37) has been divided by a factor of 2 ($\pi/\hbar$

instead of $2\pi/\hbar$), because in a scattering transition the spin usually does not change and the final spin state is therefore fixed.

The δ function potential is a reasonable model for any potential of very short range. The corresponding mobility is

$$\mu = \frac{|e|\tau}{m^*} \approx \frac{V_{ol}|e|\hbar}{N_\delta m^* 2\pi e^2 V_0^2 g(\frac{3}{2}kT)} \tag{8.39}$$

where N_δ / V_{ol} is the density of scattering centers and $\tau = \langle \tau_{tot} \rangle$.

8.3 IMPORTANT SCATTERING MECHANISMS IN SILICON AND GALLIUM ARSENIDE

Let us now discuss the important scattering mechanisms for electrons and holes in silicon and gallium arsenide.

First we consider electrons in silicon. The silicon conduction band has six equivalent ellipsoidal minima close to the Brillouin zone boundary at the connection line from Γ to X. Within each minimum we have acoustic phonon scattering, ionized impurity scattering, and perhaps some additional scattering mechanisms such as scattering by neutral impurities. At room temperature ionized impurity scattering starts to be important if the impurity density exceeds $\sim 10^{17}$ cm^{-3}. Acoustic phonons are always significant in silicon at room temperature. Optical deformation potential scattering is negligible within the particular minimum since the matrix element vanishes for reasons of symmetry. However, scattering between the minima is significant.

From energy and momentum conservation it is clear that scattering between the minima on a given axis (e.g., [−100], [100]) is different from scattering between minima on different axis (e.g., [−100], [010]). The former process is called g scattering, and the latter is called f scattering. A good approximation for these scattering processes is to use Eq. (8.32) with different coupling constants D and optical phonon energies for the particular processes. A review of all constants and phonon energies has been given by Jacoboni et al.

Table 8.1 lists these constants and additional material parameters for silicon. At higher energies a second conduction band becomes important in silicon, which introduces further complications. The total phonon scattering rate, including all scattering in between minima (intervalley) and the two conduction bands (interband) for acoustic as well as optical deformation potential scattering, has been calculated by Tang et al. and is shown in Fig. 8.7.

The scattering mechanisms in the conduction band of gallium arsenide are very different from silicon. Because of the spherical symmetry of the wave function at Γ, optical deformation potential scattering is zero (the matrix element vanishes because of the symmetry) in the lowest GaAs minimum. The

TABLE 8.1 Material Parameters for Silicon

Bulk Material Parameters		
Lattice constant	5.431	Å
Density	2.329	g/cm^3
Dielectric constant	11.7	
Sound velocity	9.04×10^5	cm/s
***X* Valley**		
Effective masses		
Transverse	0.19	m
Longitudinal	0.916	m
Nonparabolicity	0.5	eV^{-1}
Acoustic deformation potential	9.5	eV
***L* Valley**		
Effective masses		
Transverse	0.12	m
Longitudinal	1.59	m

Phonon Temperature (K)	Deformation Potential D (eV/cm)	Scattering Type
***X-X* Intervalley Scattering**		
220	3×10^7	f
550	2×10^8	f
685	2×10^8	f
140	5×10^7	g
215	8×10^7	g
720	1.1×10^9	g
***X-L* Intervalley Scattering**		
672	2×10^8	
634	2×10^8	
480	2×10^8	
197	2×10^8	

After Tang and Hess.

effective mass in this minimum is very small $m^* = 0.067\, m_0$. Therefore, acoustic phonon scattering also contributes very little. These two facts are the very reason why GaAs exhibits high electron mobility compared to silicon (at room temperature about a factor of 5 higher than silicon). This higher mobility offers advantages from the viewpoint of devices since it gives rise to highest speed and lower power dissipation.

The main scattering mechanisms in GaAs are polar optical scattering and impurity scattering if the density of impurities is significant ($\geq 10^{17}$ cm^{-3} at

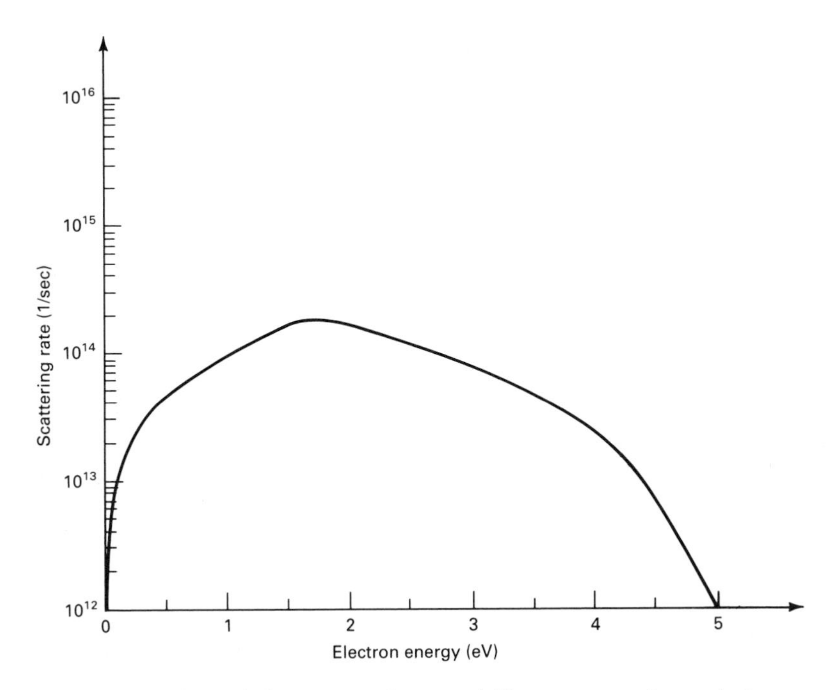

Figure 8.7 The total phonon scattering rate of silicon corresponding to the transport parameters of Table 8.1. (After Tang and Hess, Fig. 6.)

room temperature). Higher in the conduction band additional minima of ellipsoidal form appear, as shown in Fig. 4.4. At an energy of ~0.33 eV above the Γ minimum there are "germaniumlike" minima at *L* and ~0.5 eV above Γ, siliconlike minima are located close to the *X* point. Not too much is known about the details of scattering in these minima. One can assume, however, that the scattering in the *X* minima is very similar to scattering in silicon, and scattering in the *L* minima is similar to that in germanium. There is, of course, also scattering between all these minima such as *X*-*L*, Γ-*X*, Γ-*L*, and so on. A list of coupling constants that describe well a large set of experimental results (Shichijo and Hess) is given in Table 8.2, and the total scattering rate is shown in Fig. 8.8.

The maxima of the valence bands of both Si and GaAs are at Γ and the wave function does not have spherical symmetry. Therefore optical deformation potential scattering is important for holes in both GaAs and Si.

Also important are acoustic phonon scattering (especially for the heavy holes) and impurity scattering. For a detailed discussion, refer to the excellent comprehensive review of Rode.

The number and variety of scattering mechanisms discussed above for bulk semiconductors are enlarged by the presence of interfaces. Interfaces usually necessitate dealing with interface roughness scattering (Ferry et al)

TABLE 8.2 Material Parameters for Gallium Arsenide

Bulk Material Parameters			
Lattice constant	5.642		Å
Density	5.36		g/cm^3
Electron affinity	4.07		eV
Piezoelectric constant	0.16		C/m^2
LO phonon energy	0.036		eV
Longitudinal sound velocity	5.24×10^5		cm/sec
Optical dielectric constant	10.92		
Static dielectric constant	12.90		

Valley-Dependent Material Parameters			
	Γ[000]	L[111]	X[100]
Effective mass (m^*/m_0)	0.067	0.222	0.58
Nonparabolicity (eV^{-1})	0.610	0.461	0.204
Energy band gap relative to valence band (eV)	1.439 (0)	1.769 (0.33)	1.961 (0.522)
Acoustic deformation potential (eV)	7.0	9.2	9.7
Optical deformation potential (eV/cm)	0	3×10^8	0
Optical phonon energy (eV)	—	0.0343	—
Number of equivalent valleys	1	4	3
Intervalley deformation potential D (eV/cm)			
Γ	0	1×10^9	1×10^9
L	1×10^9	1×10^9	5×10^8
X	1×10^9	5×10^8	7×10^8
Intervalley phonon energy (eV)			
Γ	0	0.0278	0.0299
L	0.0278	0.0290	0.0293
X	0.0299	0.0293	0.0299

After Shichijo and Hess.

and with a host of "remote" scattering mechanisms. Remote here means that the scattering agent and the electrons are in different types of semiconductors (as at an interface). Examples are the already mentioned remote phonon scattering and remote impurity scattering.

Remote impurity scattering is an important mechanism in modulation-doped AlGaAs-GaAs layers (or other lattice-matched III–V compound systems). Typically the AlGaAs is doped with donors and the electrons move to the GaAs (which has the smaller band gap), where, because of their high density compared to the unintentional GaAs "background doping," screening is effective [Eq. (8.88)] and the remote impurities contribute only little to the scattering rate (Hess). Consequently, very low total scattering rates and high mobilities can be achieved in these materials (Chapter 12).

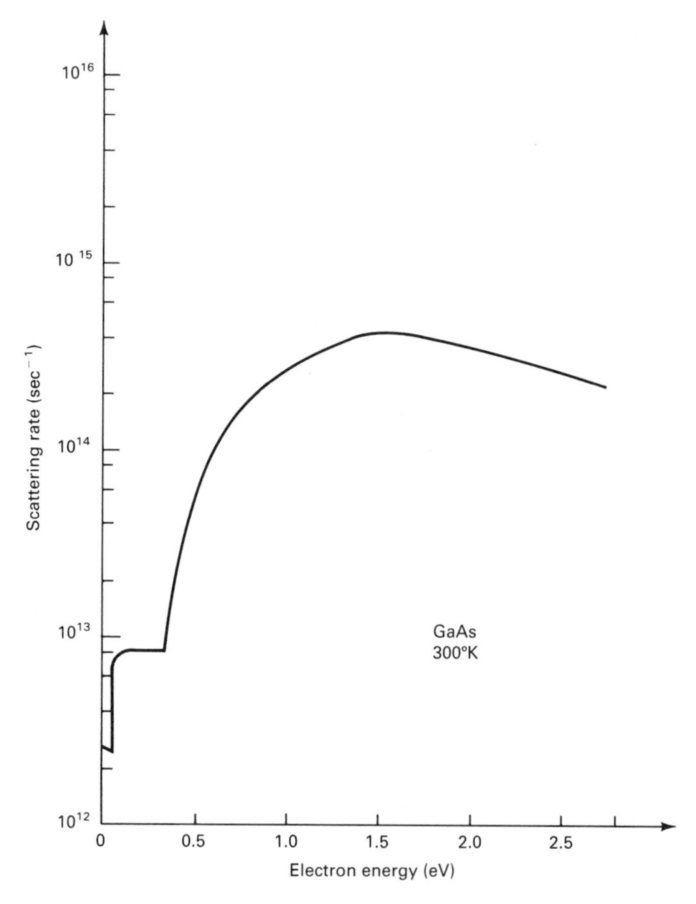

Figure 8.8 Approximate phonon scattering rate of gallium arsenide using the material parameters of Table 8.2. (After Shichijo and Hess, Fig. 7.)

8.4 SCATTERING RATES AND MOBILITY

We now return to the calculation of the mobility and conductivity. How can we calculate these quantities from the scattering mechanisms, when it appears to be very difficult already for one scattering center? This question will be answered below and in Chap. 9.

In Eq. (8.39) we had to use an approximation for the energy since the actual energy of an electron in a semiconductor varies. The electron is accelerated by an electric field; its energy therefore increases and then the electron is scattered to a different point in **k** space. It may or may not have lost energy during the scattering event, depending on the scattering agent. The new **k** wave vector could point in any direction and the electron can now again be accelerated by the electric field or decelerated by the field.

Many such events will happen to an electron as it propagates through a crystal. If we accumulate all of these events in a memory, then at the end we can average over all events and obtain the average velocity or mobility without having to invoke approximations such as Eq. (8.39).

Current computers are able to handle such a project. It turns out, however, that one does not even need to store all events in a memory. One can accumulate averages by using certain "estimators" (described below). We still have to let the electron propagate and be scattered by chance. This type of computation involving chance is called Monte Carlo simulation after the famous gambling place (Las Vegas simulation would also be a good name).

In such a calculation we need to know the probability P that the electron is not scattered in a time interval $[t_1,t_2]$. We know that the total scattering rate per unit time is $\sum_{\mathbf{k}'} S(\mathbf{k},\mathbf{k}')$ and therefore can relate the probability $P(t)$ to the probability $P(t + dt)$. Since P is the probability that the electron is not scattered, we have

$$P(t + dt) = P(t)\left(1 - dt\sum_{\mathbf{k}'} S(\mathbf{k},\mathbf{k}')\right) \tag{8.40}$$

where $dt\sum_{\mathbf{k}'} S(\mathbf{k},\mathbf{k}')$ is the probability of an electron being scattered in the time interval dt.

From the definition of differentiation, Eq. (8.40) is equivalent to

$$\frac{dP}{dt} = -P\sum_{\mathbf{k}'} S(\mathbf{k},\mathbf{k}') \tag{8.41}$$

Solving this differential equation for P in a time interval $[t_1,t_2]$, we have

$$P = \exp\left\{-\int_{t_1}^{t_2} \sum_{\mathbf{k}'} S(\mathbf{k},\mathbf{k}')dt\right\} \tag{8.42}$$

$\sum_{\mathbf{k}'} S(\mathbf{k},\mathbf{k}')$ is generally a function of time since $\mathbf{k}$ can be a function of time, as is seen from the equations of motion, Eqs. (4.18) and (4.19).

We can now design the following numerical scheme to calculate all the transport effects for the electrons (writing $1/\tau_{tot}$ for $\sum_{\mathbf{k}'} S(\mathbf{k},\mathbf{k}')$):

1. We calculate $\tau_{tot}(\mathbf{k})$ for all important scattering mechanisms. If these mechanisms are independent, then the probabilities for an electron being scattered add, which means

$$\frac{1}{\tau_{tot}^{all}(\mathbf{k})} = \frac{1}{\tau_{tot}^{I}(\mathbf{k})} + \frac{1}{\tau_{tot}^{Ph}(\mathbf{k})} + \ldots \tag{8.43}$$

where the superscript I denotes impurity and Ph phonon scattering, respectively.

2. We calculate the time that an electron is not scattered by solving the integral equation, (8.42). This is accomplished in the following way.

Since P is a probability, it will assume random numbers between 0 and 1. We can therefore replace P by such a random number r_d to obtain

$$r_d = \exp\left[-\int_{t_1}^{t_2} \frac{dt}{\tau_{\mathrm{tot}}(t)}\right] \tag{8.44}$$

This equation is now solved for t_2 (t_1 is the time the electron starts).

3. Having calculated t_2, we let a particular electron accelerate freely between t_1 and t_2 and then the electron is scattered to a different $\mathbf{k}'$ vector. Of course, we have to choose the correct $\mathbf{k}'$ vector according to the correct probability given by the dependence of $1/\tau_{\mathrm{tot}}$ on $\mathbf{k}'$. If the matrix element does not depend on $\mathbf{k}'$, any $\mathbf{k}'$ that conserves energy will do.

4. We calculate again a free acceleration time period $[t_2,t_3]$ from

$$r_d = \exp\left[-\int_{t_2}^{t_3} \frac{dt}{\tau_{\mathrm{tot}}(t)}\right] \tag{8.45}$$

by solving Eq. (8.45) for t_3. Again we let the electron accelerate freely in this time period and then scatter, and we continue this whole process for many time periods, typically $\sim 10^5$. At each instance we calculate the important physical quantities—velocity, energy, and so on.

As mentioned before, one does not need to store all the information at all times. Instead, one can accumulate information by using certain estimators that sum and average velocities, energies, and other important physical quantities.

Assume, for example, that we start a certain number of electrons at one end of a semiconductor, as shown in Fig. 8.9. We can then average over physical quantities, such as velocity, in the "bins" shown in Fig. 8.9 and obtain the space- and time-dependent macroscopic results corresponding to the averages.

It is important to note that the simultaneous simulation of many electrons (ensemble Monte Carlo) is not always necessary. If the electric field accelerating the electrons is given (and does not depend on the electron

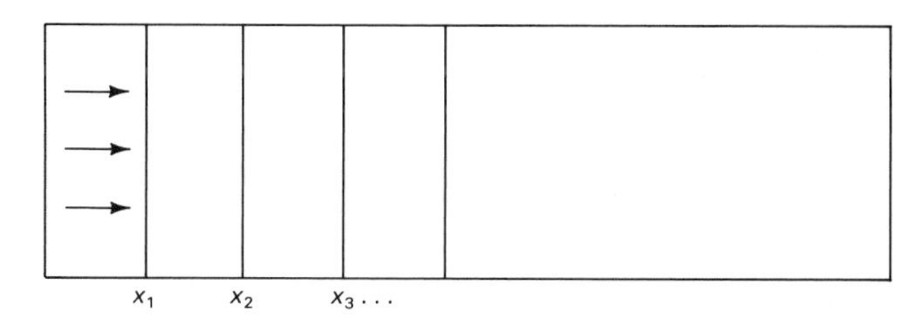

Figure 8.9 Semiconductor sample divided into "bins" for Monte Carlo simulations.

density itself), one can follow the simulation of a single electron over a long time and assume that the time average is equal to the average over the ensemble of electrons that propagates in the real semiconductor.

All the input needed for a Monte Carlo simulation that describes the transport properties are the scattering rate (especially the total scattering rate) and the equations of motion in between scattering, Eqs. (4.18) and (4.19). Both Eqs. (4.18) and (4.19) and the scattering rate depend on band structure.

Monte Carlo simulations, including a complete band structure, have been performed and have been reviewed by Shichijo and Hess. The complete band structure is especially important at high electron energies (>300 meV above E_c in GaAs, >500 meV in Si). If the electrons mainly reside close to the conduction band edge, the effective mass equations can be used, which greatly simplifies the expressions for acceleration and scattering.

With increasing speed and capacity of computers, the Monte Carlo simulation gains increasing importance, and one can foresee the day when all accurate semiconductor device models are performed using this technique.

Device models do not include only electronic transport but also make necessary a detailed knowledge of fabrication processes such as ion implantation, diffusion, and oxidation. The principal method as described above applies also for these processes. An excellent detailed review of Monte Carlo methods has been given by Price. The inclusion of band structure in Monte Carlo simulations has been discussed by Shichijo and Hess.

From the viewpoint of theoretical visualization and simple estimates, an approach different from Monte Carlo is more gratifying. This approach is based on a differential equation (the Boltzmann equation) for the probability distribution f of electrons (in real and $\mathbf{k}$ space).

After the distribution function (nonequilibrium) is known, average values $\langle \mathbf{v} \rangle$ of the velocity $\mathbf{v}$ can be calculated from

$$\langle \mathbf{v} \rangle = \int_{-\infty}^{+\infty} f\mathbf{v}\,d\mathbf{v} \Big/ \int_{-\infty}^{+\infty} f\,d\mathbf{v} \tag{8.46}$$

A similar equation applies for other averages.

A detailed discussion is given in Chap. 9.

RECOMMENDED READINGS

CONWELL, E. M., "High Field Transport in Semiconductors," in *Solid State Physics,* ed. F. Seitz, D. Turnbull, and H. Ehrenreich, Supplement 9. New York: Academic Press, 1967, (a) pp. 112–116; (b) p. 108; (c) pp. 124–127; (d) pp. 155–57.

ESAKI, L., and R. TSU, *IBM Journal of Research and Development,* 14 (1970), 61.

FERRY, D. K., K. Hess, and P. VOGL, "Physics and Modeling of Submicron Insulated-Gate Field-Effect Transistors II," in *VLSI Electronics 2,* ed. N. Einspruch. New York: Academic Press, 1981, p. 72.

HARRISON, W. A., *Physical Review,* 104 (1956), 1281.

HESS, K., *Applied Physics Letters,* 35 (1979), 484–86.

HESS, K., and J. D. DOW, *Solid State Communications,* 40 (1981), 371.

HESS, K., and P. VOGL, *Solid State Communications,* 30 (1979), 807–09.

KUCHAR, F., E. FANTNER, and K. HESS, *Journal of Physics C: Solid State Physics,* 9 (1976), 3165–71.

MADELUNG, O., *Introduction to Solid State Theory,* Springer Series in Solid-State Sciences 2, ed. M. Cardona, P. Fulde, and H. J. Queisser. New York: Springer-Verlag, 1978, pp. 183–87.

PRICE, P. J., "Monte Carlo Calculation of Electron Transport in Solids," in *Semiconductors and Semimetals,* ed. R. K. Willardson and A. C. Beer, vol. 14. New York: Academic Press, 1979, pp. 249–305.

RODE, D. L., "Low-Field Electron Transport," in *Semiconductors and Semimentals,* ed. R. K. Willardson and A. C. Beer, vol. 10. New York: Academic Press, 1975, pp. 1–89.

SHICHIJO, H., and K. HESS, *Physical Review B,* 23 (1981), 4197–4207.

TANG, J. Y., and K. HESS, *Journal of Applied Physics,* 54 (1983), 5139–43.

ZIMAN, J. M., *Elements of Advanced Quantum Theory,* Cambridge: Cambridge University Press, 1969.

PROBLEMS

8.1. Calculate the scattering probability $S(\mathbf{k},\mathbf{k}')$ for a scattering potential $U(\mathbf{r}) = U_0 \dfrac{e^{-|\mathbf{r}|/\lambda}}{|\mathbf{r}|^2}$ (use spherical coordinates relative to $\mathbf{q} = \mathbf{k} - \mathbf{k}'$).

8.2. Show that random numbers x with a given distribution $f(x)$ in an interval $a \leq x \leq b$ can be obtained (in principle) starting with random numbers r uniformly distributed in the interval (0,1). Assume that $f(x)$ is normalized on (a,b). [*Hint:* Let $F(x') = \int_a^{x'} f(x)dx$, and use the fact that $dF = f(x)dx$.] Check your result for $f(x) = \text{constant} = 1/(b - a)$.

8.3. Use Eq. (7.14) to derive the screened (by free electrons) electron-phonon interaction of the deformation potential for absorption of acoustic phonons. Discuss limiting cases $q_s \rightarrow 0,\infty$.

9

THE BOLTZMANN TRANSPORT EQUATION AND ITS APPROXIMATE SOLUTIONS

9.1 DERIVATION OF THE BOLTZMANN EQUATION

The Boltzmann transport equation (BTE) describes the nonequilibrium distribution $f(\mathbf{v},\mathbf{r},t)$ of electrons as a function of velocity $\mathbf{v}$, space coordinate $\mathbf{r}$, and time t. The quantity $f(\mathbf{v},\mathbf{r},t)d\mathbf{v}d\mathbf{r}$ defines the probability of finding an electron (hole) in a velocity range between $\mathbf{v}$ and $\mathbf{v}+d\mathbf{v}$ having a space coordinate between $\mathbf{r}$ and $\mathbf{r}+d\mathbf{r}$. Therefore, the BTE is not a quantum mechanical equation as it specifies $\mathbf{v}$ and $\mathbf{r}$ simultaneously, ignoring Heisenberg's uncertainty relations. It is, however, sufficient for many semiconductor transport problems.

The BTE is most easily derived by considering the rate of change of particles in phase space that can be treated by considering separately a cube in real space (Fig. 9.1) and one in velocity space (Fig. 9.2).

We first calculate how many electrons arrive from the left, enter the cube in the left $dydz$ plane, and how many leave at the corresponding plane to the right in a time period dt. Since the x direction travel distance of electrons with velocity $\mathbf{v}$ is $v_x\, dt$ we have

$$\text{incoming:} \quad f(\mathbf{v},\mathbf{r},t)d\mathbf{v}dydzv_x\, dt \tag{9.1}$$

$$\text{outgoing:} \quad f(\mathbf{v}, x + dx/y/z/, t)d\mathbf{v}dydzv_x\, dt \tag{9.2}$$

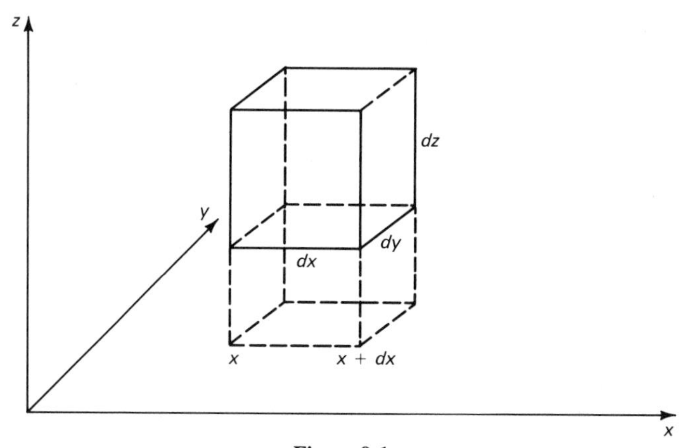

Figure 9.1

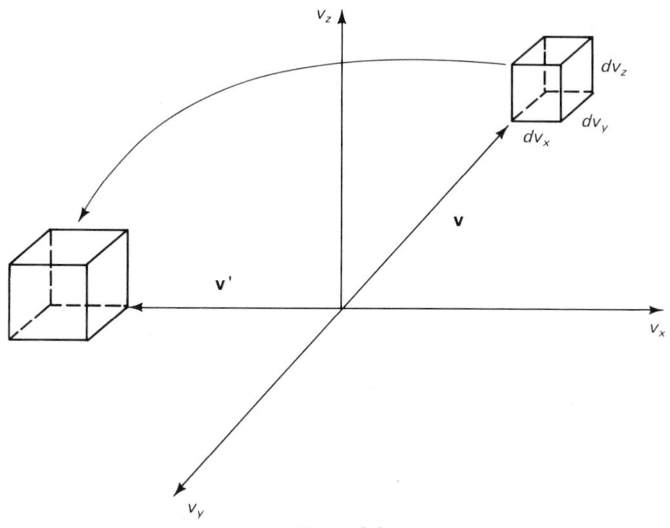

Figure 9.2

and the net particle gain, therefore, is

$$-v_x[f(\mathbf{v}, x + dx/y/z, t) - f(\mathbf{v}, \mathbf{r}, t)]dydzd\mathbf{v}dt$$

$$= - v_x \frac{\partial f}{\partial x} dxdydzd\mathbf{v}dt \tag{9.3}$$

$$\text{or} \quad = - \mathbf{v} \cdot \nabla f d\mathbf{v} d\mathbf{r} dt \quad \text{in three dimensions}$$

In a totally analogous manner, we obtain the change of electrons at **v** because of accelerations (**v** changes in velocity space):

$$-\dot{\mathbf{v}} \cdot \nabla_{\mathbf{v}} f d\mathbf{v} d\mathbf{r} dt \tag{9.4}$$

where $\dot{\mathbf{v}} = -\frac{1}{m}eF$ and F is the electric field.

There is still another possibility to change the number of electrons with velocity $\mathbf{v}$ at $\mathbf{r}$. The electrons can be scattered and change their velocity from $\mathbf{v}$ to $\mathbf{v}'$ at a given point $\mathbf{r}$ in space. Figure 9.2 shows the two infinitesimal volumes in $\mathbf{v}$ space to illustrate the scattering events. The outgoing (out of state $\mathbf{v}$) electrons are

$$\text{out} = -\sum_{\mathbf{v}'} S(\mathbf{v},\mathbf{v}')f(\mathbf{v},\mathbf{r},t)d\mathbf{v}d\mathbf{r}dt \tag{9.5}$$

The factor $f(\mathbf{v})$ is necessary because an electron has to be in the $\mathbf{v}$ state to be scattered out. In degenerate systems (Fermi statistics), a factor $1 - f(\mathbf{v}')$ arises from the Pauli principle. Notice that this treatment implies that the collision processes are *completed locally in space and time.* The incoming (into the $\mathbf{v}$ state) electrons are

$$\text{in} = \sum_{\mathbf{v}'} S(\mathbf{v}',\mathbf{v})f(\mathbf{v}',\mathbf{r},t)d\mathbf{v}d\mathbf{r}dt \tag{9.6}$$

for obvious reasons. Again, the Pauli principle will call for a factor $1 - f(\mathbf{v})$. The Boltzmann equation is obtained by balancing the particle numbers and the change in f given by the net change of incoming and outgoing electrons. Therefore, we have

$$\begin{aligned}\frac{\partial f(\mathbf{v},\mathbf{r},t)}{\partial t} = &-\mathbf{v}\cdot\nabla f(\mathbf{v},\mathbf{r},t) - \frac{1}{m}\mathbf{F}_0\cdot\nabla_{\mathbf{v}} f(\mathbf{v},\mathbf{r},t) \\ &+ \sum_{\mathbf{v}'}\{f(\mathbf{v}',\mathbf{r},t)S(\mathbf{v}',\mathbf{v}) - f(\mathbf{v},\mathbf{r},t)S(\mathbf{v},\mathbf{v}')\}\end{aligned} \tag{9.7}$$

where F_0 is the force ($-eF$ for an electric field F).

This equation is now more than 150 years old. However, because of the complicated integral form (sum) of the scattering term, explicit solutions are possible only in very special cases. Numerous approaches are known from the transport theory of gases. Some of them can be directly applied to semiconductors. It is usually demonstrated that the complicated collision sum can be replaced under certain conditions by a relaxation time if the collisions are elastic.

For most practical cases in semiconductors, however, the following approach is more appropriate. We partition the distribution functions into a part f_0, which is even in the velocity, and a contribution f_1, which is odd; that is

$$f_0(\mathbf{v}) = f_0(-\mathbf{v}), f_1(-\mathbf{v}) = -f_1(\mathbf{v}) \tag{9.8}$$

Furthermore, we assume

$$S(v_i,v_i') = S(v_i,-v_i') = S(-v_i,v_i') = S(-v_i,-v_i') \tag{9.9}$$

Here $i = x,y,z$. In other words, the transition probability is even in all velocity components.

Equation (9.9) is exact for optical deformation potential scattering and a good approximation for acoustic phonon scattering. Polar optical scattering cannot be treated this way because the squared matrix element is proportional to $|\mathbf{k}-\mathbf{k}'|^{-2}$. With the above assumptions, the collision term in the Boltzmann equation reduces to

$$\sum_{\mathbf{v}'}\{[f_0(\mathbf{v}',\mathbf{r},t)S(\mathbf{v}',\mathbf{v}) - f_0(\mathbf{v},\mathbf{r},t)S(\mathbf{v},\mathbf{v}')] - f_1(\mathbf{v},\mathbf{r},t)S(\mathbf{v},\mathbf{v}')\} \tag{9.10}$$

The term containing $f_1(\mathbf{v}',\mathbf{r}'t)$ vanishes because f_1 is odd and S is even in $\mathbf{v}'$. In equilibrium the terms containing f_0 cancel each other since f_0 is the Maxwell-Boltzmann distribution and we have detailed balance—that is, an equal number of particles is scattered in and out. Under nonequilibrium conditions, this is not true anymore, and f_0 is different from the Maxwell-Boltzmann distribution. Then we denote the terms containing f_0 symbolically by

$$\left.\frac{\partial f_0}{\partial t}\right|_{\text{coll}}$$

and the Boltzmann equation can be written as

$$\frac{\partial f(\mathbf{v},\mathbf{r},t)}{\partial t} = -\mathbf{v}\cdot\nabla f(\mathbf{v},\mathbf{r},t) - \frac{1}{m}\mathbf{F}_0\cdot\nabla_{\mathbf{v}} f(\mathbf{v},\mathbf{r},t) + \left.\frac{\partial f_0}{\partial t}\right|_{\text{coll}} - f_1(\mathbf{v},\mathbf{r},t)\sum_{\mathbf{v}'}S(\mathbf{v},\mathbf{v}') \tag{9.11}$$

As we will see below, for the case of small electric fields, $\mathbf{F}$, the term $\left.\frac{\partial f_0}{\partial t}\right|_{\text{coll}}$ vanishes (as F^2) because f_0 approaches the Maxwellian distribution function and the Boltzmann equation reduces to a partial differential equation of first order which is easy to solve, especially if the coefficients are constant. In case the term $\left.\frac{\partial f_0}{\partial t}\right|_{\text{coll}}$ is not negligible, explicit solutions can still be obtained as long as one can separate the BTE into separate equations for f_0 and f_1. Such a solution will be derived below. The case of polar optical scattering requires special treatment (Conwell, a).

In a semiconductor the equations of motion are close to the free electron equation only at the minimum of the conduction band (maximum of the valence band) where the mass m of the electron has to be replaced by an effective mass m^*. For the more general case of the one band approximation, we can use Eqs. (4.18) and (4.19) to obtain

$$\frac{\partial f(\mathbf{k},\mathbf{r},t)}{\partial t} = -\frac{1}{\hbar}\nabla_{\mathbf{k}}\mathbf{E}(\mathbf{k})\cdot\nabla f(\mathbf{k},\mathbf{r},t) - \frac{\mathbf{F}_0}{\hbar}\cdot\nabla_{\mathbf{k}} f(\mathbf{k},\mathbf{r},t) - f_1\frac{V_{\text{ol}}}{8\pi^3}\int d\mathbf{k}' S(\mathbf{k},\mathbf{k}') + \left.\frac{\partial f_0}{\partial t}\right|_{\text{coll}} \tag{9.12}$$

Notice that the term proportional to f_1 is just $1/\tau_{\text{tot}}$. Therefore, Eq. (9.12) is called the relaxation time approximation of the Boltzmann equation (τ_{tot} is the total relaxation time).

If Eq. (9.9) does not hold, the scattering is not randomizing; that is, the electron can be scattered in certain preferable directions. Both ionized impurity scattering and polar optical scattering scatter mainly in the forward direction since the matrix element is inversely proportional to $|\mathbf{k} - \mathbf{k}'|$ and becomes largest for $|\mathbf{k} - \mathbf{k}'| \to 0$. A relaxation time, then, in general cannot be defined, although it may still be a good approximation for certain energy ranges.

For nonrandomizing elastic scattering processes, detailed balance

$$\left.\frac{\partial f_0}{\partial t}\right|_{coll} = 0$$

and isotropic effective mass, it turns out that $1/\tau_{tot}(\mathbf{k})$ must be replaced by $1/\tau_m(\mathbf{k})$ with

$$\frac{1}{\tau_m} = \frac{V_{ol}}{8\pi^3} \int d\mathbf{k}' S(\mathbf{k},\mathbf{k}')(1 - \cos\theta) \tag{9.13}$$

which replaces Eq. (8.19) and differs from Eq. (8.19) by the factor $(1 - \cos\theta)$.

Where θ is again the angle between $\mathbf{k}$ and $\mathbf{k}'$. $1/\tau_m$ is called the momentum scattering rate (τ_m is the momentum relaxation time), since scattering events in the direction of the original momentum are not counted as $(1 - \cos\theta) = 0$. Only if the momentum direction changes do we have contributions to $1/\tau_m$.

It seems natural that only these processes should be counted, and indeed, for elastic scattering nothing happens at all if $\mathbf{k}$ does not change direction. For inelastic processes this is not the case. Imagine an electron at an energy of the optical phonon energy and emitting a phonon. This electron loses all its energy in the scattering process even if $\mathbf{k}$ and $\mathbf{k}'$ are parallel.

9.2 SOLUTIONS OF THE BOLTZMANN EQUATION

A simple solution of Eq. (9.12) is obtained for time- and space-independent distribution functions for the limiting case of small fields. Then the BTE reads in the relaxation time approximation

$$\frac{\mathbf{F}_0}{\hbar} \cdot \nabla_{\mathbf{k}} f(\mathbf{k},\mathbf{r},t) = \frac{-f_1}{\tau_{tot}} \tag{9.14}$$

Since $f = f_0 + f_1$ with the assumption that $f_1 \ll f_0$ and that f_0 approaches the Maxwell-Boltzmann distribution for small forces $\mathbf{F}_0$, we obtain

$$f_1 = -\tau_{tot} \frac{\mathbf{F}_0}{\hbar} \cdot \nabla_{\mathbf{k}} e^{(E_F - E)/kT} \tag{9.15}$$

For higher electric fields $\left.\frac{\partial f_0}{\partial t}\right|_{coll}$ does not usually vanish, and the solution of the BTE is much more involved.

Assuming again steady state ($\partial f/\partial t = 0$) and spatial homogeneity ($\nabla f = 0$), the BTE reads

$$\frac{\mathbf{F}_0}{\hbar} \cdot (\nabla_{\mathbf{k}} f_0(\mathbf{k}) + \nabla_{\mathbf{k}} f_1(\mathbf{k})) = -\frac{f_1}{\tau_{tot}(\mathbf{k})} + \left.\frac{\partial f_0}{\partial t}\right|_{coll} \tag{9.16}$$

This equation contains even and odd functions: f_1 *and* $\nabla_{\mathbf{k}} f_0$ are odd; $\nabla_{\mathbf{k}} f_1$ and $\left.\frac{\partial f_0}{\partial t}\right|_{coll}$ are even. The even and odd functions must cancel separately, and we therefore obtain two equations:

$$\frac{\mathbf{F}_0}{\hbar} \cdot \nabla_{\mathbf{k}} f_0(\mathbf{k}) = -\frac{f_1}{\tau_{tot}(\mathbf{k})} \tag{9.17a}$$

and

$$\frac{\mathbf{F}_0}{\hbar} \cdot \nabla_{\mathbf{k}} f_1(\mathbf{k}) = -\left.\frac{\partial f_0}{\partial t}\right|_{coll} \tag{9.17b}$$

The solution of this coupled system of equations has been treated by Hess and Sah for a two-dimensional electron gas and is reviewed by Conwell for three-dimensional systems. We follow Hess and Sah and derive the collision term (operator).

For optical deformation potential scattering we have from Eq. (8.28)

$$S(\mathbf{k},\mathbf{k}') = \overline{C}\left(N_q + \frac{1}{2} \pm \frac{1}{2}\right)\delta(E - E' \mp \hbar\omega_0) \tag{9.18}$$

where $\overline{C}$ is a constant. Therefore

$$-\left.\frac{\partial f_0}{\partial t}\right|_{coll}^{opt} = \int d\mathbf{k}'\overline{C}\left[f_0(E)\left(N_q + \frac{1}{2} \pm \frac{1}{2}\right)\delta(E - E' \mp \hbar\omega_0) \right.$$
$$\left. - f_0(E')N_q\delta(E - E' - \hbar\omega_0) - f_0(E')(N_q + 1)\delta(E - E' + \hbar\omega_0)\right] \tag{9.19}$$

The integration over $\mathbf{k}'$ is easily converted in an integration over energy by introducing the density of states $g(E)$. In three dimensions $g(E) \propto \sqrt{E}$, which introduces cumbersome square roots. In two dimensions $g(E)$ is a constant, as can easily be seen from our treatment in Chap. 6.

Therefore, for a two-dimensional system

$$-\left.\frac{\partial f_0}{\partial t}\right|_{coll}^{opt} = \overline{C}\frac{g}{2}[H(E - \hbar\omega)f_0(E)(N_q + 1) + f_0(E)N_q$$
$$- f_0(E + \hbar\omega_0)(N_q + 1) + H(E - \hbar\omega_0)f_0(E - \hbar\omega_0)N_q] \tag{9.20}$$

The step function $H(E - \hbar\omega_0)$ (defined as $H = 0$ for $E \leq \hbar\omega_0$ and 1 otherwise) has to be introduced because there can be no electrons scattered to negative energy values. The corresponding three-dimensional operator is almost the same. Only the step functions are replaced by square roots.

The collision operator for two-dimensional acoustic phonon scattering can be derived in a similar fashion, although the integration over $\mathbf{k}'$ is not as simple since the phonon energy depends on $\mathbf{k}'$. After rather tedious algebra (Hess and Sah), one obtains

$$-\left.\frac{\partial f_0}{\partial t}\right|_{\text{coll}}^{\text{ac}} = S_A \frac{\partial}{\partial \bar{x}} \bar{x} \left(\frac{\partial f_0}{\partial \bar{x}} + f_0\right) \tag{9.21}$$

where $\bar{x} = E/kT$ and

$$S_A = 2(m^*)^2 Z_A^2/\hbar^3\rho \tag{9.22}$$

We proceed to solve the Boltzmann equation, assuming acoustic phonon scattering only. Notice that in this case τ_{ac} is energy independent since the two-dimensional density of states is constant.

Using the chain rule, we obtain from Eq. (9.17a)

$$f_1 = -\tau_{\text{ac}} \frac{F_0}{\hbar} \cdot \nabla_{\mathbf{k}} E(\mathbf{k}) \frac{\partial f_0}{\partial E} \tag{9.23}$$

and Eq. (9.17b) becomes, by using Eq. (9.21)

$$\frac{\mathbf{F}_0}{\hbar} \cdot \nabla_{\mathbf{k}} f_1 = +S_A \frac{\partial}{\partial \bar{x}} \bar{x} \left(\frac{\partial f_0}{\partial \bar{x}} + f_0\right) \tag{9.24}$$

Therefore

$$\frac{\mathbf{F}_0}{kT\hbar} \cdot \nabla_{\mathbf{k}} \left[\tau_{\text{ac}} \frac{\mathbf{F}_0}{\hbar} \cdot \nabla_{\mathbf{k}} E(\mathbf{k}) \frac{\partial f_0}{\partial \bar{x}}\right] = -S_A \frac{\partial}{\partial \bar{x}} x \left(\frac{\partial f_0}{\partial \bar{x}} + f_0\right) \tag{9.25}$$

Using $E = \hbar^2 k^2/2m^*$ or Eq. (4.18) one obtains

$$\frac{\mathbf{F}_0}{kT\hbar} \tau_{\text{ac}} \cdot \nabla_{\mathbf{k}} \left[\mathbf{F}_0 \cdot \mathbf{v} \frac{\partial f_0}{\partial \bar{x}}\right] = -S_A \frac{\partial}{\partial \bar{x}} \bar{x} \left(\frac{\partial f_0}{\partial x} + f_0\right) \tag{9.26}$$

and by applying the chain rule again (the only variable is $\bar{x}$ since we have excluded space dependences)

$$\frac{F_0^2 \tau_{\text{ac}}}{kTm^*} \frac{\partial f_0}{\partial \bar{x}} + \frac{\tau_{\text{ac}}(\mathbf{F}_0 \cdot \mathbf{v})^2}{(kT)^2} \frac{\partial^2 f_0}{\partial \bar{x}^2} = -S_A \frac{\partial}{\partial \bar{x}} \bar{x} \left(\frac{\partial f_0}{\partial \bar{x}} + f_0\right) \tag{9.27}$$

The vector product $(\mathbf{F}_0 \cdot \mathbf{v})^2$ is still an obstacle for an explicit solution and is usually (Conwell, b) replaced by an average value $\langle|\mathbf{F}_0|^2|\mathbf{v}|^2 \cos^2\psi\rangle = |\mathbf{F}_0|^2|\mathbf{v}|^2/2 = F_0^2 E/m^*$. Then Eq. (9.27) becomes

$$\frac{F_0^2 \tau_{\text{ac}}}{m^* kT} \frac{\partial}{\partial \bar{x}} \bar{x} \frac{\partial f_0}{\partial \bar{x}} = -S_A \frac{\partial}{\partial \bar{x}} \bar{x} \left(\frac{\partial f_0}{\partial \bar{x}} + f_0\right) \tag{9.28}$$

The boundary condition to solve this equation is that f_0 vanishes exponentially when $\bar{x}$ (the energy) approaches infinity. The first integration constant is therefore zero, which gives

$$\frac{\partial f_0}{\partial \bar{x}}\left(\frac{F_0^2\tau_{ac}}{kTm^*}+S_A\right)=-S_A f_0 \tag{9.29}$$

This has the solution

$$f_0 = C \exp\left[\left(\frac{-S_A}{\dfrac{F_0^2\tau_{ac}}{m^*kT}+S_A}\right)\bar{x}\right] \tag{9.30}$$

The force $\mathbf{F}_0$ can be replaced, for example, by an electric field $\mathbf{F}_0 = -e\mathbf{F}$. It is very important to note that f_0 still looks like a Maxwell-Boltzmann distribution. However, the equilibrium temperature T has to be replaced by

$$T_c = T\left(1+\frac{e^2F^2\tau_{ac}}{kTm^*S_A}\right) \tag{9.31}$$

The equilibrium temperature T is, for small fields, the temperature of the crystal lattice T. However, from here on we distinguish between the lattice temperature T_L and the carrier temperature T_c.

The constant C in Eq. (9.30) can be determined from the boundary condition of constant carrier concentration:

$$n = \int_0^\infty f_0 g(E)dE$$

Since g is constant in two-dimensional systems, we have

$$n = g\int_0^\infty Ce^{-E/kT_c}\,dE$$

which gives

$$C = \frac{n}{gkT_c}$$

We can then write Eq. (9.30) exactly in the form of a Maxwell-Boltzmann distribution and therefore define

$$C = \exp\,(E_{QF}/kT_c) \tag{9.32}$$

E_{QF} is different from the Fermi energy (we do not have equilibrium) and is called the quasi-Fermi energy (the term imref, which is Fermi backwards, is sometimes used).

Equation (9.32) gives

$$E_{QF} = kT_c \ln\,(n/gkT_c) \tag{9.33}$$

Our approximations have been such that the value of T_c calculated from Eq. (9.31) is not very realistic for real semiconductors (we neglected optical phonons in the collision operator). However, the derivation reveals an important idea: If electric fields are applied to semiconductors, the even part of the distribution function may still be Maxwell-Boltzmann–like. The temperature has to be replaced by a carrier temperature T_c and the quantity corresponding to the Fermi energy becomes also a function of T_c, as can be seen from Eq. (9.33).

Later in this chapter we will see that even under less stringent approximations T_c assumes the form

$$T_c = T_L + S(F) \tag{9.34}$$

where the function $S(F)$ vanishes at least as F^2 for $F \to 0$ and T_L is the temperature of the crystal lattice, that is, the temperature without applied field.

As emphasized above, we have to distinguish from now on between electron and lattice temperatures since the two are not necessarily the same. The reason is simply that electrons gain energy from the field and can lose it only if their temperature is raised above the equilibrium value.

9.3 DISTRIBUTION FUNCTION AND CURRENT DENSITY

Having calculated the distribution function, we can obtain the current density from the definition, Eq. (4.29). The integral over $\mathbf{v}f_0$ vanishes since f_0 is even, and we have

$$\mathbf{j} = -\frac{e}{4\pi^3}\int \mathbf{v}f_1\,d\mathbf{k} \tag{9.35}$$

From Eq. (9.17a) $\mathbf{j}$ becomes

$$\mathbf{j} = -\frac{e^2}{4\pi^3}\int \tau_{\text{tot}}(\mathbf{k})\,\frac{\partial f_0}{\partial E}\,\mathbf{v}(\mathbf{v}\cdot\mathbf{F})\,d\mathbf{k} \tag{9.36}$$

In components Eq. (9.36) reads

$$\sigma_{li} = -\frac{e^2}{4\pi^3}\int \tau_{\text{tot}}(\mathbf{k})\,\frac{\partial f_0}{\partial E}\,v_l v_i\,d\mathbf{k} \tag{9.37}$$

where σ_{li} is a component of the conductivity matrix as defined by Eq. (3.3).

If $\tau(\mathbf{k})$ depends only on the absolute value of $\mathbf{k}$ (or the energy) and if the band structure is isotropic, σ becomes diagonal—that is, $\sigma = \sigma_{li}\delta_{l,i}$.

From the definition of the mobility ($\sigma = en\mu$), one obtains

$$en\mu = -\frac{e^2}{4\pi^3}\int \tau_{\text{tot}}(\mathbf{k})\,\frac{\partial f_0}{\partial E}\,v_i^2\,d\mathbf{k} \tag{9.38}$$

where $i = x$, y, or z. Equation (6.7) then gives

$$\mu = -\frac{e}{4\pi^3} \frac{\int \tau_{tot}(\mathbf{k}) \frac{\partial f_0}{\partial E} v_i^2 \, d\mathbf{k}}{\frac{1}{4\pi^3} \int f_0 \, d\mathbf{k}} \tag{9.39}$$

Assuming that f_0 is a Maxwellian distribution at carrier temperature T_c and writing a specific velocity component v_i in polar coordinates (e.g., $v_z = v \cos \theta$), we obtain in the effective mass approximation

$$\mu = \frac{e\langle \tau_{tot} \rangle}{m^*} = \frac{\frac{2}{3} e \int_0^\infty \tau_{tot}(\mathbf{k}) e^{-E/kT_c} \frac{E}{kT_c m^*} g(E) \, dE}{\int_0^\infty e^{-E/kT_c} g(E) \, dE} \tag{9.40}$$

Here we have made use of the facts that in the numerator $\int_0^\pi \cos^2 \theta \sin \theta \, d\theta = 2/3$ (in the denominator $\int_0^\pi \sin \theta \, d\theta = 2$) and $m^* v^2/2 = E$.

For the average value $\langle \tau_{tot} \rangle$ of τ_{tot}, one therefore obtains the equation

$$\langle \tau_{tot} \rangle = \frac{\int_0^\infty \tau_{tot}(\mathbf{k}) e^{-\bar{x}} \bar{x}^{3/2} d\bar{x}}{\frac{3}{2} \int_0^\infty e^{-\bar{x}} \bar{x}^{1/2} d\bar{x}} \tag{9.41}$$

which is equivalent to

$$\langle \tau_{tot} \rangle = \frac{1}{\Gamma(5/2)} \int_0^\infty \tau_{tot}(\mathbf{k}) e^{-\bar{x}} \bar{x}^{3/2} d\bar{x} \tag{9.42}$$

where $\bar{x} = E/kT_c$.

In some cases $\tau_{tot}(\mathbf{k})$ follows a power law in energy:

$$\tau_{tot}(\mathbf{k}) = \tau_0 \bar{x}^P \tag{9.43}$$

Then

$$\langle \tau_{tot} \rangle = \frac{\tau_0 \Gamma(p + 5/2)}{\Gamma(5/2)} \tag{9.44}$$

In the following, we discuss the temperature-dependent mobility for several scattering mechanisms. It is important to realize that τ_{tot} may depend on the carrier temperature and on the lattice temperature at the same time. The carrier temperature dependence is introduced by the energy dependence of τ_{tot}, and the dependence on lattice temperature comes from the phonon occupation number N_q.

For acoustic phonon scattering we have, from Eq. (8.33):

$$1/\tau_{ac} \propto \sqrt{E} \, kT_L \tag{9.45}$$

Therefore,

$$\tau_{ac} \propto \frac{(\overline{x})^{-1/2}}{kT_L\sqrt{kT_c}} \tag{9.46}$$

For optical phonons no simple expression can be found, as can be seen from Eqs. (9.20) and (8.32). At very high electron temperatures, however, optical phonons behave similarly to acoustic phonons [see Eq. (8.32)] and

$$\tau_{op} \propto \frac{\overline{x}^{-1/2}}{\sqrt{kT_c}} \tag{9.47}$$

The lattice temperature dependence is still complicated because of the exponents in N_q.

For the case $T_c = T_L$, τ_{op} can be approximated by $\tau_{op} \propto T_L^p$, where p is close to ~2 in silicon and GaAs.

Impurity scattering does, of course, not depend on the lattice temperature (N_q is not involved).

There are two limiting cases of impurity scattering, depending on the strength of screening. For very strong screening, we have

$$1/\tau_{imp} \propto \sqrt{E} \tag{9.48}$$

This square root of energy comes from the density of states, while the matrix element is independent of energy since the screened potential is of short range [see the matrix element of the δ function potential, and Eq. (8.18) with large q_s].

For weak screening (small q_s), on the other hand,

$$\tau_{imp} \propto E^{3/2} \tag{9.49}$$

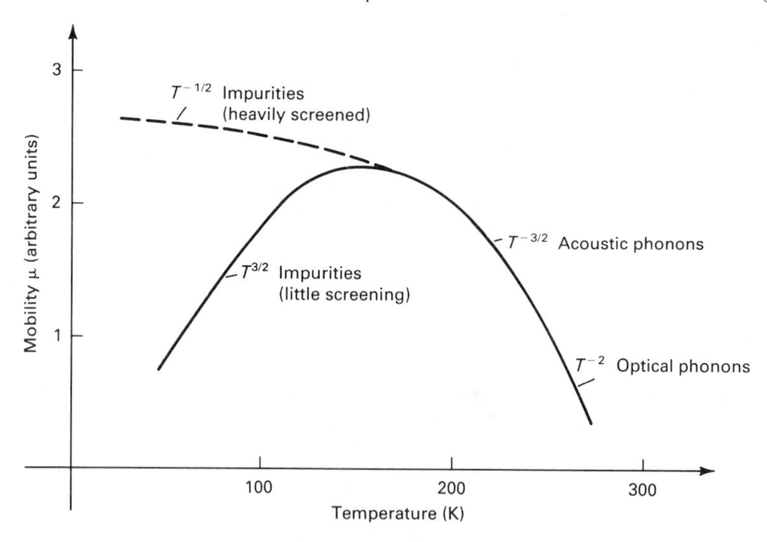

Figure 9.3 Typical temperature dependence of the mobility in semiconductors.

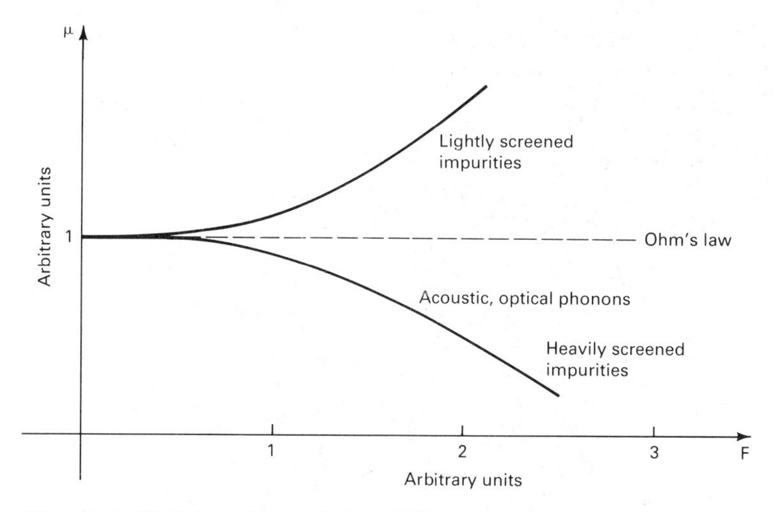

Figure 9.4 Field dependences of the mobility for various scattering mechanisms.

This proportionality is only approximate and can be verified from Eq. (8.20) for weak screening.

Ionized impurity scattering can therefore be proportional to both $T_c^{-1/2}$ and to $T_c^{3/2}$, depending on screening. This gives the typical temperature dependence of the mobility in semiconductors as shown in Fig. (9.3).

The field dependence follows from the proportionalities to T_c as outlined above. T_c increases with the electric field resulting in the schematic field dependences shown in Fig. 9.4.

RECOMMENDED READINGS

CONWELL, E. M., "High Field Transport in Semiconductors," in *Solid State Physics,* ed. F. Seitz, D. Turnbull, and H. Ehrenreich, Supplement 9. New York: Academic Press, 1967.

HESS, K., and C. T. SAH, *Physical Review,* B10 (1974), 3375–86.

PROBLEMS

9.1. Solve the Boltzmann equation for constant relaxation time τ_{tot} in a time-independent, uniform electric field F in one dimension:

$$\frac{-eF}{m^*}\frac{\partial f(v)}{\partial v} = -\frac{(f(v) - f_0(v))}{\tau_{tot}}$$

(v = velocity) where $f_0(v) = n\left(\frac{2\pi kT}{m^*}\right)^{-1/2} \exp\left(-\frac{m^* v^2}{2kT}\right)$ is the equilibrium distribution function (in this case, Maxwell-Boltzmann). (*Hint:* Integrate from v to ∞; your answer will contain an error function.)

9.2. Using the effective mass theorem and Eqs. (4.21) and (9.37), derive the conductivity mass, Eq. (4.22), by considering a two-dimensional system with four equivalent constant energy ellipses along the [1,1], [−1,1], [1,−1], and [−1,−1] directions. Each ellipse is characterized in the coordinate system of the main axes by two masses: m_l^* = longitudinal mass, m_t^* = transverse mass.

10

GENERATION-RECOMBINATION

Generation-recombination (GR) processes are scattering processes similar to those described in Chap. 8 and can usually be calculated in a similar fashion—that is, by the use of the Golden Rule. The reason why another chapter is devoted to these processes is that the number of particles is not conserved in a particular band in the GR process. Typically, an electron from the conduction band will recombine with a hole in the valence band or an electron-hole pair will be generated. The involvement of two bands renders the one band approximation invalid and also makes a simplistic use of the effective mass theorem impossible. We will, therefore, not be able to calculate explicitly the matrix elements in this chapter, but rather will point out their proportionalities to quantities such as carrier concentration and density of states.

10.1 IMPORTANT MATRIX ELEMENTS

In principle, we can numerically calculate all matrix elements by using wave functions computed, for example, by the method of Chap. 4. Notice that GR gives a richness to the theory of transport in semiconductors (not present in the case of metals), which also forms the basis of the theory of *p–n* junctions (Chap. 13). The GR processes also usually involve a considerable change in energy. This change is due to interactions with photons, phonons, and the like, and is treated separately below. We describe mainly recombination processes; the generation processes are the exact inverse.

10.1.1 Radiative Recombination

Electrons can lose their energy by emitting a photon, as shown in Fig. 10.1. The photon can be emitted by a band to band transition or by a transition involving impurity states or exciton states. An exciton is a hydrogen atom–like entity where the nucleus is replaced by a hole, that is, an electron-hole pair bound to each other. If the effective mass of the hole is large compared to the electron, the theory of the exciton proceeds like the theory of impurity levels. If the electron and hole masses are comparable, a more complicated approach is in order.

We treat here only band to band transitions; band to impurity transitions can be treated in an analogous fashion. The transition probability is again obtained from the Golden Rule. The only problem is the derivation of the matrix element $M_{c,v}$ for the electron-photon interaction. Since this derivation is given in many textbooks (see, e.g., Casey and Panish, a), we just repeat the result.

The so-called momentum matrix element is given by

$$M_{c,v} = \int_{V_{ol}} \psi_v^* \hat{\mathbf{p}} \psi_c \, d\mathbf{r} \tag{10.1}$$

where ψ_v and ψ_c are the valence and conduction band wave functions, respectively, and $\hat{\mathbf{p}}$ is the momentum operator:

$$\hat{\mathbf{p}} = \frac{\hbar}{i} \nabla \tag{10.2}$$

We now write the wave function using Bloch's theorem

$$\psi_c = e^{i\mathbf{k}_c \cdot \mathbf{r}} u_c(\mathbf{r}) \tag{10.3}$$

and

$$\psi_v = e^{i\mathbf{k}_v \cdot \mathbf{r}} u_v(\mathbf{r}) \tag{10.4}$$

where $\mathbf{k}_c$ and $\mathbf{k}_v$ are the wave vectors of electrons in the conduction and

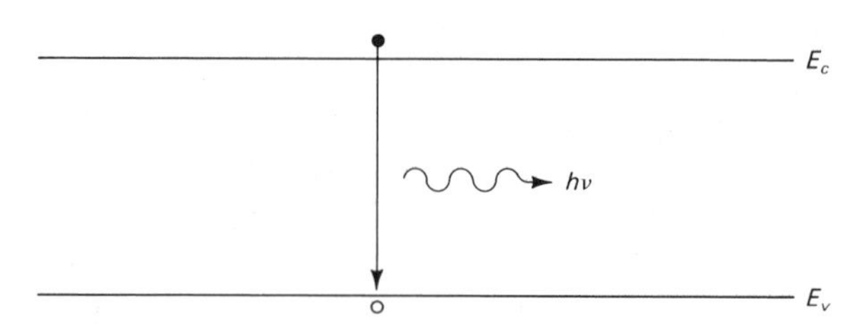

Figure 10.1 Schematic diagram for radiative recombination.

valence bands, respectively, and $u_{c,v}$ are the periodic parts of the wave functions, Eq. (3.20b).

Using Eqs. (10.2), (10.3), and (10.4) in Eq. (10.1), and differentiating, we get

$$\begin{aligned} M_{c,v} &= \frac{\hbar}{i} \int_{V_{\text{ol}}} e^{-i\mathbf{k}_v \cdot \mathbf{r}} u_v^*(\mathbf{r}) \nabla e^{i\mathbf{k}_c \cdot \mathbf{r}} u_c(\mathbf{r})\, d\mathbf{r} \\ &= \frac{\hbar}{i} \left[\int_{V_{\text{ol}}} e^{i(\mathbf{k}_c - \mathbf{k}_v)\cdot \mathbf{r}} u_v^*(\mathbf{r}) \nabla u_c(\mathbf{r})\, d\mathbf{r} \right. \\ &\quad \left. + \int_{V_{\text{ol}}} e^{-i\mathbf{k}_v \cdot \mathbf{r}} u_v^*(\mathbf{r}) u_c(\mathbf{r}) \nabla e^{i\mathbf{k}_c \cdot \mathbf{r}}\, d\mathbf{r} \right] \end{aligned} \tag{10.5}$$

Because $u_c(\mathbf{r})$ and $u_v(\mathbf{r})$ are orthogonal and the exponential components vary slowly, the second integral is usually neglected. The first integral can be evaluated in the following way.

The product $u_v^*(\mathbf{r})\nabla u_c(\mathbf{r})$ is the same for each unit cell and is multiplied by a slowly varying function that can be taken out of the integral. We then have

$$\begin{aligned} M_{c,v} &= \frac{\hbar}{i} \sum_{\text{all cells}} e^{i(\mathbf{k}_c - \mathbf{k}_v)\cdot \mathbf{r}} \int_{\Omega} u_v^*(\mathbf{r}) \nabla u_c(\mathbf{r})\, d\mathbf{r} \\ &= \frac{\hbar}{i} \int_{V_{\text{ol}}} e^{i(\mathbf{k}_c - \mathbf{k}_v)\cdot \mathbf{r}}\, d\mathbf{r} \cdot \frac{1}{\Omega} \int_{\Omega} u_v^*(\mathbf{r}) \nabla u_c(\mathbf{r})\, d\mathbf{r} \end{aligned} \tag{10.6}$$

And using the definition

$$\langle v|\hat{\mathbf{p}}|c\rangle \equiv \frac{1}{\Omega} \int_{\Omega} u_v^*(\mathbf{r}) \nabla u_c(\mathbf{r})\, d\mathbf{r}$$

we have

$$M_{c,v} = \langle v|\hat{\mathbf{p}}|c\rangle \int_{V_{\text{ol}}} e^{i(\mathbf{k}_c - \mathbf{k}_v)\cdot \mathbf{r}}\, d\mathbf{r} \tag{10.7}$$

We can see that $M_{c,v}$ is then finite only if

$$\mathbf{k}_c - \mathbf{k}_v = 0 \tag{10.8}$$

This means geometrically that only vertical transitions can happen in **k** space, as shown in Fig. 10.2. This, of course, is important in optical devices (lasers, LEDs, and so on).

Transitions that involve changes in **k** cannot happen in first order. However, second-order transitions involving a photon and a phonon are possible. The photon is then responsible for the energy change (its momentum is very small, which is the reason for the vertical transition), and the phonon supplies most of the momentum and contributes only little to the energy change. Detailed calculations and discussion have been given by Bebb and Williams.

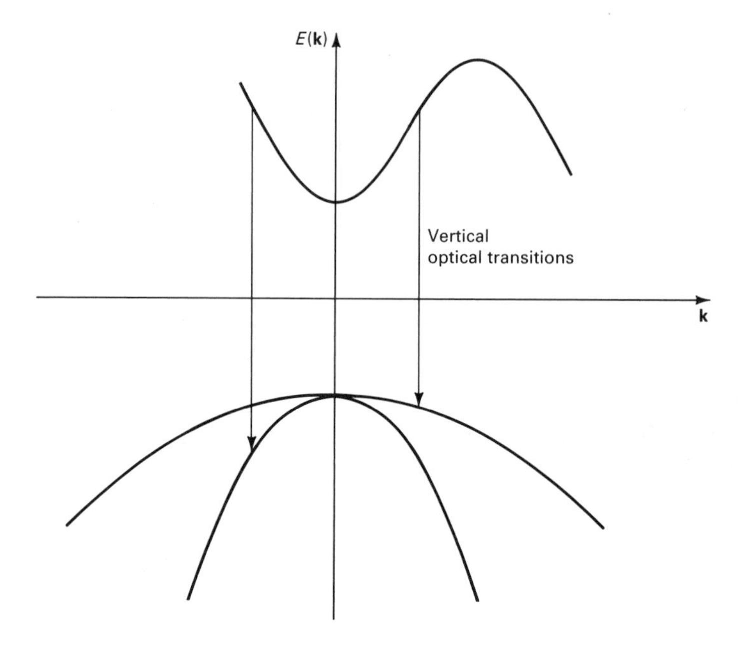

Figure 10.2 Possible transitions in **k** space caused by photon emission.

The treatment discussed above is very useful and shows immediately how the matrix element has to be modified if external potentials are present. If we have, for example, a quantum well heterolayer structure, then the envelope (effective mass) wave function is replaced by [see Eq. (1.11)]

$$e^{i(\mathbf{k}_c - \mathbf{k}_v)\cdot \mathbf{r}} \rightarrow e^{i(\mathbf{k}_{c\|} - \mathbf{k}_{v\|})\mathbf{r}_\|} \sin \mathbf{k}_{c\perp} z \sin \mathbf{k}_{v\perp} z \tag{10.9}$$

where $\|$ indicates that the vector is parallel to the layers and $\perp$ indicates the direction perpendicular to the layers.

For infinite square wells of width L, we have $\mathbf{k}_{c\perp} = \mathbf{k}_{v\perp} = n\pi/L$, where n is an integer. We see then that the matrix element is only finite if the transition occurs between levels with the same index n; in other words, we have a selection rule $\Delta n = 0$ (see Fig. 10.3). The term $\langle v|\hat{\mathbf{p}}|c\rangle$ is more difficult to evaluate; approximations have been published (see, e.g., Casey and Panish, b).

The emission of phonons in band to band transitions is a rather rare event since the phonon energy is typically much smaller than the energy of the band gap. However, if impurity states are involved, the emission of phonons becomes easier for two reasons: First, the energy that must be overcome is smaller since the electron can recombine in cascades, as shown in Fig. 10.4. Second, the electron-phonon coupling is enhanced by the impurity since the momentum conservation is relaxed, as the impurity potential can "supply" any necessary component **q**. The matrix element for electron capture by an impurity has been reviewed in detail by Ridley and will not be derived here.

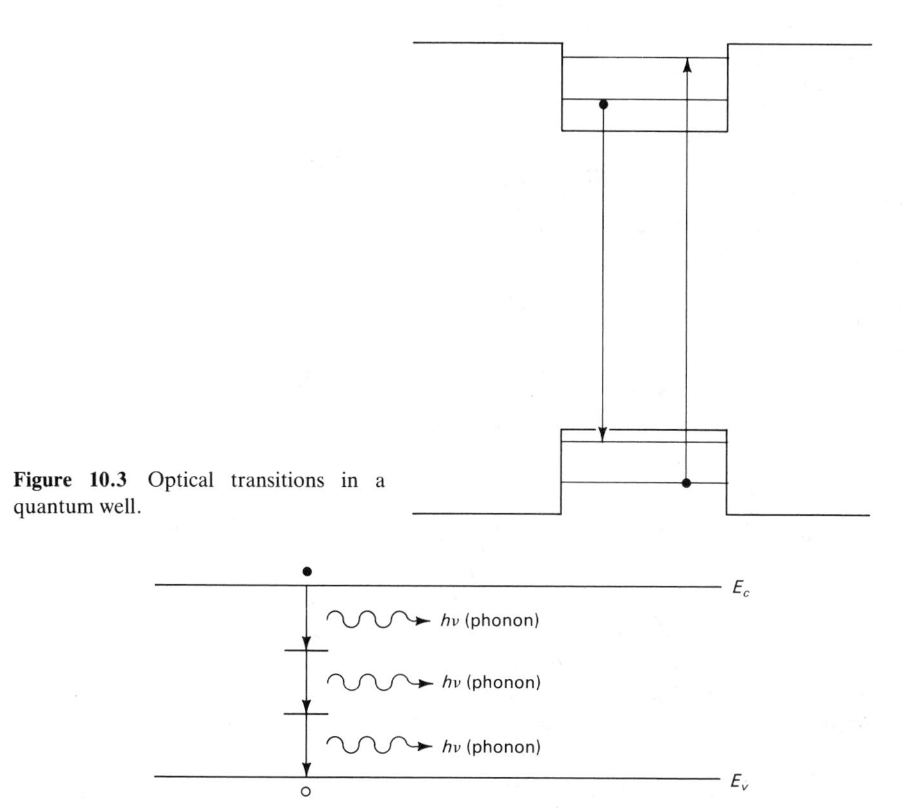

Figure 10.3 Optical transitions in a quantum well.

Figure 10.4 Recombination involving phonons and impurities.

The matrix element for optical transitions involving impurities has been treated by Casey and Panish (c).

10.1.2 Auger Recombination

This recombination process involves more than one electron but no radiation. As shown in Fig. 10.5, an electron can transfer its energy to another electron and recombine with a hole. The exact inverse of this process is called impact ionization and is described in Chap. 13. Here it is sufficient to understand that the Auger process is only important if the electron density is high (one needs electron-electron interaction) and therefore is only important in heavily doped semiconductors as the rate increases with the square of the carrier concentration. The interested reader is referred to Ridley's treatment.

Before we proceed with generation-recombination mechanisms, an important concept must be explained: *quasi-Fermi energy* (sometimes also quasi-Fermi level).

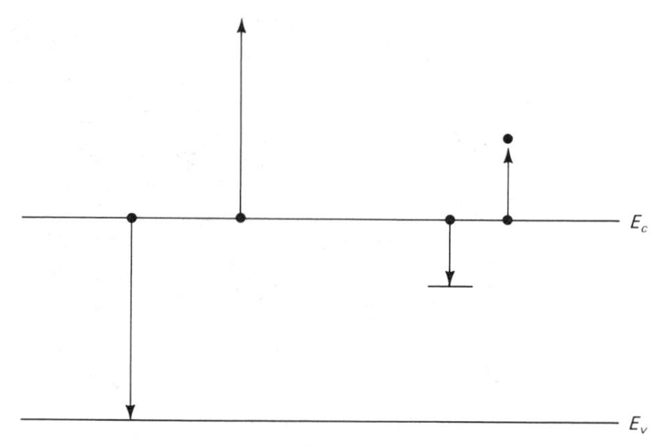

Figure 10.5 Auger recombination of an electron.

10.2 QUASI-FERMI LEVELS (IMREFS)

We have discussed quasi-Fermi energy (imref) in connection with high field solutions of the Boltzmann equation. If the equilibrium is disturbed optically or by electrical injection of charge carriers (as we will see when developing the theory of p–n junctions), the imref concept also applies. This is especially easy to see for the case of optical excitation.

Assume that we radiate light onto a semiconductor and generate a large number of electron-hole pairs. The recombination of an electron with a hole over the band gap by light emission takes place typically in 10^{-9} s. The electron-electron interaction, hole-hole interaction as well as electron (hole)-phonon interactions have typically a much shorter time constant (of the order of 10^{-13} s). Therefore the electrons and holes will be in equilibrium with each other in a time of the order of several 10^{-13} s. Then electrons and holes obey Fermi-like distributions separately.

To understand this in more detail, consider the following analogy. In Fig. 10.6 two gas containers connected with a very thin tube are shown. The total particle density, which in equilibrium is a measure of the Fermi energy, is different in the two containers ($n_1 \neq n_2$) and so is the pressure ($p_1 \neq p_2$). The

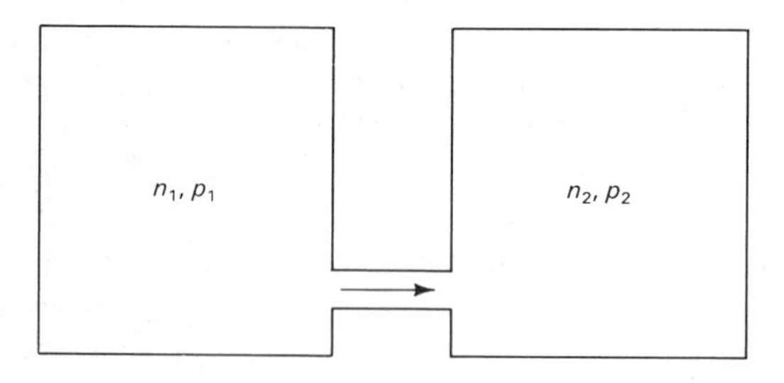

Figure 10.6 Hydrodynamic analogy for the explanation of imrefs.

system is not in equilibrium since there is a steady stream from one container to the other. However, if the tube is just small enough, the particles in each container separately will be in equilibrium with themselves; that is, the even part of the distribution function in each container will have the Fermi shape:

$$f_0(E) = \frac{1}{e^{(E-E_{QF})/kT} + 1} \tag{10.10}$$

10.3 GENERATION-RECOMBINATION RATES

Using the distribution function of Eq. (10.10) we can calculate the transition rates for most of the GR processes.

We know from the treatment of the Boltzmann equation that we have to determine the transition rate by using Eq. (8.36). To calculate this term properly, we must include the $(1-f)$ terms required by the Pauli principle if the distribution function cannot be approximated by a Maxwell-Boltzmann distribution. This is frequently the case for generation-recombination problems since they involve two bands. Therefore

$$\left.\frac{\partial f_0}{\partial t}\right|_{\text{coll}} = \frac{V_{\text{ol}}}{8\pi^3}\int d\mathbf{k}'\,[S(\mathbf{k},\mathbf{k}')f_0(\mathbf{k})(1-f_0(\mathbf{k}')) - S(\mathbf{k}',\mathbf{k})f_0(\mathbf{k}')(1-f_0(\mathbf{k}))] \tag{10.11}$$

To proceed with the calculation, we assume that all functions are functions of energy instead of the **k** vector, which is usually a good approximation, and we introduce the density of states in the calculation. Generation and recombination are treated separately. That is, we separate the two terms in Eq. (10.11) and take the difference at the end of the calculation. Since we have to deal mainly with two bands, we apply again the indices c, v to energy and **k** vector. The edges of the conduction and valence bands are now denoted by E_c^0 and E_v^0, respectively.

Suppose, then, that a conduction band state of energy E_c is full and a valence band state of energy E_v is empty and the transition probability per unit time is $S(E_c,E_v)$. The density of full states at energy E_c between E_c and $E_c + dE_c$ is given by

$$f_c(E_c)g_c(E_c)dE_c \tag{10.12}$$

Here f_c is the Fermi distribution in the conduction band with quasi-Fermi energy E_{QF}^c (we assume that E_{QF}^c is well defined). $g_c(E_c)$ is the density of states in the conduction band. The density of empty valence band states around E_v is given by

$$[1-f_v(E_v)]\,g_v(E_v)\,dE_v \tag{10.13}$$

The meaning of the symbols is the same as above, only now they are applied to the valence band instead of the conduction band.

The "recombination" term of Eq. (10.11) is then

$$\int_{-\infty}^{E_v^0} dE_v f_c(E_c)(1-f_v(E_v))g_v(E_v)S(E_c,E_v) \tag{10.14}$$

Notice that, again, as in the case of phonon scattering, the spin of the electron may not change in a particular transition, depending on the mechanism. As described in connection with Eq. (8.30), the final density of states has to be replaced by one-half of its actual value. Here, however, we do not need to be concerned about this factor because it will cancel out in the final results.

In the following discussion we will need the total rate of recombination R, which is obtained by integrating Eq. (10.14) over all conduction band energies.

This total rate can be brought into the following form: The product of electron density in the conduction band n and hole density in the valence band p is given by (see derivation of Eq. (6.27)):

$$n \cdot p = \int_{E_c^0}^{\infty} dE_c \int_{-\infty}^{E_v^0} dE_v f_c(E_c)[1-f_v(E_v)]g_c(E_c)g_v(E_v) \tag{10.15}$$

It therefore follows that

$$R = n \cdot p \langle S(E_c,E_v) \rangle$$

where $S(E_c,E_v)$ is the average defined by

$$\langle S(E_c,E_v) \rangle = \frac{\int_{E_c^0}^{\infty} dE_c \int_{-\infty}^{E_v^0} dE_v f_c(E_c)(1-f_v(E_v))g_c(E_c)g_v(E_v)S(E_c,E_v)}{\int_{E_c^0}^{\infty} dE_c \int_{-\infty}^{E_v^0} dE_v f_c(E_c)(1-f_v(E_v))g_c(E_c)g_v(E_v)} \tag{10.16}$$

Inserting for f_c and f_v and assuming a Boltzmann distribution, we can see that the quasi-Fermi levels cancel in Eq. (10.16) and $\langle S(E_c,E_v) \rangle$ is a number independent of the density of electrons and holes [except for Auger processes, where $S(E_c,E_v)$ depends on E_{QF}]. This is very fortunate and helps give simple rate equations.

Let us now consider the inverse process—the generation of an electron-hole pair by the same mechanism. To obtain the total generation rate G we only have to exchange E_c and E_v in Eqs. (10.14)–(10.16), which gives

$$G = \int_{E_c^0}^{\infty} dE_c \int_{-\infty}^{E_v^0} dE_v \, (1-f_c(E_c))f_v(E_v)g_c(E_c)g_v(E_v)S(E_v,E_c) \tag{10.17}$$

In many cases one can neglect f_c vs. 1 and approximate $f_v(E_v) \approx 1$. (If this does not hold, the generation undergoes a "Moss-Burstein" shift to higher energies.)

With the above approximations, the generation rate is given by

$$G = \int_{E_c^0}^{\infty} dE_c \int_{-\infty}^{E_v^0} dE_v g_c(E_c) g_v(E_v) S(E_v, E_c) \tag{10.18}$$

which is a constant independent of the carrier densities (imref's) as long as $S(E_v, E_c)$ is not the probability of an Auger process the latter depending on the density. G for equilibrium can be determined from the requirement of detailed balance:

$$G = R$$

Therefore

$$G = n_i^2 \langle S(E_c, E_v) \rangle \tag{10.19}$$

where n_i is the intrinsic carrier concentration. This equation is also used for situations close to equilibrium but not exactly in equilibrium, which is to be justified in each case. Auger processes can be treated in a similar manner; however, additional dependences on n and p are then introduced.

Some of the most important recombination processes involve a trap for electrons or holes (impurity level), and we therefore have to consider these processes (Shockley-Hall-Read).

Assume for simplicity that we have a total density N_{TT} of identical non-interacting trapping centers with levels at energy E_T in the band gap. Let f_T be the probability of finding a trap full and n_T the number of full traps. We then have

$$n_T = N_{TT} f_T \tag{10.20}$$

f_T is a Fermi distribution in equilibrium but may be very different far away from equilibrium. In some instances the introduction of a quasi-Fermi energy may be appropriate. However, traps usually do not "communicate" with each other, and therefore long time constants will be needed to establish a Fermi distribution in some instances. Especially for the case of electron injection into traps, f_T may be very different from Fermi-like distributions.

Using Eq. (10.20) and the above definition, we have as first term of the collision operator

$$\sum_{T_{\text{raps}}} f_c(E_c)(1 - f_T) S(E_c, E_T) = N_{TT} f_c(E_c)(1 - f_T) S(E_c, E_T) \tag{10.21}$$

We will need later the sum over all possible recombination events, labeled the recombination term R:

$$R = N_{TT} \int_{E_c^0}^{\infty} dE_c f_c(E_c)(1 - f_T) g_c(E_c) S(E_c, E_T)$$

As before, we can rewrite this as

$$R = N_{TT} (1 - f_T) n \langle S(E_c, E_T) \rangle \tag{10.22}$$

with the average $\langle S(E_c, E_T)\rangle$ being given by

$$c_n \equiv \langle S(E_c, E_T)\rangle = \frac{\int_{E_c^0}^{\infty} dE_c f_c(E_c) g_c(E_c) S(E_c, E_T)}{\int_{E_c^0}^{\infty} dE_c f_c(E_c) g_c(E_c)} \tag{10.23}$$

The recombination rate is then rewritten as

$$R = (N_{TT} - n_T) n c_n \tag{10.24}$$

To calculate the generation process we have to exchange E_T and E_c in Eq. (10.21), which gives

$$G = N_{TT} \int_{E_c^0}^{\infty} dE_c f_T(E_T)(1 - f_c(E_c)) S(E_T, E_c) g_c(E_c) \tag{10.25}$$

Using a Maxwell-Boltzmann distribution for f_c and assuming $f_c \ll 1$, we arrive similarly as in the above calculation for R at

$$G = n_T e_n \tag{10.26a}$$

where e_n is independent of the trap and electron densities. Note, however, that e_n does depend on the energy of the trap (Ridley) as

$$e_n \propto \exp E_T / kT \tag{10.26b}$$

since the electrons need to be activated to get out of the trap (escape) (Ridley). Here we measure E_T from the conduction band edge (i.e., it is negative).

The capture coefficient c_n is often independent of E_T, since the probability of falling into a trap (or a deep well) usually does not depend on the depth of the trap (well). There are some exceptions to this rule, however, which have been reviewed by Ridley.

10.4 RATE EQUATIONS

The net rate GR of electron loss (or gain) becomes

$$-GR = c_n n (N_{TT} - n_T) - n_T e_n = R(n) - G(n) \tag{10.27}$$

The subscript n in the rates stands for electrons. We can derive similar rates for hole capture and emission at a trapping center. The total change in the density of full traps is then

$$\frac{\partial n_T}{\partial t} = R(n) - G(n) - R(\mathrm{h}) + G(\mathrm{h}) \tag{10.28}$$

where again n and h stand for electrons and holes, respectively. If we consider both electron and hole recombination, we can have a situation where GR is equal to zero while the system, in fact, is not in equilibrium. There can be a steady capture of electrons and holes in the traps, which results in the destruction of electron-hole pairs. Of course, to make the flow steady we have to replenish the electron-hole pairs, for example, by shining light onto the semiconductor. In this case ($\partial n_T / \partial t = 0$), which is very important for the steady state theory of p–n junctions, we have

$$R(n) - G(n) = R(\mathrm{h}) - G(\mathrm{h}) \tag{10.29}$$

Equation (10.29) gives

$$c_n n(N_{TT} - n_T) - n_T e_n = c_p p n_T - e_p(N_{TT} - n_T) \tag{10.30}$$

The rates for the holes differ from the rates for the electrons since a capture of a hole means that the *trap is full* while an electron capture means that the *trap is empty*. Therefore the factor of $c_p p$ is n_T, while the factor of $c_n n$ is $(N_{TT} - n_T)$.

We can deduce from Eq. (10.30) n_T, which is

$$n_T = \frac{N_{TT}(c_n n + e_p)}{c_n n + e_n + c_p p + e_p} \tag{10.31}$$

We can use this expression for n_T to calculate the net electron recombination in steady state U_s:

$$U_s = R(n) - G(n) = \frac{c_n c_p N_{TT} n \cdot p - e_n e_p N_{TT}}{c_n n + e_n + c_p p + e_p} \tag{10.32}$$

The total GR term for electrons is then the generation by light (or in other instances by an electric field) $\mathrm{GR}_{\mathrm{light}}$ minus U_s:

$$\mathrm{GR} = \mathrm{GR}_{\mathrm{light}} - U_s$$

Close to equilibrium U_s takes a simple form. Since in equilibrium $U_s = 0$, we have

$$c_n c_p n \cdot p = e_n e_p$$

It is customary to assume that the rates e_n, e_p, c_n, and c_p do not depend on the perturbation and retain their equilibrium value. This assumption is not always valid [see Eq. (10.26b) and Fig. 13.15] and should be carefully examined before it is used. Nevertheless, we use it in the following to simplify U_s. This is appropriate for not very high electric fields in the junction (Fig. 13.15).

Therefore, using Eq. (6.27), we have

$$\frac{e_n e_p}{c_n c_p} = n_i^2$$

and

$$U_s = c_n c_p N_{TT} \frac{n \cdot p - n_i^2}{c_n n + e_n + c_p p + e_p} \tag{10.33}$$

Denoting the equilibrium electron density by n_0 and hole density by p_0 and approximating in the denominator $n \approx n_0$, $p \approx p_o$ (but not in the numerator, which would vanish), we obtain

$$U_s = \bar{c}(n \cdot p - n_i^2) \tag{10.34}$$

where $\bar{c}$ is a constant given by

$$\bar{c} = c_n c_p N_{TT} \frac{1}{c_n n_0 + e_n + c_p p_0 + e_p} \tag{10.35}$$

If we generate a small perturbation of the electron density with $n = n_0 + \delta n$ and $\delta n \ll n_0$, then

$$n \cdot p_0 = (n_0 + \delta n)p_0 = n_0 p_0 + p_0 \delta n$$

and

$$U_s = \bar{c} p_0 \delta n = \frac{\delta n}{\tau_n} \tag{10.36}$$

τ_n is called the lifetime of electrons. This time constant is in its nature quite different from the phonon scattering times since the electrons vanish from the conduction band. There also exists a lifetime for holes that is derived in an analogous fashion.

RECOMMENDED READINGS

BEBB, H., and E. WILLIAMS, in *Semiconductors and Semimetals,* vol. 8, ed. R. K. Willardson and A. C. Beer. New York: Academic Press, 1972, pp. 243–53.

CASEY, JR., H. C., and M. B. PANISH, *Heterostructure Lasers. Part A. Fundamental Principals.* New York: Academic Press, 1978, (a) pp. 122–27; (b) p. 146; (c) pp. 144–50.

LANDSBERG, P. T., *Solid State Theory,* New York: Wiley/Interscience, 1969, pp. 279–94.

RIDLEY, B. K., *Quantum Processes in Semiconductors.* Oxford: Clarendon Press, 1982, pp. 251–63.

PROBLEMS

10.1. Show that in a semiconductor with only one type of acceptor trapping center the minimum possible electron lifetime is

$$\tau_{no} = \frac{1}{c_n N_{TT}}$$

Hint: Assume here that $\delta n \simeq n$. (Likewise, $\tau_{po} = 1/c_p N_{TT}$.)

10.2. Consider the same situation as in problem 10.1 and assume a disturbed electron and hole population as $n_0 + \delta_n$ and $p_0 + \delta_p$. If $\delta_n = \delta_p$ is very small and the trap concentration is also small, show that n and p return to equilibrium with a lifetime

$$\tau = \frac{\tau_{p0}(n_0 + n_1) + \tau_{n0}(p_0 + p_1)}{n_0 + p_0}$$

where $n_1 = e_n/c_n$ and $p_1 = e_p/c_p$ (the rates are equilibrium rates).

10.3. Using the expression in problem 10.2, try to find out whether or not a trap close to the conduction bandedge (shallow) is a more efficient recombination center than one at the middle of the gap (deep).

11

THE DEVICE EQUATIONS

To calculate the electronic current in a semiconductor device, we need to solve the Boltzmann equation subject to (usually very complicated) boundary conditions. From the Boltzmann equation we can obtain n, Eq. (6.7), and the current density $\mathbf{j}$, Eq. (9.35), and all other quantities we may need in order to characterize a device. There is, however, a complication which we have discussed (quantum mechanically) already in Chap. 7. The potentials and therefore the electric fields depend on the charge density via the Poisson equation and have to be determined self-consistently. That is, we have to solve the Boltzmann equation coupled with the Poisson equation, Eq. (7.9).

This is, of course, very complicated, and therefore simpler ways have been sought. It turns out that with very few approximations simpler formulas can be found for simple band structures. If these approximations are not good enough (generally they are not for fine-line III–V semiconductor devices), the previously discussed Monte Carlo approach is more appropriate.

In this chapter we assume in many instances $E = \hbar^2 k^2/2m^*$.

11.1 THE METHOD OF MOMENTS

The way simpler equations than the Boltzmann equation can then be found is called the *method of moments*. To illustrate this method (which is described in more detail in Appendix E), we multiply the Boltzmann equation by $1/4\pi^3$ and integrate it over all $\mathbf{k}$ space. This is done below for each term of the BTE:

1.
$$\frac{1}{4\pi^3}\int_{-\infty}^{+\infty}\frac{\partial f}{\partial t}d\mathbf{k}=\frac{\partial}{\partial t}\frac{1}{4\pi^3}\int_{-\infty}^{+\infty}fd\mathbf{k}=\frac{\partial n}{\partial t} \tag{11.1}$$

2.
$$\frac{1}{4\pi^3}\int_{-\infty}^{\infty}\mathbf{v}\cdot\nabla fd\mathbf{k}=-\frac{1}{e}\nabla\mathbf{j} \tag{11.2}$$

which follows from the definition of $\mathbf{j}$, Eq. (9.35).

3.
$$\frac{1}{4\pi^3}\int_{-\infty}^{+\infty}e\mathbf{F}\cdot\nabla_{\mathbf{k}}fd\mathbf{k}=0 \tag{11.3}$$

as can be found by using the fact that f_0 is an even and $\nabla_{\mathbf{k}}f_0$ is an odd function of $\mathbf{k}$. Inserting Eq. (9.15) into Eq. (11.3), one can also show that the integral over f_1 vanishes. (One also has to use partial integration to get rid of $\nabla_{\mathbf{k}}$, and the fact that f_0 vanishes exponentially as $|\mathbf{k}|\to\infty$.)

4. We have calculated the integral over the collision operator in Chaps. 9 and 10. The integral over the f_1 terms vanishes as f_1 is odd. The rest of the scattering operator can be written as

$$\left.\frac{\partial f_0}{\partial t}\right|_{\text{coll}}^{\text{ph}}+\left.\frac{\partial f_0}{\partial t}\right|_{\text{coll}}^{\text{GR}}$$

where ph denotes phonon scattering and GR denotes scattering due to generation-recombination. For special cases, it is easy to show [using, e.g., Eq. (9.21)] that the integral over the phonon term vanishes and only the generation-recombination terms, which we calculated in Chap. 10, are left.

For electrons recombining via a trap we have, therefore

$$\frac{\partial n}{\partial t}-\frac{1}{e}\nabla\mathbf{j}=G(n)-R(n) \tag{11.4}$$

This is the well-known *equation of continuity.*

We still need additional equations to determine the electric fields, current densities, carrier concentration, and electron temperatures. These equations can be obtained by multiplying the BTE with appropriate functions of $\mathbf{k}$ and integrating over all $\mathbf{k}$. Using the same methods as described in Eqs. (11.1)–(11.3) and multiplying the Boltzmann equation with a generic function $Q(\mathbf{k})$, one obtains, with the definition

$$\langle Q(\mathbf{k})\rangle=\int_{-\infty}^{+\infty}Q(\mathbf{k})f(\mathbf{k})\,d\mathbf{k}\Big/\int_{-\infty}^{+\infty}f(\mathbf{k})\,d\mathbf{k} \tag{11.5a}$$

a general equation that is valid for any scalar "moment" $Q(\mathbf{k})$. Accordingly

$$\frac{\partial}{\partial t}(n\langle Q(\mathbf{k})\rangle) + \frac{\hbar}{m^*}\nabla(n\langle Q(\mathbf{k})\mathbf{k}\rangle) - \frac{e}{\hbar}\mathbf{F}n\left\langle Q(\mathbf{k})\frac{\nabla_{\mathbf{k}}f}{f}\right\rangle$$
$$= \frac{1}{4\pi^3}\int_{-\infty}^{+\infty} d\mathbf{k}\, Q(\mathbf{k})\left.\frac{\partial f_0}{\partial t}\right|_{\text{coll}} - \frac{1}{4\pi^3}\int_{-\infty}^{+\infty} d\mathbf{k}\frac{f_1}{\tau_{\text{tot}}(\mathbf{k})}Q(\mathbf{k}) \tag{11.5b}$$

It can be shown by partial integration that the term $\langle Q(\mathbf{k})\nabla_{\mathbf{k}}f/f\rangle$ is equal to $-\langle\nabla_{\mathbf{k}}Q(\mathbf{k})\rangle$. On occasion, this second form is more convenient to use. For some purposes it is also more appropriate to rewrite Eqs. (11.5a) and (11.5b) with respect to the appearance of $\tau_{\text{tot}}(\mathbf{k})$. Since Eq. (11.5b) is derived from the Boltzmann equation and subsequent integration, it is possible first to multiply the equation by $\tau_{\text{tot}}(\mathbf{k})$ and to integrate subsequently. Then $\tau_{\text{tot}}(\mathbf{k})$ will appear as a multiplicative factor for each $Q(\mathbf{k})$ except for the last term on the right-hand side, where it cancels out.

In order to illustrate the intricacies that are inevitable if one wants to derive device equations from Eq. (11.5b), we discuss the case of $Q(\mathbf{k}) = k_z$. Consider, for the moment, a distribution function that does not explicitly depend on time. We can then rewrite Eq. (11.5b) to read

$$\frac{\hbar}{m^*}\nabla(\langle\mathbf{k}k_z\tau_{\text{tot}}(\mathbf{k})\rangle n) - \frac{e}{\hbar}Fn\left\langle k_z\tau_{\text{tot}}(\mathbf{k})\frac{\nabla_k f}{f}\right\rangle = -\frac{1}{4\pi^3}\int_{-\infty}^{+\infty} d\mathbf{k}k_z f_1 \tag{11.6a}$$

The term containing $\partial f_0/\partial t|_{\text{coll}}$ vanishes since f_0 is even in $\mathbf{k}$. It is important at this point to note that the average defined by Eq. (11.5a) is different from the average defined in Eq. (9.40), which contains an additional energy. To simplify Eq. (11.6a) further it is convenient to neglect f_1 compared to f_0 in the second term and in all the averages. Precise solutions of the Boltzmann equation show that this approximation is good for the scattering mechanisms which are important in silicon (even at high electric fields) but is basically invalid in GaAs because of the peculiarity of polar optical phonon scattering to prefer small angles and thus enhances the streaming terms (terms that are odd in $\mathbf{k}$).

The integrations (averages) can then be performed as in Eqs. (9.38)–(9.40), and the right-hand side of Eq. (11.6a) is equal to the z component of the current density multiplied by a factor $m^*/(\hbar e)$.

One then obtains

$$\mathbf{j} = en\mu\mathbf{F} + e\nabla Dn \tag{11.6b}$$

with

$$D = \mu kT_c/e \tag{11.7}$$

Here we have assumed that f_0 is a Maxwell-Boltzmann distribution at electron temperature T_c and f_1 is given by Eq. (9.17a). We also have assumed that $\tau_{\text{tot}}(\mathbf{k})$ does not explicitly depend on the space coordinate. (It may depend on T_c, which may in turn depend on the space coordinate.)

Equation (11.7) is the Einstein relation. Notice, however, that the diffusion constant D is a function of T_c and therefore belongs after the gradient in Eq. (11.6), as T_c may be a function of the space coordinate. The second term in Eq. (11.6) is the well-known diffusion current. If the time derivative is not neglected in Eq. (11.5), a term $\bar{\tau}\frac{\partial \mathbf{j}}{\partial t}$ arises in Eq. (11.6) in addition to $\mathbf{j}$. $\bar{\tau}$ is a certain average of τ_{tot} which cannot be determined within our formalism since we do not know the precise form of f. Only if τ_{tot} is independent of $\mathbf{k}$ do we have $\bar{\tau} = \tau_{tot}$.

The equation obtained from the Boltzmann equation for the average energy is (neglecting spatial gradients, which are derived in Appendix E)

$$\frac{\partial\langle E\rangle}{\partial t} = \frac{1}{n}\mathbf{F}\cdot\mathbf{j} - \left.\frac{\partial\langle E\rangle}{\partial t}\right|_{coll} \tag{11.8a}$$

where

$$\left.\frac{\partial\langle E\rangle}{\partial t}\right|_{coll} \equiv \frac{1}{8\pi^3}\int d\mathbf{k}E\left.\frac{\partial f_0}{\partial t}\right|_{coll} \tag{11.8b}$$

The term $\left.\frac{\partial f_0}{\partial t}\right|_{coll}$ gives an important contribution. Without this term the average energy diverges with time, since

$$\langle E\rangle = \frac{t}{n}\mathbf{F}\cdot\mathbf{j} + \text{const} \tag{11.9}$$

It is therefore the term $\left.\frac{\partial\langle E\rangle}{\partial t}\right|_{coll}$ that keeps the energy low, and it therefore represents the heat (Joule's heat) given to the crystal lattice.

One can prove in general (for Boltzmann statistics) that for small $|\mathbf{F}|$

$$\left.\frac{\partial\langle E\rangle}{\partial t}\right|_{coll} = \left(\langle E\rangle - \frac{3}{2}kT_L\right)\Big/\tau_E \tag{11.10}$$

where τ_E is a time constant called energy relaxation time and T_L is the temperature of the crystal lattice. Often one defines an electron temperature independent of the actual form of the nonequilibrium distribution function by

$$\langle E\rangle = \frac{3}{2}kT_c \tag{11.11}$$

Then

$$\left.\frac{\partial\frac{3}{2}kT_c}{\partial t}\right|_{coll} = \frac{3}{2}k(T_c - T_L)/\tau_E \tag{11.12}$$

This gives

$$\frac{\partial T_c}{\partial t} = \frac{2}{3k}\frac{1}{n}\mathbf{F}\cdot\mathbf{j} - (T_c - T_L)/\tau_E \tag{11.13a}$$

from which one can see that the temperature of the electron gas is always raised when a finite electric field is applied. If $T_c \gg T_L$, one calls the electrons "hot electrons."

For silicon at room temperature ($T_L = 300$ K), the energy relaxation time is almost entirely due to the interaction with optical phonons and can roughly be approximated by

$$\tau_E \approx 4 \cdot 10^{-12} \sqrt{\frac{T_c}{T_L}} \text{ sec} \tag{11.13b}$$

The general calculation of $\partial\langle E\rangle/\partial t|_{\text{coll}}$, including all inelastic scattering mechanisms, is usually connected with tedious algebra and has been explained in detail by Conwell.

For the special case of optical deformation potential scattering, the calculation is straightforward, at least in principle: For any absorption event, the energy $\hbar\omega_0$ is gained, while for emission the same energy is lost. The net energy gain or loss per unit time $\partial E/\partial t|_{\text{coll}}$ is therefore given by the difference of absorption and emission rates. Using Eq. (8.32) one obtains

$$\left.\frac{\partial E}{\partial t}\right|_{\text{coll}} = \frac{Z_0^2 m^{*3/2} \omega_0^2}{\sqrt{2}\pi\hbar^2 \rho v_s^2}[N_q\sqrt{E + \hbar\omega_0} - (N_q + 1)\sqrt{E - \hbar\omega_0}] \tag{11.14}$$

To obtain $\partial\langle E\rangle/\partial t|_{\text{coll}}$ we have to average Eq. (11.14) using Eq. (11.5a) and recognize that the integral over f_1 vanishes. The integration involves modified Bessel functions and the result will not be given here (see Conwell). For large energies, however, we can neglect $\hbar\omega_0$ in the square roots of Eq. (11.14) and obtain

$$\left.\frac{\partial E}{\partial t}\right|_{\text{coll}} = -\frac{Z_0^2 m^{*3/2} \omega_0^2}{\sqrt{2}\pi\hbar^2 \rho v_s^2}\sqrt{E} \tag{11.15}$$

which gives

$$\left.\frac{\partial\langle E\rangle}{\partial t}\right|_{\text{coll}} = -\frac{Z_0^2 m^{*3/2} \omega_0^2 \sqrt{2}}{\pi\sqrt{\pi}\hbar^2 \rho v_s^2}\sqrt{kT_c} \tag{11.16}$$

Comparing this equation with Eq. (11.12) for large T_c (and neglecting T_L vs. T_c), we can obtain the energy relaxation time τ_E. It is seen that τ_E still depends (although weakly) on the average energy and is therefore not a relaxation time in the strict sense. For T_c approaching T_L, however, τ_E approaches a constant, as can be seen from the exact integration of Eq. (11.14) (Conwell).

The derivation of $\partial\langle E\rangle/\partial t|_{\text{coll}}$ we gave above is a more intuitive one. Actually, the definition Eq. (11.8b) involves the complicated collision operator $\partial f_0/\partial t|_{\text{coll}}$. One can derive the energy loss also from this definition and get the same result, Eq. (11.16), if the same approximations ($E \gg \hbar\omega_0$) are made.

11.2 SIMPLE APPLICATIONS OF THE METHOD OF MOMENTS

11.2.1 Hot Electrons, Velocity Transients

We have developed a rather complete set of semiconductor transport equations. It is clear that these equations, which are all coupled, cannot be solved explicitly for a semiconductor device without further simplifying assumptions. The explicit techniques to solve these equations for special cases are discussed in the chapters on devices.

To illustrate some implications of these equations, we focus our attention on the current density in a homogeneous semiconductor. From Eqs. (11.6) and (11.13), it is clear that the current density is a nonlinear function of the applied voltage (electric field), since the mobility μ is a function of T_c (we neglect diffusion in this discussion). Using Eqs. (11.13a) and (11.13b) and putting $\partial T_c / \partial t = 0$ (steady state), we have

$$e\mu F^2 \approx \frac{C}{(300\text{ K})}(T_c - T_L)\sqrt{\frac{T_L}{T_c}} \tag{11.17}$$

where $C \approx 10^{-9}$ watts (the approximation holds for silicon at room temperature).

The mobility μ can often be approximated by

$$\mu = \mu_0 \sqrt{\frac{T_L}{T_c}}$$

as has been derived for scattering by acoustical phonons (also by optical phonons at high energy) in Eqs. (9.46) and (9.47). Therefore we obtain

$$T_c = T_L + e\mu_0 F^2 \cdot (300\text{ K})/C \tag{11.18}$$

Inserting this value of T_c into the equation for the mobility and using $\mathbf{j} = en\mu\mathbf{F}$, we have

$$\mathbf{j} = en\mu_0 \mathbf{F} \sqrt{\frac{1}{1 + e\mu_0 F^2 \left(\frac{300\text{ K}}{T_L C}\right)}} \tag{11.19}$$

As can be seen, $\mathbf{j}$ follows Ohm's law for small F since the term proportional to F^2 can be neglected. For large F, however, $\mathbf{j}$ is constant (it saturates), since then the term proportional to F^2 dominates and F cancels, giving

$$j_{\text{sat}} = \frac{en\mu_0}{\sqrt{e\mu_0\left(\frac{300\text{ K}}{T_L C}\right)}} \tag{11.20}$$

Dividing j_{sat} by en, we obtain the saturation velocity v_{sat} of the electrons.

The physical explanation of this phenomenon is, of course, that electrons are scattered more frequently as they are accelerated to higher velocities and therefore the mobility decreases. The fact that the mobility decreases proportional to just $1/F$ at high fields F is more general than suggested by the above derivation. The term $\sqrt{T_c}$ in the above expressions arises from the density of states $g(E)$, which is proportional to $\sqrt{E}$. In general, $g(E)$ increases much more steeply than $\sqrt{E}$ at higher energies in the bands (see Fig. 6.2). Even for steep increases in the density of states, the proportionality of the mobility to $1/F$ holds as is shown below.

Assuming that $g(E) \propto E^P$, and assuming $E \gg \hbar\omega_0$ and $T_c \gg T_L$, one obtains [see Eqs. (11.15) and (11.16)]

$$e\mu F^2 \propto \left(\frac{T_c}{T_L}\right)^p \tag{11.21}$$

This is because the scattering rates are roughly proportional to the final density of states, and $\sqrt{E}$ in Eq. (11.15) becomes E^p. Note, however, that this is only exact for the deformation potential interaction with optical phonons, which are the most important scattering agents for the energy loss in silicon.

For the same reasons one obtains approximately

$$\mu \propto \left(\frac{T_c}{T_L}\right)^{-p} \tag{11.22}$$

and thus using Eq. (11.21)

$$\frac{T_c}{T_L} \propto F^{1/p} \tag{11.23}$$

Inserting Eq. (11.23) into Eq. (11.22), we have

$$\mu \propto 1/F \tag{11.24}$$

It has been tacitly assumed in this derivation that $E \gg \hbar\omega_0$. This is only the case if the electrons are very hot (i.e., T_c is very high), which implies high electric fields. We therefore have shown that $\mu \propto 1/F$ (which is equivalent to velocity saturation) under rather general conditions.

The electron velocity saturates at high fields in silicon, germanium, and many other semiconductors. However, some III–V compounds, having the conduction band minimum at Γ, are special in certain respects. At Γ, polar optical scattering is predominant and the effective mass is small. This means that the mobility for small electron energies and low electric fields will be high. As the electron energy increases with increasing field, the L minima (germaniumlike) and the X minima (siliconlike) are also populated by intervalley scattering. Electrons in L and X valleys experience deformation-potential scattering and have a large mass. Thus, their mobility is much smaller compared to the Γ valley. This leads to a drop in the average velocity of the carriers above a certain field, as shown in Fig. 11.1. At very high fields the velocity

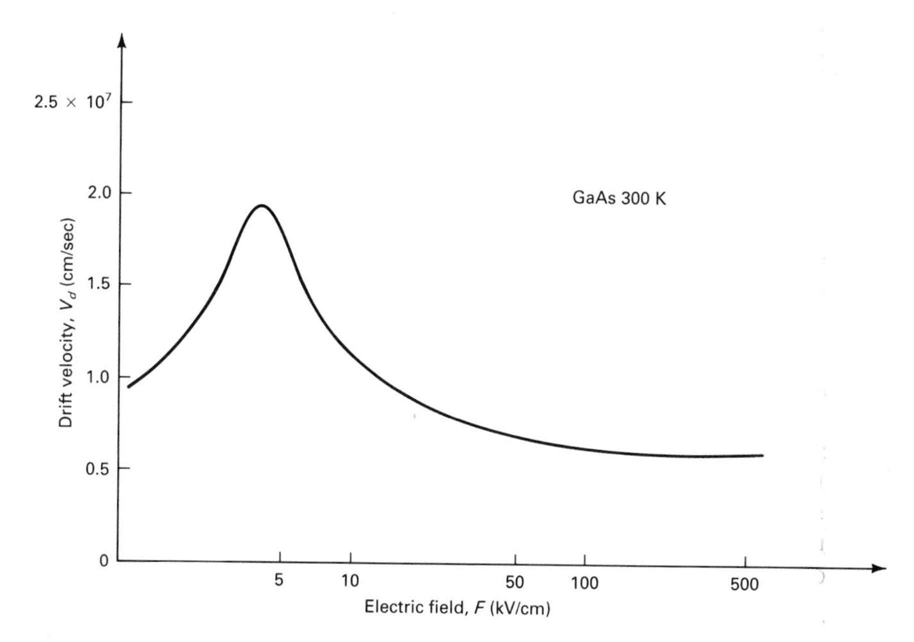

Figure 11.1 Calculated electron drift velocity in GaAs at room temperature. The calculations have been performed by Shichijo and Hess.

saturates approximately. This effect is the basis for the Gunn effect, which is discussed in Chap. 13.

In the above derivations it has been assumed that $\partial T_c / \partial t = 0$. If this is not the case, the situation becomes more complicated, as another differential equation needs to be solved, Eq. (11.13a), to obtain the electric current. A very important effect connected to the time development of the electron temperature is *velocity overshoot.* To understand this effect, assume that all electrons start at $\mathbf{k} = 0$ at time $t = 0$ (at which we apply a high electric field). The electrons are then accelerated and for a short time period do not scatter ($T_c = 0$). Since all electrons go in the forward direction, very high average velocities can be reached. As time goes on scattering events occur, and the electron energy is partially randomized, which also means that the average velocity decreases.

As shown by Ruch, this leads to the velocity vs. time curve shown in Fig. 11.2. In some instances (e.g., electrons in GaAs) this effect is very much enhanced because electrons start in the Γ valley and are accelerated beyond the L and X valleys. Then (after, typically, 10^{-13} sec or less) they scatter into the L and X valleys, where they have a large mass. This effect decreases the velocity further (as it does for the steady state, Fig. 11.1) and therefore leads to a significant overshoot effect, as shown in Fig. 11.2.

The physical mechanism leading to the overshoot is also illustrated in Fig. 11.2. In this figure a distinction is made between the drift velocity and the

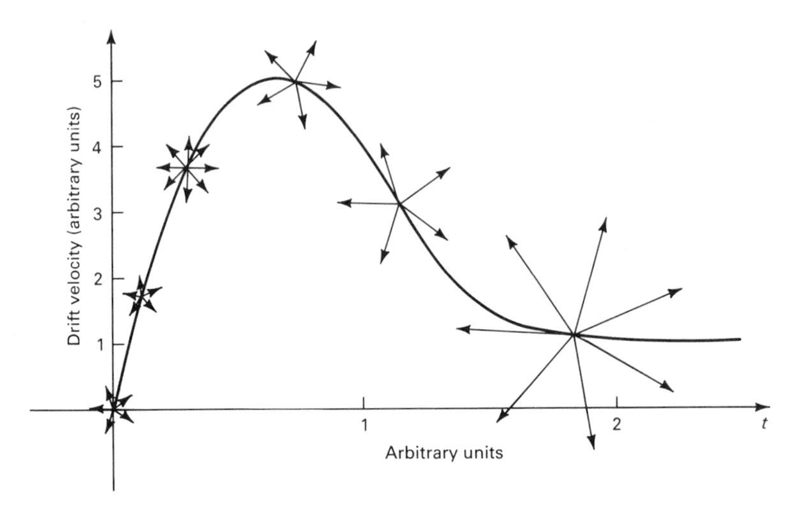

Figure 11.2 Velocity overshoot in semiconductors. (The time scale is typically in picoseconds, the velocity scale typically in 10^7 cm/sec.)

instantaneous random velocity of the electron gas. The drift velocity is the average over all random velocities of the electrons which point in forward and backward direction. The random velocity is indicated by the arrows plotted at a number of points. At $t = 0$, the electrons are in equilibrium and the drift velocity is zero. After a short time electrons are accelerated in the forward direction and very few scattering events occur. The drift velocity therefore increases while the random velocity still stays small. As time goes on the electrons are scattered and their forward velocity is randomized, leading to a large random velocity and to a smaller saturated drift velocity. Typical values in this range are 10^8 cm/s for the random velocity and 10^7 cm/s for the drift velocity. Velocities much higher than 10^8 cm/s do not usually occur in semiconductors because of Bragg refraction (the mass becomes negative).

Let us note in passing that Eq. (11.13a) is valid only if ∇f (or ∇n) can be neglected. In the presence of spatial gradients additional terms need to be added (see Appendix E). These terms need to be included when overshoot phenomena are possible, since overshoot implies automatically space-dependent velocities. Electrons start with small drift velocities in the contact areas and once accelerated overshoot their steady state velocity and finally approach steady state typically after a distance of the order of 500–5000 Å.

11.2.2 Space-Dependent Carrier Distributions

The patterns of space-dependent carrier distribution that can be connected to the dependence of the drift velocity on the space coordinate are best demonstrated by the example of the simplest possible semiconductor device, a

n^+nn^+ junction (semiconductor sandwiched between semiconductor contacts, the heavily doped n^+, or p^+ for holes, regions).

If the n region of the semiconductor is very long, then the equations for the current that have been developed above for a constant carrier concentration apply. For very short lengths of the n region, a new effect becomes important. Electrons can be pulled out of the contacts, and the electron concentration in the semiconductor increases independent of the doping of the semiconductor. The current associated with this effect is called the space charge limited current, since the electron concentration is in excess of the concentration of positively charged donors. The electric field is then given by the Poisson equation [see Eq. (7.9)]:

$$\frac{\partial F}{\partial x} = \frac{\delta\rho}{\epsilon\epsilon_0} \tag{11.25}$$

Here $\delta\rho$ is the excess charge that can be approximated by the total free carrier concentration n if the donor (acceptor) charge is small. If the electric field is high and diffusion currents are also neglected, the current density is $j = en\mu F$ and n becomes

$$n = \frac{j}{e\mu F} \tag{11.26}$$

Below we discuss three separate cases for the solution of Eqs. (11.25) and (11.26).

First we assume that the mobility is constant $\mu = \mu_0$. Then

$$\frac{\partial F}{\partial x} = \frac{j}{\epsilon\epsilon_0 \mu_0 F} \tag{11.27}$$

and

$$F^2 = \frac{2jx}{\epsilon\epsilon_0 \mu_0} \tag{11.28}$$

Here we have also assumed that j = constant (steady state) and that $F = 0$ at $x = 0$. This boundary condition is appropriate if we place the end of the n^+ contact at $x = 0$. The field is necessarily small in the contact region since the resistance of this region and, therefore, the external voltage drop over this region are small.

Assuming a length L for the n region and using

$$-\int_0^L F dx = V_{ext} \tag{11.29}$$

where V_{ext} is the external voltage, and using Eq. (11.28), one obtains the Mott-Gurney law:

$$j = \frac{9}{8}\frac{\epsilon\epsilon_0 \mu_0}{L^3} V_{ext}^2 \tag{11.30}$$

As discussed at the start of this calculation, the carrier concentration is space dependent. This can be seen from Eqs. (11.30), (11.28) and (11.26). Furthermore, a current will flow independently of the doping of the semiconductor. Even if there are no carriers at all in the semiconductor, for $V_{ext} = 0$ there will be a current for $V_{ext} \neq 0$, the reason, of course, being that electrons are supplied by the contacts. This is a strange result because it seems to contradict the existence of thin insulating films in nature, at least if they are sandwiched between contacts. However, close to a conducting contact made out of the same material, and differing only in doping, an insulator does not exist. Even if we do not apply an external voltage, electrons will spill over to the lesser doped semiconductor from the highly doped contacts. The spilling length is equal to the Debye length, which was derived in Chap. 7. The Debye length decreases with increasing carrier concentration. This does not mean, however, that a contact with infinite carrier concentration will not spill electrons into the neighboring material. On the contrary, it will spill the most. The Debye length represents in this case the penetration length for any given carrier concentration.

Second, we assume that the velocity is saturated, that is, $\mu F = v_{sat}$. Equation (11.27) then becomes

$$\frac{\partial F}{\partial x} = \frac{j}{\epsilon\epsilon_0 v_{sat}} \tag{11.31}$$

Integrating this equation (the right-hand side is now constant) and using Eq. (11.30), one obtains for the current density j

$$j = 2\epsilon\epsilon_0 v_{sat} V_{ext}/L^2 \tag{11.32}$$

It is important to note that in this case the current obeys Ohm's law. However, the velocity of the electrons is saturated and the increase of j with V_{ext} arises from an increase in charge density.

If μF does not follow a simple law as assumed for the derivations above, a numerical integration is necessary to obtain the current density j. Especially when velocity overshoot is involved, the problem can be quite complicated and a precise solution can only be achieved by a many-particle Monte Carlo calculation.

There is still one more important case that can be solved explicitly. In the case of negligible energy loss of the electrons, the velocity can be calculated from the kinetic energy and the potential energy $eV(x)$

$$m^* v^2/2 = eV(x) \tag{11.33}$$

and we have

$$\mu F = v = \sqrt{2eV(x)/m^*} \tag{11.34}$$

In this case we obtain for the current density

$$j = \frac{4}{9}\left(\frac{2e}{m^*}\right)^{1/2}\frac{\epsilon_0\,\epsilon}{L^2}V_{\text{ext}}^{3/2} \tag{11.35}$$

which is known as Child's law and was known for electrons emitted into a vacuum as early as about 1900.

Equations (11.25)–(11.36), together with Eqs. (11.17)–(11.24), demonstrate how complicated the simplest semiconductor "device" (a piece of semiconductor between two ohmic contacts) is and prepares us for further complications in the next chapters.

RECOMMENDED READINGS

CONWELL, E. M., "High Field Transport in Semiconductors," in *Solid State Physics,* ed. F. Seitz, D. Turnbull, and H. Ehrenreich. New York: Academic Press, 1967, pp. 155–60.

SHICHIJO, H., and K. HESS, *Physical Review,* B23 (1981), 4197–4207.

RUCH, J. G., *IEEE Transactions on Electron Devices,* ED-19 (1972), 652–59; see also K. Hess, *IEEE Transactions on Electron Devices,* ED-28 (1981), 937–40.

PROBLEMS

11.1. Prove Eq. (11.3).

11.2. Use Eq. (9.21) to show that $\int d^3\mathbf{k}\,\frac{\partial f_0}{\partial t}\Big|_{\text{coll}}^{\text{ph}}$ vanishes. Be sure to state explicitly any assumptions you make.

11.3. Derive the term corresponding to $\frac{h}{m^*}\nabla(n\langle Q(\mathbf{k})\mathbf{k}\rangle)$ in Eq. (11.5) for $Q(\mathbf{k}) = E$.

11.4 Assume that the n-type region of a n^+nn^+ diode is doped with N_D donors. Calculate the range of V_{ext} for which the current is space charge limited [i.e., follow Eqs. (11.30), (11.32), and (11.35)] and determine for which range the current is given by $j = \mu V_{\text{ext}}/L$. Distinguish carefully between the different laws for μF.

12

THE HETEROJUNCTION BARRIER AND RELATED TRANSPORT PROBLEMS

Up to now we have considered mainly the equations of motion, the scattering and the statistics of electrons in the conduction and valence bands of semiconductors, that have led us to device equations and to the understanding of some transport effects that are important in semiconductor devices. In all cases we had assumed homogeneous doping and one type of semiconductor throughout. Devices are usually based on doping inhomogeneities (homojunctions) or the combination of more than one semiconductor (heterojunctions). Obviously the Poisson equation will play an important role, as we already have seen, for the case of space charge limited current. There are, however, also several new transport effects that we need to know in order to understand what makes devices tick. This chapter is intended to give the transition between bulk (homogeneous) transport and electronic transport in devices. We describe here the transport of electrons close to the interface of two different neighboring semiconductors. Two of the most important devices—the metal-oxide-silicon transistor and the Schottky barrier diode—work on the basis of the principles that are developed below.

12.1 THE THERMIONIC EMISSION OF ELECTRONS OVER BARRIERS

Consider two different neighboring semiconductors with a conduction band discontinuity ΔE_c such as that of $Al_xGa_{1-x}As - GaAs$, which has been described in Chap. 4. We assume that the bands are in a fixed position, as shown in Fig. 12.1, and we attempt to calculate the current from GaAs to AlGaAs, and vice versa. The calculation can easily be performed by using an approximation introduced by Bethe.

Bethe assumed that the distribution function is equal to the Fermi distribution (or Maxwell-Boltzmann distribution) having a quasi-Fermi level E_{QF} which is different in the two materials but constant on each side. This means that a strong electron-electron interaction must be present, which causes the distribution function to look Fermi-like independent of the fact that electrons can be lost at a high rate to the neighboring material. In a Schottky barrier diode this loss of electrons to the neighboring material can be so substantial that it is hard to believe that the electron distribution still stays Fermi- or Maxwell-like. Exact calculations, however, necessitate an ensemble Monte Carlo method. The following explicit treatment is instructive and probably correct within a factor of 2 or so for most practical cases. Classically, all the electrons with a velocity component in the positive z direction (perpendicular to the interface) large enough to overcome the band edge discontinuity ΔE_c will propagate into the $Al_xGa_{1-x}As$ and all the electrons in the $Al_xGa_{1-x}As$ having a component in the negative z direction will propagate to the GaAs. The current density from the left (GaAs) to the right ($Al_xGa_{1-x}As$) is then

$$j_{LR} = \frac{e}{4\pi^3}\int_{k_xk_y} dk_x dk_y \int_{k_z>k_{z0}} dk_z v_z f_0(k) \tag{12.1a}$$

where k_{z0} is the minimum $\mathbf{k}$ vector component which is necessary to overcome the barrier. Classically, we can calculate k_{z0} from

$$\frac{m^*(v_{z0})^2}{2} = \Delta E_c \quad \text{and} \quad m^* v_z = \hbar k_z \tag{12.1b}$$

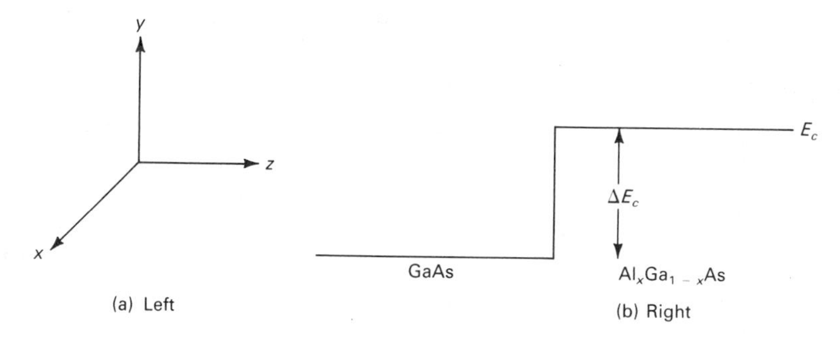

Figure 12.1 The conduction band edge at a heterojunction interface.

Here v_{z0} is the minimum velocity necessary to overcome ΔE_c. There are several quantum corrections (the quantum-transmission coefficient, deviations from the effective mass approximation, and tunneling, which will be discussed later). Assuming, then, for $f_0(k)$ a Maxwellian distribution at a quasi-Fermi level, we may rewrite the current as

$$j_{LR} = \left(\frac{m^*}{\hbar}\right)^3 \frac{e}{4\pi^3} e^{E_F^L/kT} \int_{-\infty}^{\infty} dv_x dv_y \int_{v_{z0}}^{\infty} dv_z v_z \times \exp\left[-\frac{m^*}{2kT}(v_x^2 + v_y^2 + v_z^2)\right] \tag{12.2}$$

Here E_F^L is the quasi-Fermi level on the left side measured from the left-side conduction band edge. It is easy to perform the x-y integration in cylindrical coordinates. Transforming $dv_x dv_y$ to $v dv d\phi (v = \sqrt{v_x^2 + v_y^2})$ and denoting $(v_x^2 + v_y^2)\frac{m^*}{2}$ by $\overline{E}$, one obtains by using $d\overline{E} = m^* v dv$

$$\int_{-\infty}^{+\infty} dv_x dv_y \exp\left(-\frac{m^*}{2kT}(v_x^2 + v_y^2)\right) = \frac{2\pi}{m^*} kT \tag{12.3}$$

From (12.3) we reduce (12.2) to

$$j_{LR} = \frac{e}{2\pi^2} \frac{m^{*2}}{\hbar^3} kT e^{E_F^L/kT} \int_{v_{z0}}^{\infty} v_z e^{-m^* v_z^2/2kT} dv_z \tag{12.4}$$

which gives

$$j_{LR} = A^* T^2 e^{(E_F^L - E_c^L - |\Delta E_c|)/kT} \tag{12.5}$$

j_{LR} as given by Eq. (12.5) is called the thermionic emission current, and $A^* \equiv \frac{e}{2\pi^2}\frac{m^*}{\hbar^3} k^2$ is the Richardson constant, which for the free electron $(m^* = m)$ is 120 amp/(cm$^2K^2$), and T is the temperature of the charge carriers. If the electrons are heated to a temperature T_c different from the lattice temperature T_L, then T_c has to appear in the exponent.

The current from the $Al_xGa_{1-x}As$ toward the GaAs, j_{RL} is given by

$$j_{RL} = A^* T^2 e^{E_F^R/kT} \tag{12.6}$$

where E_F^R is the quasi-Fermi level in the $Al_xGa_{1-x}As$ measured from the conduction band edge in the $Al_xGa_{1-x}As$.

Assume now that the $Al_xGa_{1-x}As$ is doped with a density N_D of donors and the GaAs is undoped. The quasi-Fermi levels then have a different distance from the conduction band edge of the two materials and consequently $j_{LR} \neq j_{RL}$, and a net current can flow. This, of course, cannot be true in equilibrium. We, therefore, have to conclude that the diagram in Fig. 12.1 changes shape as equilibrium is approached. The reason is easy to understand. If we start out at a certain time $t = 0$ with the circumstance of Fig. 12.1, then, of course, electrons start to flow from the $Al_xGa_{1-x}As$ (doped) to the GaAs (undoped) where their potential energy is lowest. We can estimate roughly the carrier density as a function of time in the $Al_xGa_{1-x}As$ close to the interface by using the equation of continuity, Eq. (11.4), generation-recombination neglected:

$$e\frac{\partial n}{\partial t} = \frac{\partial j}{\partial z} \tag{12.7}$$

and by assuming that current flows from $Al_xGa_{1-x}As$ to GaAs only. We also assume that this thermionic emission current flows only within a certain region which is of the order of the mean free path L_m of the electrons. (Beyond this region the average velocity is much smaller because of collisons.) Then, using $\partial j/\partial z \approx j/L_m$, we have

$$e\frac{\partial n(t)}{\partial t} \approx -A^*T^2 e^{E_F^R(t)/kT}/L_m \tag{12.8}$$

Since from Eq. (6.20), by using a time-dependent quasi-Fermi level

$$n(t) = \int_0^\infty e^{(E_F^R(t)-E)/kT} g(E)dE$$

we obtain

$$e^{E_F^R(t)/kT} = n(t)4\left(\frac{\hbar^2}{2m^*\pi kT}\right)^{3/2} \equiv n(t)C_1 \tag{12.9}$$

Equations (12.8) and (12.9) give

$$\frac{\partial n(t)}{\partial t} = -\frac{A^*T^2}{eL_m}C_1 n(t)$$

and

$$n(t) = n_c \exp\left[-\left(\frac{A^*T^2C_1}{eL_m}\,t\right)\right] \tag{12.10}$$

where n_c is the concentration in the $Al_xGa_{1-x}As$ at $t = 0$. The time constant $eL_m/A^*T^2C_1$ is, near room temperature, of the order of picoseconds; that is, the $Al_xGa_{1-x}As$ loses initially its electrons in picoseconds, and positively charged donors remain behind.

These positively charged donors give rise to a potential barrier on the $Al_xGa_{1-x}As$ side which slows down the electron transfer to the GaAs and finally prevents electrons from leaving $Al_xGa_{1-x}As$ at a higher rate than electrons returning from the GaAs. Then equilibrium has been reached; the corresponding equilibrium band diagram is shown in Fig. 12.2.

The theory of transport over a heterobarrier as shown in Fig. 12.2 is basic to the theory of semiconductor devices. A rigorous solution to the heterobarrier problem can in general be obtained only by ensemble Monte

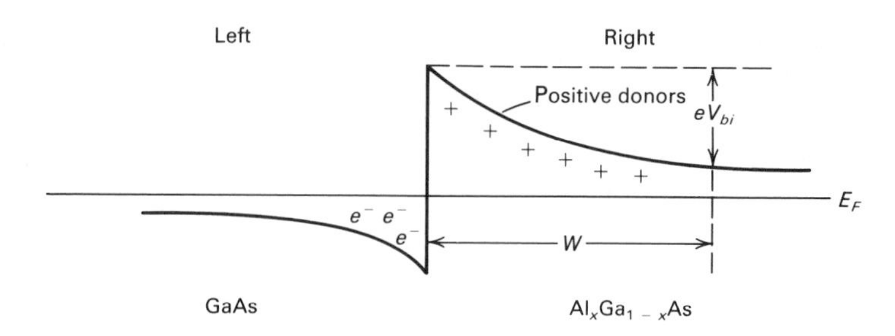

Figure 12.2 As Fig. 12.1 but with band bending due to a charge distribution included.

Carlo simulations. However, good approximations can be achieved by using the moment equations of Chap. 11 and solving them together with the Poisson equation. Because of the inhomogeneous charge distribution, Poisson's equation plays a significant role. We show the approximate solution in two steps. First we treat the right-hand side of Fig. 12.2, which is almost depleted of mobile electrons. Next we discuss the left-hand side where electrons accumulate, and the matching boundary conditions that connect the two sides.

12.2 FREE CARRIER DEPLETION OF SEMICONDUCTOR LAYERS

As can be seen from Fig. 12.2, the $Al_xGa_{1-x}As$ depletes a distance $\approx W$. In other words, the electron concentration is considerably smaller in this region than it is far away to the right-hand side. There is a "fuzzy" region where the depletion region goes over into the normal semiconductor. This transition occurs within the Debye length. We will return to this point below. The calculation of W and the potential drop V_{bi} (Fig. 12.2) necessitates, of course, the complete solution of the Poisson equation (also at the left side of Fig. 12.2). However, an important relation between W and V_{bi} can be derived without this knowledge, and this relation will be the subject of this section. The Poisson equation, Eq. (7.9), reads in one dimension (including the fixed charge of positive donors N_D^+ and negative acceptors N_A^-, as well as electrons of density n and holes of density p):

$$\frac{\partial F}{\partial z} = -\frac{e}{\epsilon\epsilon_0}(p - n + N_D^+ - N_A^-) \tag{12.11}$$

with F being the electric field.

We now solve the Poisson equation by using a very general trick that is based on two steps: (1) partial integration and (2) the depletion approximation.

It is convenient to choose the zero of the coordinate system at the interface. We have then

$$V_{bi} = -\int_0^W F\,dz$$

and

$$V_{bi} = -zF\Big|_0^W + \int_0^W z\frac{\partial F}{\partial z}\,dz \tag{12.12}$$

The depletion approximation can be stated in the following way: The semiconductor is depleted (i.e., free of *mobile* electrons or holes) over a distance W at the end of which (at $z = W$) the electric field is zero. The merit of this approximation can be seen directly from Fig. 12.2. One assumes that the electric field is screened out by the free carriers after the width W. However, to screen a field we need a Debye or Thomas-Fermi length or the general length $2\pi/q_s$ as given in Eq. (7.15). Within this length there is a transition region containing mobile charge and a finite electric field. We therefore can expect the depletion approximation to work only if

$$W \gg \frac{2\pi}{q_s} \tag{12.13}$$

Then we obtain from Eq. (12.12)

$$V_{bi} \approx \int_0^W z\frac{\partial F}{\partial z}\,dz \tag{12.14}$$

This equation is very useful for estimates of depletion voltages and can be regarded as one of the basic tools to calculate explicitly the properties of semiconductor devices. Neglecting n and p in the depleted region and assuming that only charged donors are present, we have

$$V_{bi} \approx -\int_0^W z\frac{e}{\epsilon\epsilon_0}N_D^+\,dz \tag{12.15}$$

For a constant donor concentration N_D^+ this gives

$$V_{bi} \approx -\frac{1}{2}\frac{W^2 e}{\epsilon\epsilon_0}N_D^+ \tag{12.16}$$

This equation relates V_{bi} to W but does not give us the value of either. As outlined above, to obtain both we need to solve the Poisson equation also on the left side and connect the two solutions. This will be done later, although it is pretty complicated since we cannot use the depletion approximation on the other side. For the time being we approximate the solution on the left side by assuming that the GaAs is much heavier doped than the $Al_xGa_{1-x}As$, and therefore the bands do not distort substantially in the GaAs.

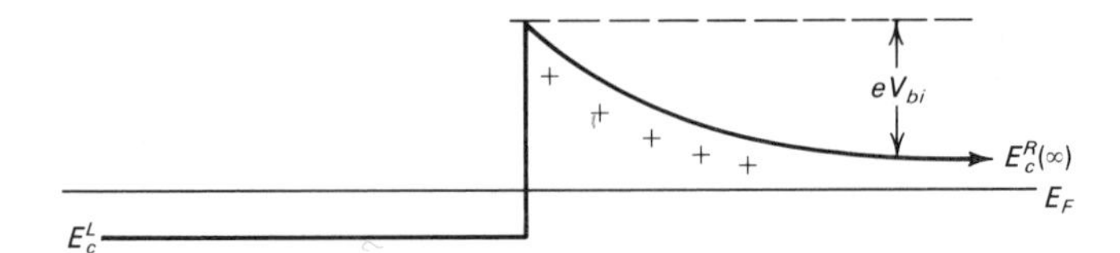

Figure 12.3 Band structure close to a band edge discontinuity. The band bending at the undepleted (left) side is neglected. However, it is important to note that negative charge must accumulate in the GaAs since the semiconductor as a whole must be neutral.

Within this approximation, we write (see Fig. 12.3)

$$|eV_{bi}| \approx \Delta E_c + E_c^L - E_c^R(\infty) \tag{12.17}$$

For j_{RL} one obtains [see Eq. (12.5)]

$$j_{RL} = A^*T^2 \exp\,(E_F^R - E_c^R(\infty) - |eV_{bi}|)kT \tag{12.18}$$

Of course in equilibrium $E_F^R = E_F^L$ and $j_{LR} = j_{LR}$, that is, no net current is flowing.

For W one obtains

$$W \approx \left[\frac{2\epsilon\epsilon_0(\Delta E_c + E_c^L - E_c^R(\infty))}{e^2 N_D^+}\right]^{1/2} \tag{12.19}$$

For rough estimates, the term $E_F^R - E_c^R(\infty)$ can be neglected.

12.3 CONNECTION RULES FOR THE POTENTIAL AT AN INTERFACE

Up to now we have treated mainly the right side of the junction and assumed that the other side (GaAs) is heavily doped and exhibits a flat conduction band edge (Fig. 12.1). We are now going to solve the Poisson equation for the GaAs side.

The connection rules for the solutions of the Poisson equation at the interface of two dielectrics are that the potential ϕ is continuous and the normal field (in absence of interface charge) changes as the ratio of the dielectric constant does. The parallel field is also continuous. To write down the equations, we need to remember that we solve Poisson's equation for the additional charges in the crystal, that is, for donors, acceptors, electrons, and holes, while the band structure (the $E(\vec{k})$ relation) of the crystal is predetermined. Also predetermined are the band edge discontinuities between one crystal and another. The external potential just shifts these energy bands rigidly as long as it is not too strong [see the effective mass theorem and Eqs. (4.23) and (4.24)]. It is often convenient to introduce a vacuum reference energy, which is the electron energy (at rest) outside the semiconductor (metal), and to measure the band edge energies from there. We will do this in several instances.

Given these facts, we can connect the left and right sides of the interface potential ϕ_i

$$\phi_i^R - \phi_i^L = \Delta E_c \tag{12.20}$$

and

$$\epsilon_L \partial\phi_i^L|\partial z = \epsilon_R \partial\phi_i^R/\partial z \tag{12.21}$$

In a homogeneous material the potential connects, of course, continuously and

$$\phi_i^R = \phi_i^L \tag{12.22}$$

This condition must also be used in cases when ϕ is calculated from Poisson's equation including only the charge of dopants and free carriers. The band edge discontinuity involves the crystal atoms and does not appear in such a solution. It has to be added subsequently if necessary. Notice that Eq. (12.22) applies only in the absence of interface dipoles. To proceed with the solution of Poisson's equation on the left-hand side, we can therefore assume ϕ_i^L and $\partial\phi_i^L/\partial z$ as given boundary values and perform the connection later.

12.4 SOLUTION OF POISSON'S EQUATION IN THE PRESENCE OF MOBILE (FREE) CHARGE CARRIERS: CLASSICAL CASE

We discuss now the solution of Poisson's equation at the left side of the heterojunction. (The index L, for left, is used only in cases of possible confusion.) Figure 12.4 shows a blowup of a possible form of the conduction band edge at the GaAs (left) side of the junction. To proceed with the calculation of

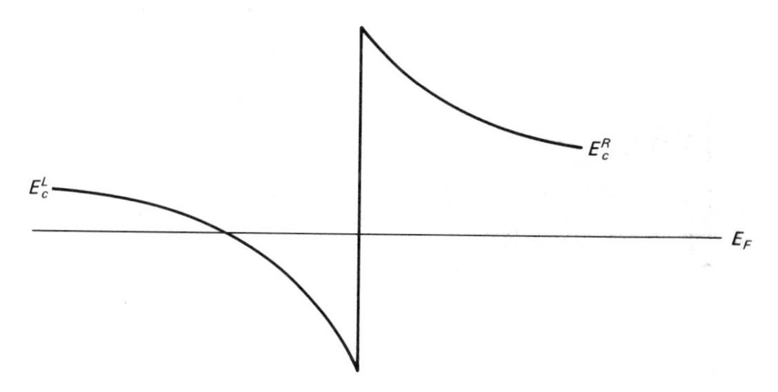

Figure 12.4 Conduction band edge for the case of electron accumulation at the heterojunction.

the potential ϕ from Poisson's equation, we need to express the carrier concentration as a function of ϕ. Since n follows the classical Boltzmann's law—(to convince yourself solve Eq. (11.6b) for $\mathbf{j} = 0$)—we have

$$n = n_c(-\infty) \exp(e\phi/kT_c) \tag{12.23}$$

Here $n_c(-\infty)$ is the equilibrium electron concentration far away to the left from the junction. Equation (12.33) is derived in the Problems section.

Denoting $e\phi/kT_c$ by Φ, Poisson's equation reads (in the absence of holes or acceptors):

$$\frac{\partial^2 \Phi}{\partial z^2} = -\frac{e^2}{\epsilon\epsilon_0 kT_c}(N_D^+ - n_c(-\infty)e^{\Phi}) \tag{12.24}$$

If $n_c(-\infty) \approx N_D^+$, as is the case for constant doping, we have

$$\frac{\partial^2 \Phi}{\partial z^2} = +\frac{e^2 n_c(-\infty)}{\epsilon\epsilon_0 kT_c}(e^{\Phi} - 1) \tag{12.25}$$

There is no explicit solution of this differential equation and we therefore cannot get $\Phi(z)$ explicitly. We can, however, obtain the total concentration of excess charge Q_{tot} at the left side of the junction (GaAs) as a function of interface potential ϕ_i and field F_i. This information is sufficient for some approximate considerations in the theory of devices. A special technique, worthwhile to be remembered, is still necessary to calculate Q_{tot} as a function of ϕ_i and F_i. We multiply Eq. (12.25) by the normalized electric field $\overline{F} = -(\partial\Phi/\partial z)$ and integrate. The left-hand side of Eq. (12.25) gives

$$-\int \frac{\partial^2 \Phi}{\partial z^2}\frac{\partial \Phi}{\partial z}\,dz = \frac{1}{2}\left[\frac{\partial \Phi}{\partial z}\right]^2 \tag{12.26}$$

The right-hand side becomes

$$\begin{aligned} -\frac{e^2 n_c(-\infty)}{\epsilon\epsilon_0 kT_c}\int (e^{\Phi} - 1)\frac{\partial \Phi}{\partial z}\,dz &= -\frac{e n_c(-\infty)}{\epsilon\epsilon_0}\int (e^{\Phi} - 1)d\Phi \\ &= -\frac{e^2 n_c(-\infty)}{\epsilon\epsilon_0 kT_c}(e^{\Phi} - \Phi + \text{constant}) \end{aligned} \tag{12.27}$$

Therefore we have the relation

$$\overline{F}^2 = -\frac{2e^2 n_c(-\infty)}{\epsilon\epsilon_0 kT_c}(e^{\Phi} - \Phi + \text{constant})$$

The constant has to be equal to -1 if we choose $\Phi = 0$ and $\overline{F} = 0$ for $z = -\infty$. One then obtains

$$\overline{F}^2 = -\frac{2e^2 n_c(-\infty)}{\epsilon\epsilon_0 kT_c}(e^{\Phi} - \Phi - 1) \tag{12.28}$$

Integrating Eq. (12.24) between $-\infty$ and the location $z = 0$ (the interface), one has

$$\epsilon\epsilon_0 F_i = -Q_{\text{tot}} \tag{12.29}$$

where Q_{tot} is the total excess charge (per unit area) between $z = -\infty$ and 0.

Equation (12.28) with $\phi = \phi_i$ and Eq. (12.29) can be used together with the connection rules to obtain the complete solution of Poisson's equation. Note, however, that we obtained only the overall charge (integral) and not the charge as a function of z. We will return to this point after the quantum mechanical solution, as we will discuss the consequences of these solutions for semiconductor devices. For large values of ϕ, the exponent dominates Eq. (12.28), and with the help of Eq. (12.29) one then obtains $Q_{\text{tot}} \propto e^{\phi_i/2}$.

The combination of Eqs. (12.28) and (12.29) gives us Q_{tot} as a function of the interface potential ϕ_i, which is schematically shown in Fig. 12.5. This figure demonstrates that, with increasing interface potential, the net carrier concentration increases exponentially on the GaAs side. In other words, electrons accumulate according to Boltzmann's law in the GaAs. They are located mostly at the interface and one speaks of an accumulation layer.

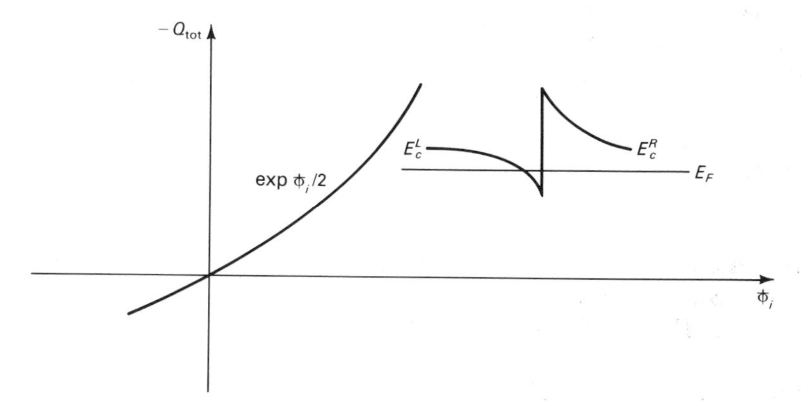

Figure 12.5 Schematic dependence of total charge Q_{tot} on the interface potential. The inset shows a typical band-structure diagram for accumulation.

If the GaAs were doped with acceptors at a density N_A, the derivation above would proceed in almost precisely the same way. However, since we then would have to permit the presence of holes, electrons initially accumulating in the GaAs would recombine with holes, and for small ϕ_i a depletion layer would form before electrons are further acccumulated. Since the depletion width W changes as $\sqrt{\phi_i}$ [see Eq. (12.19)], Q_{tot} changes initially only as $\sqrt{\phi_i}$. At higher values of ϕ_i electrons start to accumulate in the p-type semiconductor, and one says that an inversion layer forms and again $Q_{\text{tot}} \approx \exp \phi_i/2$. This is shown graphically in Fig. 12.6.

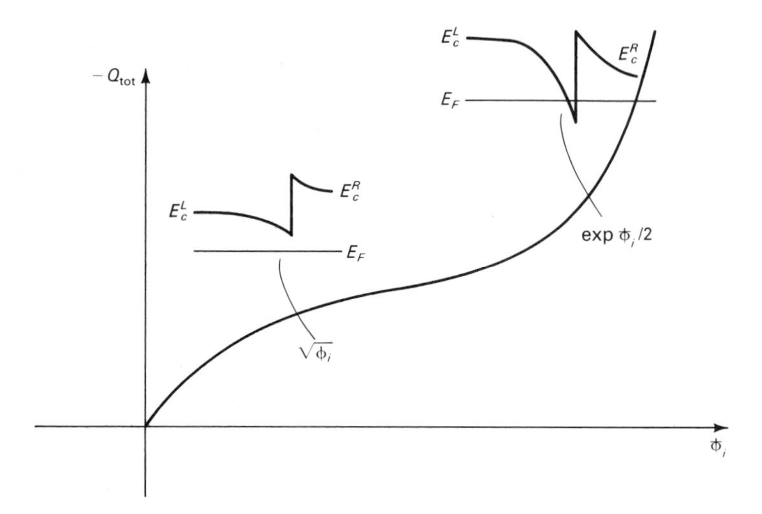

Figure 12.6 Schematic dependence of total charge Q_{tot} on the interface potential. The inserts show typical band diagrams for depletion and inversion, respectively.

The onset of strong inversion in Fig. 12.6 is not defined rigorously. However, inspection of Eqs. (12.28) and (12.29) shows that it occurs approximately at an interface potential

$$\phi_s = 2kT_c \ln (N_A / n_i)/e \tag{12.30}$$

where n_i is the intrinsic concentration at $-\infty$, that is, in bulk GaAs.

12.5 SOLUTION TO POISSON'S EQUATION IN THE PRESENCE OF MOBILE (FREE) CHARGE CARRIERS: QUANTUM MECHANICAL CASE

The potential well in Fig. 12.4 can be very narrow. For inversion layers with an electron density around 10^{18} cm^{-3} the typical well width is of the order of 10–100 Å for a given material. A typical value for the de Broglie wavelength of conduction electrons is 50 Å. Therefore, quantum effects can be very important, and we have to solve Schrödinger's equation self-consistently with Poisson's equation, much in the way as was done in Chap. 7. However, the well cannot be treated as a small perturbation, and therefore a numerical procedure has to be used to obtain a solution.

As shown in Eq. (1.12), the electrons in a quantum well populate only discrete energy levels due to size quantization. However, our well is not purely one dimensional. Parallel to the interface the electrons can move uninhibited, and in the effective mass approximation we have for the parallel kinetic energy of the electrons

$$E_{\parallel} = \hbar^2 k_{\parallel}^2 / 2m^* \tag{12.31}$$

Thus, the elecrons are still in a "band." However, since their perpendicular energy is quantized into levels with energy E_n ($n = 0, 1, 2, \ldots$), one speaks about subbands. The energies E_n are the solutions of the one-dimensional Schrödinger equation

$$\left(-\frac{\hbar^2 d^2}{2m^* dz^2} + \phi_{\text{tot}}(z)\right)\zeta_n(z) = E_n \zeta_n(z) \tag{12.32}$$

The potential $\phi(z)$ that confines the electrons consists of several contributions:

$$\phi_{\text{tot}}(z) = \phi(z) + \Delta E_c H(z) + \phi_{\text{im}}(z) + \phi_{\text{ex}}(z) \tag{12.33}$$

$\phi_{\text{ex}}(z)$ is a "many-body" contribution and comes from the fact that electrons are Fermions and are correlated due to their spin and the Pauli principle. The functional form of $\phi_{\text{ex}}(z)$ is known (see the review of Stern and Das Sarma), and it can easily be included in the calculation; for our purpose here we disregard it. $\phi_{\text{im}}(z)$ represents the contribution of the image force (i.e., the image potential energy), which can be significant since we are discussing the boundary of two dielectrics. For the AlGaAs-GaAs system the image force is negligible since the dielectric constants of both materials are very similar. Values for $\phi_{\text{im}}(z)$ are also given by Stern and Das Sarma. $H(z)$ is the step function with $H(z) = 1$ for $z \geq 0$ and $H(z) = 0$ otherwise, and $\phi(z)$ is the electrostatic potential, which in turn is a function of charge density and therefore of E_m and $\xi_n(z)$. Analogous to Eq. (7.5) we can write the charge density ρ due to the inverted (accumulated) free carriers as

$$\rho_{\text{inv}} = e\sum_{k_\parallel}\sum_{n} |\zeta_n(z)|^2 f(k_\parallel)/A \tag{12.34}$$

The sum over $k_\parallel$ can be replaced by a twofold integral analogous to Eq. (6.9) as

$$\sum_{k_\parallel} \rightarrow \frac{2A}{(2\pi)^2}\int dk_x\, dk_y \tag{12.35}$$

Here A is the inteference area. If $f(k_\parallel)$ is simply the Fermi distribution, which depends only on energy, the integration in polar coordinates is over the polar angle and gives 2π. The **k** integration is easily transformed to an energy integration, which gives for a spherical parabolical band

$$\sum_{k_\parallel} \rightarrow \frac{m^*}{\pi\hbar^2}\int_{E_n}^{\infty} dE \tag{12.36}$$

The integration of Eq. (12.34) gives then

$$\rho_{\text{inv}} = e\sum_{n} N_n |\zeta_n(z)|^2 \tag{12.37a}$$

with

$$N_n = \frac{m^* k T_c}{\pi\hbar^2} \ln\left[1 + \exp\left(\frac{E_F - E_n}{kT_c}\right)\right] \tag{12.37b}$$

Therefore, Poisson's equation reads

$$\frac{\partial^2 \phi}{\partial z^2} = + \frac{e}{\epsilon\epsilon_0} \left(N_A^- + e \sum_n N_n |\zeta_n(z)|^2 \right) \tag{12.38}$$

Here we have assumed a background of acceptors instead of donors, that is, we have treated the inversion case. Holes have again been neglected, although they can easily be included in Eq. (12.38).

This equation shows that the carrier concentration is not a simple function of the potential but rather a functional; that is, the charge density does not depend only on the potential $\phi(z)$ at a certain point z but depends on the potential as a whole. Each different function $\phi(z)$ will result in different $\zeta(z)$ and therefore in different ρ_{inv}. Therefore we cannot find an equation corresponding to Eq. (12.28), which completed the classical solution. What we have to do instead is solve Eqs. (12.32) and (12.38) iteratively. This proceeds as follows.

We start with a guessed potential $\phi_{\text{tot}}(z)$ in Eq. (12.32) and solve it to obtain E_n and $\zeta_n(z)$ by using, for example, the Numerov method. Then, using these results, we integrate Eq. (12.38) twice to obtain $\phi(z)$. A numerical procedure that accomplishes just that is listed in Appendix F.

Although it is relatively easy to develop a numerical code, a few analytical considerations are in order. A useful analytical approximation to $\zeta(z)^2$ has been given by Stern and Howard:

$$\zeta(z)^2 = \frac{1}{2} b^3 z^2 \exp(-bz) \tag{12.39}$$

with $3/b$ being the average extension of the inversion layer in the z direction. The expression for b will not be given here. It can be found in the work of Stern and Howard. A typical value for $3/b$ is 5° Å. Eq. (12.39) is only for the lowest subband and we have omitted the subscript n.

Equation (12.39) permits analytical solution of the Poisson equation and also an approximate solution for E_n. There is, however, little use for these solutions from our point of view, since for many device applications the classical solution is sufficient, and for calculations of the electron mobility, which will be discussed in the next section, one prefers either a simple $\delta(z)$ function approximation to $\zeta(z)^2$ or the complete numerical solution as described in Appendix F. This appendix also deals with the complications that arise from the many-valley character of the silicon conduction band.

The classical solution is sufficient for most device applications connected to the calculation of Q_{tot}. The reason is the following. Equation (12.29) is also valid from a quantum point of view since it merely represents one formal integration of Poisson's equation (Gauss's law). Assume now for simplicity that the AlGaAs is free of charge and that the potential at the surface of the AlGaAs is fixed. This corresponds to the situation in a metal insulator semiconductor device where on top of the insulator (AlGaAs,SiO_2) a metal is placed as a "gate" and the gate voltage V_G is fixed. A twofold integration of Poisson's equation in the charge-free AlGaAs gives then

$$\phi_R = C_1 z + C_2 \tag{12.40}$$

with C_1 and C_2 being integration constants. With the interface at $z = 0$, the boundary condition at the gate gives

$$V_G = C_1 d + C_2 \tag{12.41}$$

where d is the AlGaAs thickness. The connection rules Eq. (12.20) and (12.21) give

$$\epsilon_L F_i^L = \epsilon_R C_1 \tag{12.42a}$$

and, since our solutions for ϕ have been obtained from doping and free carrier charge, only Eq. (12.22) applies (we can disregard ΔE_c), which gives

$$C_2 = \phi_i^R \tag{12.42b}$$

Equations (12.42) together with Eq. (12.41) and Eq. (12.29) yield

$$V_G = -\frac{Q_{\text{tot}} d}{\epsilon_0 \epsilon_R} + \phi_i^R \tag{12.43}$$

where $\epsilon_0 \epsilon_R / d \equiv C_{\text{ins}}$ is the (insulator) capacitance of the AlGaAs layer and ϕ_i^R the interface potential. We remember that for the classical case we have derived an equation for $\phi_i^L = \phi_i^R$ as a function of $F_i^L = -Q_{\text{tot}}/\epsilon_0 \epsilon_L$, and therefore we can calculate Q_{tot} as a function of V_G.

A complete quantum treatment is more involved. However, the classical treatment is approximately valid for the following reason.

The interface potential rises first linearly with Q_{tot} and then saturates at the onset of strong inversion (accumulation) of free charge at the interface as shown in Fig. 12.7. The reason is that the free charge screens the potential at the interface and prevents any further rise. The potential at the onset of strong inversion, ϕ_s^R, has been given in Eq. (12.30). This equation is also valid from a quantum point of view since before strong inversion occurs, the potential well is not very deep and the potential varies slowly enough to make size quantization unimportant. Beyond the onset of strong inversion quantization occurs. However, ϕ_s^R stays approximately constant and one therefore obtains from Eq. (12.43)

$$-Q_{\text{tot}} = C_{\text{ins}}(V_G - \phi_s^R) \tag{12.44}$$

which is approximately valid from both a classical and a quantum point of view. This equation is basic to the operation of the important metal-oxide-silicon semiconductor (MOS) field effect transistor, discussed in Chap. 14.

The general validity of Eq. (12.44) is the precise reason why the inclusion of quantum effects is mostly unnecessary in order to obtain a semi-quantitative picture of MOS transistor operation.

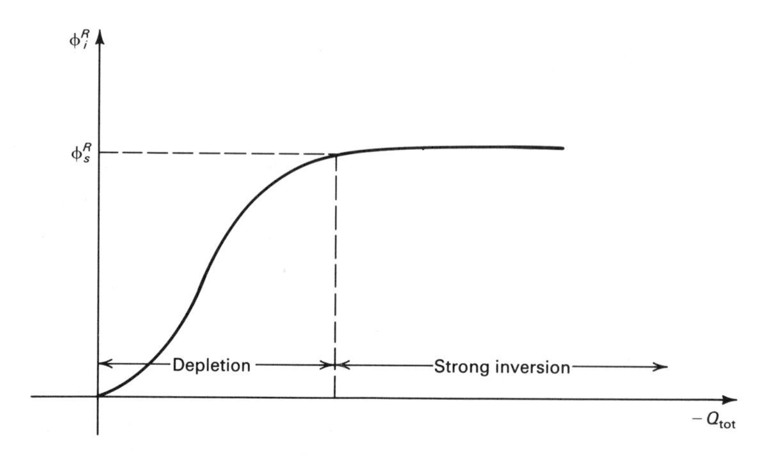

Figure 12.7 Interface potential as a function of total charge Q_{tot}. This figure corresponds precisely to Fig. 12.6.

Quantitatively, quantum effects are important for two reasons. First, it is not the total charge that one needs to know to characterize a transistor completely but rather the spatial distribution of free and fixed charge. This distribution differs somewhat classically and quantum mechanically. From Eq. (12.39) it can be seen that the wave function vanishes at the interface. This approximation is only good for the MOS system and not quite as good for AlGaAs-GaAs. The reason is that in the latter system the band edge discontinuity is rather small and the wave function penetrates the AlGaAs somewhat. Therefore, the probability to find free interface charge Q_i directly at the interface is quantum mechanically much smaller than one would expect from the classical theory [Eqs. (12.28) and (12.29)]. Fig. 12.8 shows this effect.

The quantum distribution of charge leads to slightly different device characteristics than expected from classical reasoning. We will see later that the insulator can be regarded as capacitance and so can the depletion region at the left side. This is schematically shown in Fig. 12.9.

As is evident from Fig. 12.9, the quantum mechanical calculation gives an additional almost insulating range of length $d_1 - d$. The dielectric constant $\bar{\epsilon}$ in this range is approximately the average between the left and right side dielectric constants, that is

$$\bar{\epsilon} \approx (\epsilon_L + \epsilon_R)/2$$

The length $d_1 - d$ is roughly one-third of the extension of the inversion layer, which is $3/b$. This gives the quantum capacitance

$$C_{QM} \approx \epsilon_0 \bar{\epsilon} b \tag{12.45}$$

Since $1/b$ is typically 10 Å, the series capacitance C_{QM} is only significant for very thin insulator (AlGaAs or SiO_2) thicknesses.

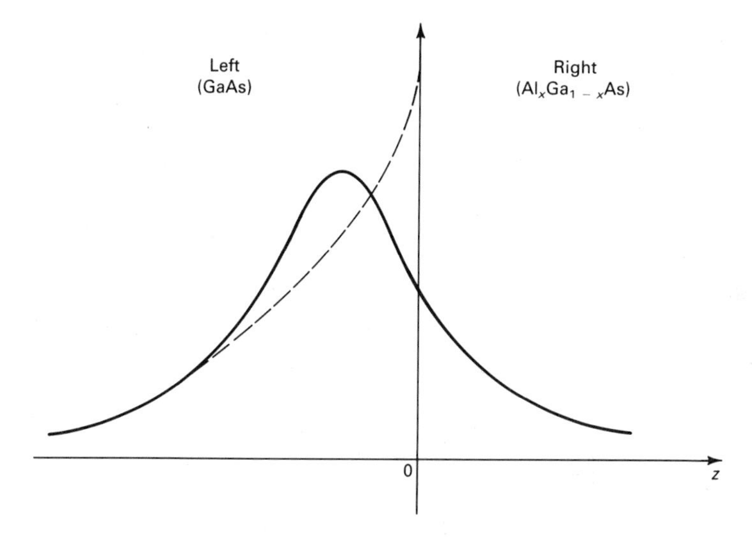

Figure 12.8 Classical (---) and quantum mechanical (—) distribution of free charges close to the heterointerface of AlGaAs-GaAs.

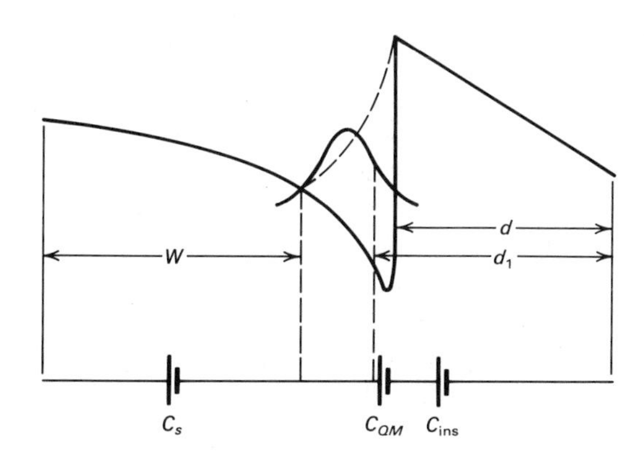

Figure 12.9 Insulating regions and corresponding capacitance for the classical and quantum mechanical calculations at a heterojunction.

Thus, concerning charge distribution, quantum effects will be typically small. The major effects of size quantization are related to the mobility of electrons parallel to the layer, which will be treated in the next section. Some of these ideas are due to H. Shichijo (unpublished).

Before discussing mobilities in size-quantized systems, a few remarks on the insulating qualities of AlGaAs should be made. We have referred to the AlGaAs as an insulator; however, the energy gap of the AlGaAs is only slightly larger than the gap of the GaAs and the intrinsic concentration of

AlGaAs is usually not negligible. Furthermore, if we apply a voltage, a thermionic emission current will flow according to Eqs. (12.5) and (12.6) (with the appropriate quasi-Fermi levels). We know already from Eqs. (11.30) and (11.32) that undoped semiconductors can carry currents that are much larger than expected from their intrinsic concentration. This is even true for different neighboring semiconductors, depending on doping of the neighbor and the band edge discontinuity. Notice also that electrons penetrate the barrier because of the extension of the wave function beyond the interface, which represents still another "influence" of a neighboring material that can change the insulating qualities (tunneling).

12.6 PRONOUNCED EFFECTS OF SIZE QUANTIZATION AND HETEROLAYER BOUNDARIES

We have seen from the derivation of Eq. (12.44) that the quantum corrections to the total charge density at interfaces will usually be small. However, the effects of size quantization on the electronic properties at interfaces can be significant. As we have seen in Chap. 9, the density of states enters into the scattering rates and collision operators. In fact the two-dimensional constant density of states has helped us solve some problems that are not tractable in three dimensions. Therefore, mobility, energy loss, diffusion constant, and other transport properties will be different at interfaces as size quantization leads to a quasi two-dimensional behavior of the electrons. These differences are more pronounced than the contributions of the capacitance of Eq. (12.45). They are not, however, order of magnitude changes. The mobility may change by a factor of 2 or so but will not change drastically except under unusual conditions.

These unusual conditions are mostly connected to a separation of scattering centers and electrons, as has been illustrated in Fig. 12.3. In this figure dopants are only on the AlGaAs (right) side, while all the electrons have left their parent donors and reside in the GaAs, where they are confined by the heterointerface and are remote from the scattering centers.

Remote and confined are the key words which indicate unusual transport effects (such as the enhanced mobility) at interfaces. It is not only the impurities that can be remote, but also phonon modes characteristic of only one medium can be separated from the electrons. We will discuss some questions related to phonons toward the end of the section. Now we will focus our attention on the detailed treatment of impurity scattering under conditions as shown in Fig. 12.3, which is the circumstance known as modulation doping (Dingle, Störmer, Gossard, and Wiegmann). Below we will derive the scattering rate for electrons interacting with remote impurities. This example gives us an excellent opportunity to repeat and illustrate the principles discussed in Chaps. 7 and 8.

Since the motion of electrons is confined in the z direction, the wave function is more complicated than the free electron wave function and we have to use the solution $\zeta_n(z)$ of Eq. (12.32) or approximations to this solution. The perturbed wave function (by the impurity charge) is still given by Eq. (7.3), but the coefficient $b_{\mathbf{k}+\mathbf{q}}$ now takes the form

$$b_{\mathbf{k}+\mathbf{q}} = \frac{\int e^{-i(\mathbf{k}_\parallel + \mathbf{q}_\parallel)\cdot \mathbf{r}_\parallel} \zeta_l^*(z) e V_{\mathrm{tr}} \zeta_m(z) e^{i\mathbf{k}_\parallel \cdot \mathbf{r}_\parallel} d r_\parallel \, dz}{E(\mathbf{k})_l - E(\mathbf{k}+\mathbf{q})_m} \tag{12.46}$$

Performing the $r_\parallel$ integration and denoting the two-dimensional Fourier transform of V_{tr} by $V_{\mathrm{tr}}^{q_\parallel}(z)$, we have

$$b_{\mathbf{k}+\mathbf{q}} = \frac{e\int_{-\infty}^{+\infty} \zeta_l^*(z)\zeta_m(z) V_{\mathrm{tr}}^{q_\parallel}(z) dz}{E(\mathbf{k})_l - E(\mathbf{k}+\mathbf{q})_m} \tag{12.47}$$

Here we have implicitly assumed that the indices l and m run over all subbands in the conduction band. If also the valence band contributes (e.g., for the calculation of the dielectric constant of the semiconductor, and not only the contribution of the free electrons), then the sum runs also over all valence band subbands.

If $V_{\mathrm{tr}}^{q_\parallel}(z)$ varies only slowly with z, we can approximate Eq. (12.47) by

$$b_{\mathbf{k}+\mathbf{q}} = \frac{e V_{\mathrm{tr}}^{q_\parallel}(0) \int_{-\infty}^{+\infty} \zeta_l^*(z)\zeta_l(z) dz}{E(\mathbf{k})_l - E(\mathbf{k}+\mathbf{q})_l} \tag{12.48}$$

since the integral vanishes for $l \neq m$ as the $\zeta_l(z)$ are orthogonal functions. Analogous to the derivation of Eq. (7.8) (with l being now the subband index), we obtain

$$\delta\rho = e^2 V_{\mathrm{tr}}^{q_\parallel}(0) e^{i\mathbf{q}_\parallel \cdot \mathbf{r}_\parallel} \int_{-\infty}^{+\infty} \zeta_l^*(z)\zeta_l(z) dz \sum_{k_\parallel, l} \frac{f^l(\mathbf{k}) - f^l(\mathbf{k}+\mathbf{q})}{E(\mathbf{k})_l - E(\mathbf{k}+\mathbf{q})_l} + \mathrm{cc} \tag{12.49}$$

where cc stands for complex conjugate and the additional ζ's are introduced as in Eq. (7.6). The sum can be converted into an integral and expanding $f(\mathbf{k}+\mathbf{q})$ and $E(\mathbf{k}+\mathbf{q})$ for small q, one obtains

$$\delta\rho = e V_{\mathrm{tr}}^{q_\parallel}(0) e^{i\mathbf{q}_\parallel \cdot \mathbf{r}_\parallel} 2 \left\{ \sum_l \int \left(-\frac{\partial f^l}{\partial E} \right) g(E) dE \right\} \int_{-\infty}^{+\infty} \zeta_l^*(z)\zeta_l(z) dz \cdot \zeta_l^*(z)\zeta_l(z) \tag{12.50}$$

To simplify further, we consider now the case when only one subband is populated and assume that the electron gas is essentially two dimensional. In this spirit we approximate

$$\zeta_l^*(z)\zeta_l(z) \sim \delta(z) \tag{12.51}$$

which will give us good results only if the electrons are confined in a very narrow well.

The potential can be written in its original form (before Fourier transformation)—that is, $V_{tr}^{q_{\|}}(0)e^{i\mathbf{q}_{\|}\cdot\mathbf{r}_{\|}} = V_{tr}(z = 0)$—and we have

$$\delta\rho = eV_{tr}(z = 0)2\int_0^{\infty}\left(-\frac{\partial f}{\partial E}\right)g(E)dE \tag{12.52}$$

where $g(E)$ is the two-dimensional density of states.

We proceed now to calculate the scattering rate of two-dimensional electrons for scattering by screened impurities at an arbitrary distance z_0 from the sheet of electronic charge. From perturbation theory, Eq. (8.13), we know that all we need is the (two-dimensional) Fourier transform of the scattering potential, that is, $V_{tr}^{q_{\|}}(0)$. To obtain $V_{tr}^{q_{\|}}(0)$, we need to Fourier transform Eq. (7.9) in two dimensions. Multiplying Eq. (7.9) by $e^{i\mathbf{q}_{\|}\mathbf{r}_{\|}}$ and integrating over $r_{\|}$ gives

$$\left(\frac{\partial^2}{\partial z^2} - q_{\|}^2\right) V_{ad}(\mathbf{q}_{\|},z) = -\frac{1}{\epsilon_0}\,\delta\rho \tag{12.53}$$

The differential operator $\dfrac{\partial^2}{\partial z^2} - q_{\|}^2$ has the Greens function

$$G(z,z') = \exp\,(-q_{\|}|z - z'|)/2q_{\|} \tag{12.54}$$

which means

$$V_{ad}(q_{\|},z) = -\frac{1}{\epsilon_0}\int_{-\infty}^{+\infty} dz'G(z,z')\delta\rho \tag{12.55}$$

as can be proven by inspection [insertion of Eq. (12.55) into (12.53)]. Since according to Eq. (7.1) $V_{ad} = V_{tr} - V_{ap}$ and since

$$\nabla^2 V_{ap} = -\frac{e}{\epsilon_0}\,\delta(r - r_0) \tag{12.56}$$

for a point charge located at $r_0 = (0, 0, z_0)$, one obtains

$$V_{tr}(q_{\|},0) = -\frac{1}{\epsilon_0}\int_{-\infty}^{+\infty} dz'G(0,z')(\delta\rho + e\delta(z - z_0)/A) \tag{12.57}$$

This equation is a linear algebraic equation for $V_{tr}(\mathbf{q}_{\|},0)$ (which appears also in $\delta\rho$) and can easily be solved.

Defining the two-dimensional screening constant by

$$q_s^{\|} = \frac{e^2}{\epsilon_0}\int_0^{\infty}\left(-\frac{\partial f}{\partial E}\right)g(E)dE \tag{12.58}$$

one arrives at the matrix element for the scattering probability as outlined in Chap. 8, Eqs. (8.13)–(8.17):

$$|M_{\mathbf{kk}'}|^2 = \frac{e^4}{4(q_\parallel^2 + q_s^{\parallel 2}/\epsilon)^2(\epsilon\epsilon_0)^2 A^2}\, e^{-2q_\parallel|z_0|} \tag{12.59}$$

where A is the area of the two-dimensional electron gas, z_0 is the distance of the impurity from the electrons in z direction, and $\mathbf{q}_\parallel = \mathbf{k}_\parallel - \mathbf{k}'_\parallel$. A corresponding relaxation time can again, as in Chap. 8, be derived explicitly only for limiting cases such as strong or weak screening (large or small $q_s^\parallel$).

The most important feature of the result is the exponent in Eq. (12.59), which tells us that the scattering matrix element decreases exponentially with the distance from the impurity. The concept of modulation doping makes use of this decrease: One dopes the higher energy gap layers such as AlGaAs neighboring to the GaAs. The electrons then leave their parent donors and accumulate at a distant place of lowest potential energy (GaAs). There the scattering is much reduced and, if phonon scattering is unimportant (low electron temperatures), enormous mobilities can be reached. In the GaAs-AlGaAs system mobilities well above 10^6 cm^2/Vs have been achieved. Phonon scattering in quasi two-dimensional systems has also been treated in detail and shows interesting features because of size quantization.

A key to understanding the existing literature is that while the dimensionality of electrons is reduced by the presence of interfaces, the phonons may (or may not) still propagate in three dimensions. One can therefore have scattering of two-dimensional electrons by three-dimensional phonons but also scattering of the electrons by the surface (fringing) fields of remote phonons, which can be understood from Fig. 2.7. For more details the reader is referred to the review by Hess and to the paper by Mason and Das Sarma.

RECOMMENDED READINGS

DINGLE, R., H. L. STÖRMER, H. C. GOSSARD, and W. WIEGMAN, *Applied Physics Letters* 33 (1978), 665.

HESS, K., "High Field Transport in Semiconductors," in *Advances in Electronics and Electron Physics,* ed. P. W. Hawkes. New York: Academic Press, 1982.

MASON, B., and S. DAS SARMA, *Physical Review* B (March 1987).

STERN, F., and S. DAS SARMA, *Physical Review* B 30 (1984), 840.

STERN, F., and W. E. HOWARD, *Physical Review* 163 (1967), 816.

PROBLEMS

12.1. Derive the Greens function of Eq. (12.54) from its definition

$$\left(\frac{\partial^2}{\partial z^2} + q_\parallel^2\right) G(z,z') = \delta(z - z')$$

12.2. Derive Eq. (12.59) from Eq. (12.57) using the methods of Chap. 8.

12.3. Show the validity of Eq. (12.23) by balancing field and diffusion current.

12.4. Solve the expression for the built-in voltage (Eq. 12.14) in the presence of exponential doping distributions, that is, $N_D^+ = N_D^0 \exp(z/L)$ with L being a fixed length and the highest dopant occurring at $z = 0$.

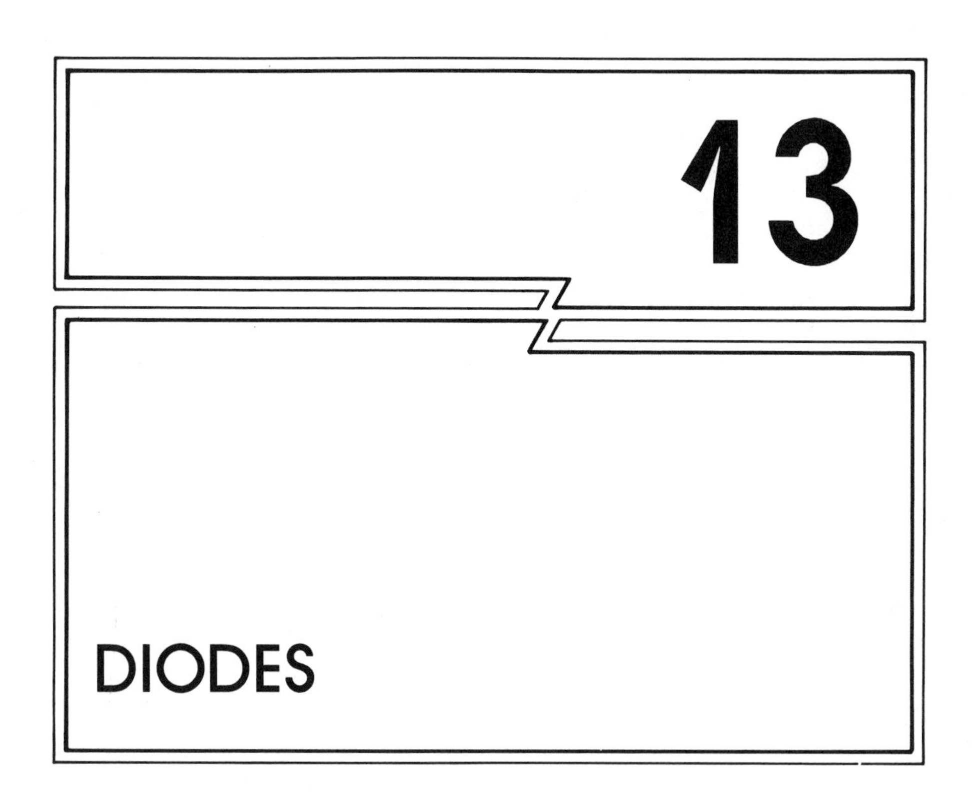

We have already discussed simple diodes (two terminal devices) in the previous two chapters and have developed some of the theoretical concepts that are important for their understanding. In this chapter we discuss several types of diodes, which represent useful semiconductor devices. The use in electronics will not be specifically emphasized here since it is outlined in many other texts (Streetman) and is also, at times, obvious (rectification, light generation, etc.). The basic physical concepts necessary for the understanding of those diodes that have not been discussed in the previous two chapters are treated here in connection with the description of diode operation.

13.1 SCHOTTKY BARRIERS—OHMIC CONTACTS

Section 12.1 dealt with the ideal lattice-matched semiconductor heterojunction. Similar techniques apply, of course, for other heterojunctions, such as metal-semiconductor contacts. In fact, historically these have been investigated first and a detailed understanding has been provided by Schottky.

The band structure of a separated semiconductor and metal is shown in Fig. 13.1. Imagine that we bring the semiconductor and metal closer together so that charge can flow from one to the other. The Fermi levels will then line up, and if we start with a Fermi level that is higher in the semiconductor, electrons will flow from the semiconductor to the metal. This will cause depletion of the semiconductor and a potential change between the semiconductor (at the end of the depletion width) and the metal. This is shown in Fig. 13.2. Finally when metal and semiconductor are in contact, the situation of Fig. 13.3 emerges.

The quantity $e(\phi_m - \chi)$ is usually called the Schottky barrier height, denoted by $e\phi_{Bn}$. The subscript n is for electronic semiconductors. If we had considered a junction with a p-type semiconductor (holes), then we would have found a barrier height

$$e\phi_{Bp} = E_G - e(\phi_m - \chi) \tag{13.1}$$

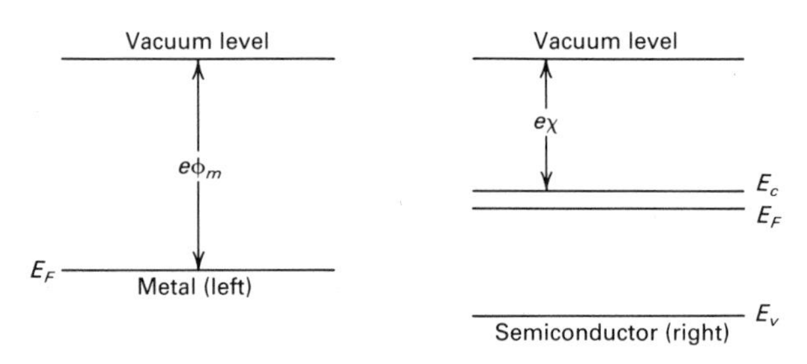

Figure 13.1 Band structure of semiconductor and metal when they are still separated, $e\phi_m$ is the metal work function and $e\chi$ the semiconductor electron affinity.

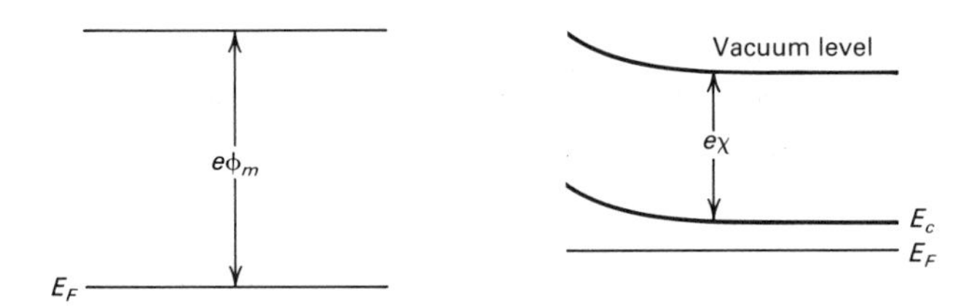

Figure 13.2 Same as Fig. 13.1, but for closer distance of metal and semiconductor.

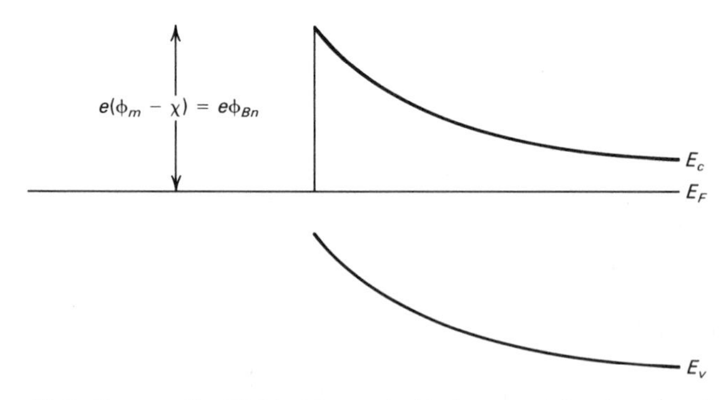

Figure 13.3 Same as Fig. 13.2 but for contact between metal and semiconductor.

which gives us

$$e(\phi_{Bp} + \phi_{Bn}) = E_G \tag{13.2}$$

This result applies only to ideal junctions.

In a real semiconductor-metal junction, a new phenomenon occurs: The Fermi level can be fixed around a certain energy, which is typically close to the middle of the gap of the semiconductor material. The reason for this "pinning" is that high densities of electrons can be trapped close to the interface. These electrons form an interface charge that determines the barrier height which corresponds then to the "pinning" energy, as shown in Fig. 13.4. J. Bardeen suggested that surface states of the kind shown in Fig. 5.4 are responsible for the pinning. However, numerous recent investigations have shown that the problem is more involved, and the explanation of the pinning depends on the actual preparation of the Schottky barrier contact.

An ideal semiconductor surface that has reconstructed (Fig. 5.5) does not necessarily pin the Fermi level at all since the bonds are saturated. Such an ideal surface can only be generated by cutting (cleaving) the semiconductor in ultrahigh vacuum. As soon as the semiconductor comes in contact with air, oxides form at the surface and oxygen remnants remain no matter how carefully the semiconductor is treated afterward. The surface contamination and subsequent metal deposition will "undo" the reconstruction, and interface (not surface) traps of various kinds will form and have to form since the metals are not lattice matched to the semiconductors as demonstrated by Woodall. This is schematically shown in Fig. 13.5.

It is not clear at present what precisely pins the Fermi level, which is understandable from the complicated variety of possibilities inherent in a picture such as Fig. 13.5. Most probably there exists no universal explanation

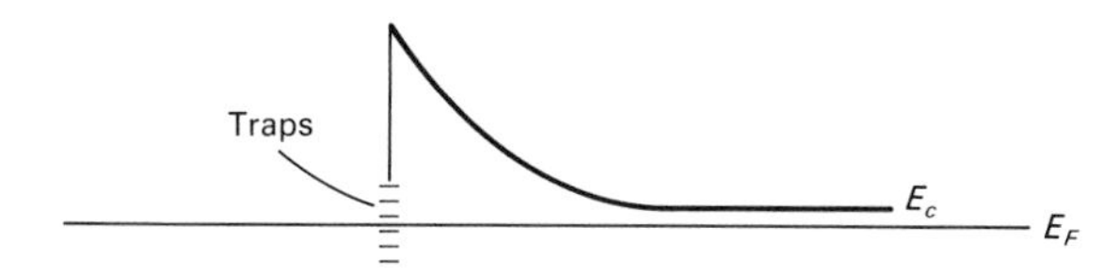

Figure 13.4 Pinning of the Fermi level by localization of electrons at the interface.

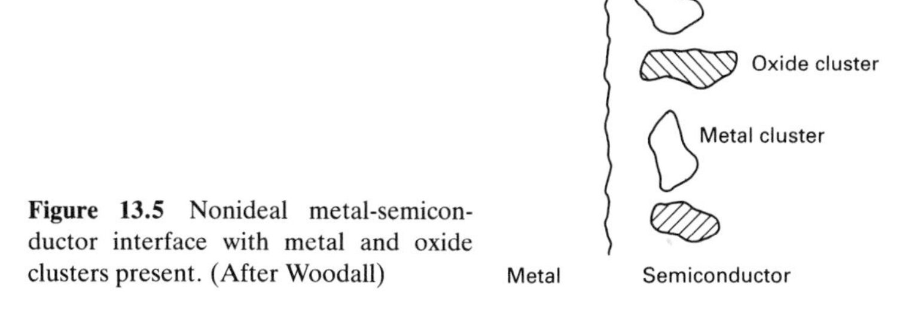

Figure 13.5 Nonideal metal-semiconductor interface with metal and oxide clusters present. (After Woodall)

of pinning energies. The experimental findings indicate that the explanation of pinning will depend on the particular technology of barrier formation. For our purpose we can accept these trapping or pinning energies as given facts and insert into our transport theory experimental values such as those shown in Table 13.1. The barrier heights in the table are approximate and typical for "clean" interfaces. Contamination of the interface may lead to widely varying results. The barrier heights also apply only to equilibrium (no applied electric field).

The addition of an external field leads to a barrier lowering (the Schottky effect) caused by the image force, which is equal to

$$e^2(16\pi\epsilon_0\epsilon z) \tag{13.3}$$

as is well known from electrostatics. The total potential energy $PE(z)$, including the applied field, is then

$$PE(z) = e^2/(16\pi\epsilon_0\epsilon z) + eFz \tag{13.4}$$

This function has a maximum that is lowered by

$$\Delta\phi_B = \sqrt{eF/4\pi\epsilon_0\epsilon} \tag{13.5}$$

from the original barrier height. The maximum is also some distance from the interface, which can be easily found by equating $dPE(z)/dz = 0$. Deducting Eq. (13.5) from the given equilibrium barrier height gives the barrier height ϕ_B^t, which is appropriate for transport calculations. The barrier lowering is illustrated in Fig. 13.6.

There is an additional term that contributes to the Schottky barrier height. This term arises from the penetration of the metal wave functions into the semiconductor and the corresponding penetration of metal charge. This term is easy to add to Eq. (13.4) and analytical expressions have been given by Dutton and coworkers. The interested reader is referred to Shenai et al.

We can now proceed to calculate the current over a Schottky barrier. The Schottky barrier from the metal to the semiconductor can be treated in the same way as we have treated the semiconductor-semiconductor junction of Fig. 12.1. If we measure the barrier height from the Fermi energy of the metal, putting $E_F^L - E_c^L$ in Eq. (12.5) equal to 0, and replace $|\Delta E_c|$ by $e\phi_B^t$, we obtain

$$j_{LR} = A^*T^2e^{-e\phi_B^t/kT} \tag{13.6}$$

TABLE 13.1 Typical Schottky Barrier Heights (volts at 300 K)

Semiconductor	Ag	Au
Si *n*	0.78	0.80
Si *p*	0.54	0.34
Ge *n*	0.54	0.59
Ge *p*	0.50	0.30
GaAs *n*	0.97	1.02
GaAs *p*	0.63	0.42

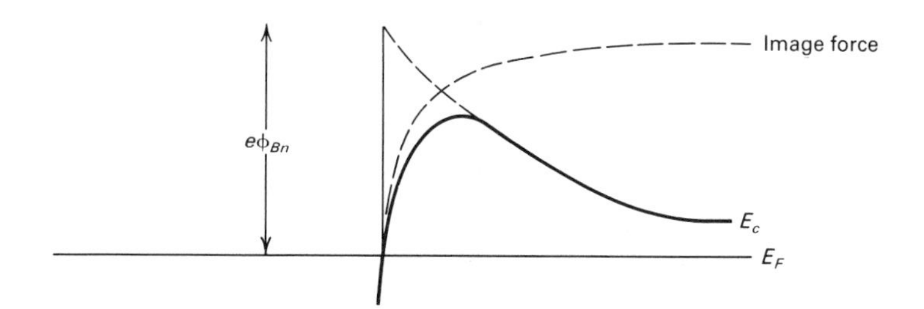

Figure 13.6 Schottky barrier lowering by applied (built-in) electric field. The dashed arrow is the semiconductor conduction band edge without and the full curve with the image force contribution. The dotted curve indicates the image potential alone.

for the equilibrium current j_{RL} from the semiconductor to the metal. The total current is the difference between j_{RL} and j_{LR}. To calculate this difference, we need to digress briefly and discuss the quasi-Fermi levels of a barrier structure when a voltage is applied.

Assume that we have a very long semiconductor that is, far to the right, essentially unperturbed by the presence of the junction with the metal. If we bring an electron from the quasi-Fermi level of the metal to the quasi-Fermi level (which is constant away from the junction) of the semiconductor, we need to perform a certain amount of work that needs to be done by external voltages. We therefore conclude with the important notion that the difference in quasi-Fermi levels is equal to eV_{ext}:

$$\boxed{|E_F^L - E_F^R| = |eV_{\text{ext}}|} \tag{13.7}$$

The applied voltage will mainly drop in the depletion region (another important notion), and therefore we will change the built-in potential V_{bi} of Eq. (12.17) just by eV_{ext}.

Figure 13.7 shows the typical form of the quasi-Fermi levels of a Schottky barrier under bias. Figure 13.7a is plotted for forward bias, the direction of voltage which causes a large current to flow [positive $\overline{V}_{\text{ext}}$ in Eq. (13.10)], while Fig. 13.7b shows reverse bias [negative $\overline{V}_{\text{ext}}$ in Eq. 13.10)].

A simple analogy of the current flow can be obtained by comparing the semiconductor side with a mountain that can be varied from flat plateau form (Fig. 13.7a) to peaked form (Fig. 13.7b). The electron flow corresponds, then, to the way the atmosphere (air) would flow if the mountain were moved. In Fig. 13.7a, we have moved the air up, and high density air at the right will flow to the low density side (left), and vice versa in Fig. 13.7b. It is important to note that this Schottky barrier current is totally different (in its origin) from the current in Eq. (9.36). This current was a field current due to the drift (odd) part f_1 of distribution function f. Here we can have a current entirely due to the even part f_0 of the distribution function. The current arises from the difference

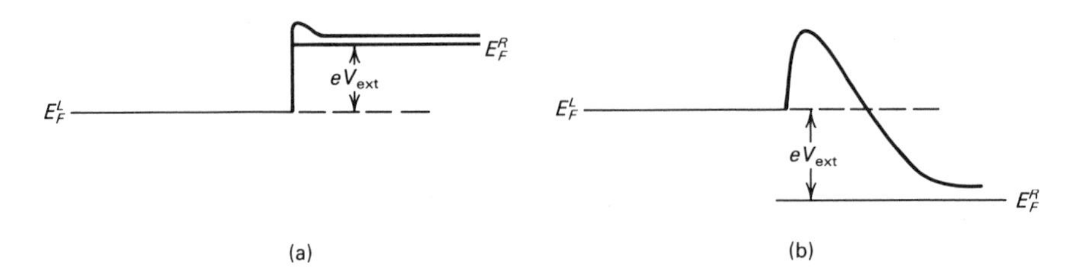

Figure 13.7 Schottky barrier under forward (a) and reverse (b) bias.

of the quasi-Fermi levels relative to the band edges of the different materials, while for a current as in Eq. (9.36) the distance of the quasi-Fermi levels from the band edge is constant (both quasi-Fermi level and band edge change by eV_{ext} in the same way). In the case of a Schottky barrier, all the external voltage drops in the depletion region of the semiconductor and modulates the built-in voltage V_{bi} to $V_{bi} \pm V_{ext}$ while the quasi-Fermi level stays approximately constant on either side.

The current density from the right to the left of our junction is therefore equal to

$$j_{RL} = A^* T^2 e^{-e\phi_B^t/kT} e^{eV_{ext}/kT} \tag{13.8}$$

with an external voltage applied [see Eq. (12.18)]. This makes the total current density $j = (j_{RL} - j_{LR})$ equal to

$$j = A^* T^2 e^{-e\phi_B^t/kT}(e^{eV_{ext}/kT} - 1) \tag{13.9}$$

In the following it is convenient to denote eV_{ext}/kT by $\overline{V}_{ext}$ and analogous $\overline{\phi}_B^t$ so that we obtain

$$j = A^* T^2 e^{-\overline{\phi}_B^t}(e^{\overline{V}_{ext}} - 1) \qquad 13.10)$$

Fig. 13.8 shows schematically this current density and demonstrates the use of Schottky barriers as rectifiers. Notice that the application of an external voltage to a Schottky barrier diode also changes the depletion width. From

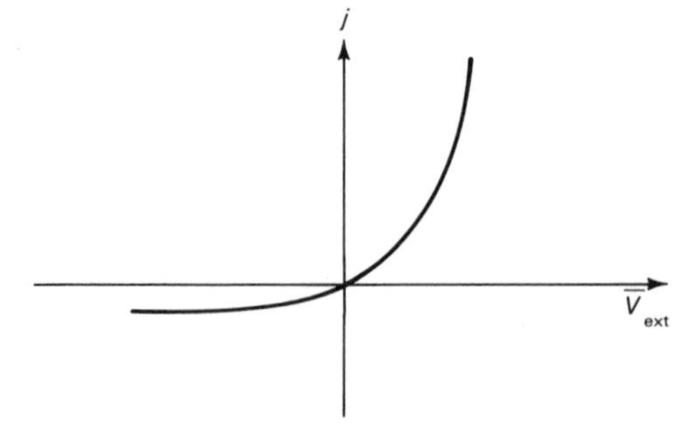

Figure 13.8 Schematic plot of the current in a Schottky barrier demonstrating the rectifying property.

Eq. (12.16), the depletion width, including external potentials [see Eq. (12.14) and its deviation], is

$$W = \left(\frac{2\epsilon\epsilon_0 (V_{\text{bi}} \pm V_{\text{ext}})}{eN_D^+} \right)^{1/2} \tag{13.11}$$

The minus sign applies for barrier lowering, that is, for negative polarity of V_{ext} at the n-type semiconductor side.

We turn now to the speed limitations of Schottky barrier diodes, which will also give us a better understanding of the assumptions involved in our discussion.

The ultimate speed for switching of a Schottky barrier diode is given by the time t_{ext}, which is the time the bulk of the carriers need to transfer from the semiconductor over the semiconductor depletion width W to the metal. Within the framework of our derivation the speed of the carriers can roughly be put equal to v_{z0}, as given in Eq. (12.1b) (a more precise evaluation would involve appropriate averaging over the velocities). Using Eq. (12.16), one obtains

$$t_{\text{ext}} = \frac{W}{v_{z0}} \approx \left(\frac{\epsilon\epsilon_0 m^*}{e^2 N_D} \right)^{1/2} \tag{13.12}$$

which is of the order of 10^{-13} sec for typical values of parameters (such as $N_D = 10^{16}\ \text{cm}^{-3}$).

It is this extraordinary high speed that makes Schottky barriers interesting.

Of course, other factors are involved in the actual speed of Schottky barrier diodes. A trivial limitation is the product of resistance and capacitance, which includes the total semiconductor resistance and the depletion width capacitance (which is given by $\epsilon\epsilon_0/W$ for current purposes). Another speed limitation would be the time the electrons need to lose energy after going over the barriers (so they cannot return), which is typically equal to the inverse scattering rate by phonons (10^{-13} sec). However, even more important (for speed limitations) is the fact that the transport of electrons has a diffusive nature. This is because of the scattering and because Eq. (12.1a) holds only if the electrons that transfer to the metal are replenished from lower energies. For this replacement of electrons it is necessary that electron-electron interactions and phonon absorption events are frequent enough to maintain a Maxwellian distribution, which we have assumed in the deviation. The electrons also must be replenished from the right side, which can represent a bottleneck if the distances are longer than the scattering length since then the transport becomes diffusive. Indeed, if $W > 1000$ Å, transport by diffusion dominates and the thermionic emission process that takes place close to the interface represents just a boundary condition for the diffusion process (the bottleneck for electron supply). We give a treatment of this diffusion process in the next section in the context of the p–n junction, and also in Appendix G.

This appendix gives also an explanation of the constancy of the quasi-Fermi levels on each side as shown in Fig. 13.7.

Before leaving this section, we need to discuss another mode of transport through Schottky barriers, which is tunneling.

Electrons can cross barriers quantum mechanically by tunneling through them. Tunneling is, as we will see, especially important for thin barriers. For thick barriers the thermionic current is larger except if the temperature becomes very small, as can be seen from Eq. (12.5). The tunneling current can be derived from Eqs. (1.30) and (A.16) as well as (A.19).

To calculate the current from the left ($z < z_a$) to the right ($z > z_a$), we proceed as follows: We multiply the probability per unit time of a tunneling transition by the probability to find an electron at the left (which is equal to the energy distribution f_L on the left side) and by the probability that the state is not occupied at the right ($1 - f_R$) and then sum over all states on the right. This gives the current from the left to the right due to one state. The total current is the sum over all left states multiplied by the elementary charge. Denoting $\mathbf{k}$ by $\mathbf{k}_L$ at the left and by $\mathbf{k}_R$ at the right and including a factor of 2 for the spin, we get

$$j_{LR} = \frac{4\pi e}{\hbar} \sum_{k_L} \sum_{k_R} \frac{|\bar{k}_{z_a}| |\bar{k}_{z_b}|}{L_a L_b} \exp\left(-2 \int_{z_a}^{z_b} |\bar{k}_z| dz\right) \delta(E_R - E_L) f_L (1 - f_R) \tag{13.13}$$

Now transforming the sums into energy integrals, one obtains

$$j_{LR} = \frac{\pi e}{\hbar} \iint \frac{|\bar{k}_{z_a}| |\bar{k}_{z_b}|}{L_a L_b} \exp\left(-2 \int_{z_a}^{z_b} |\bar{k}_z| dz\right) \delta(E_R - E_L) \times f_L (1 - f_R) g_L(E_L) g_R(E_R) dE_L dE_R \tag{13.14}$$

where $g_L(E_L)$ and $g_R(E_R)$ represent the density of states on the left and right sides, respectively. The notation is otherwise the same as in Appendix A. Since these density of states functions each include a factor of 2 for the spin, which has already been accounted for above, we have divided Eq. (13.13) by a factor of 4.

The current from the right to the left can be calculated in analogous fashion. The integral is easily evaluated. One integration involving the δ function simply replaces all energies by E_L or E_R. The other integration can be performed for special cases (triangular barrier, etc.) or can be done with a pocket calculator for general potential forms.

After the discussion of the metal-semiconductor contact and its rectifying properties, it seems difficult to understand how an ohmic contact is formed. Of course, there is the possibility that the Schottky barrier height is zero or such that electrons are transferred to the n-type semiconductor and no depletion region forms (see also Streetman). There is, however, another possibility that is of greater importance. The metal can be enriched by donor material (or acceptors) and, by alloying, a thin layer of highly doped semiconductor will form in the proximity of the metal. If this is achieved (by whatever technology), the depletion layer width W will be very thin. Using $z_b - z_a = W$ in the tunneling formula, we can see that the tunneling current will be large. In fact it can be large enough to satisfy the most stringent requirements of low contact resistance. Note that doping fluctuations will enhance the current at certain places and reduce it at others, the effect being an overall enhancement by a factor of 2–10 over the computation for homogeneous material, as has been shown by McGill and coworkers.

13.2 THE *p–n* JUNCTION

We saw in Chap. 12 the importance and significance of heterojunctions, that is, junctions in which the semiconductor and its band structure vary. However, as we know from Chap. 5, minute changes in the chemical composition can drastically alter the electronic properties of semiconductors. For this reason, the homojunction, a junction of identical semiconductor but different doping (typically, donors on one side, acceptors on the other) has its own special importance. The doping agents can be introduced by diffusion at high temperatures or by ion implantation. The simplistic approach of a theorist "to glue" two semiconductors of different types together does not work because of the interface states between glue and semiconductors. We do not attempt to discuss the experimental details and assume an ideal abrupt junction (Fig. 13.9) where the doping changes suddenly at $z = 0$ from p-type at the left and n-type at the right. Gradual junctions can be treated similarly (Streetman).

We first discuss the band diagram of such a junction in equilibrium (no external forces applied) and list the rules to graph such diagrams.

The voltages and band edge energies are measured relative to a fixed point, for example, the point c, where c is far away from the junction (far

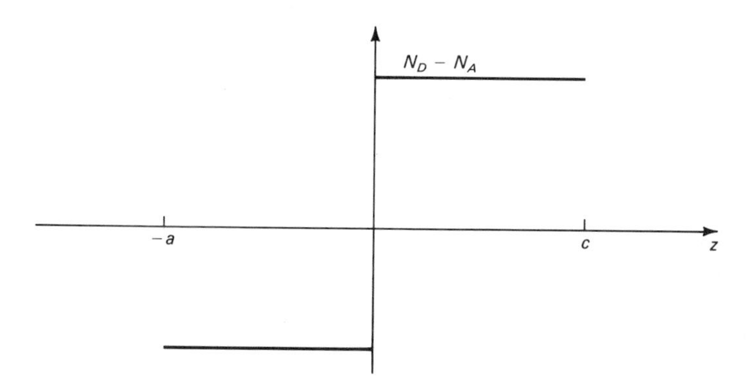

Figure 13.9 Doping concentration in an ideal abrupt homojunction. Notice that $N_D \neq N_A$ at the different sides.

enough not to be disturbed by the presence of the junction). The rules for plotting band diagrams are

1. The band edges follow the additional potential V_{tr} (Chap. 7) caused by impurity charges and external voltages. This rule is a consequence of the effective mass theorem [Eqs. (4.23) and (4.24)]. Notice that the potential energy varies with the opposite sign of the electrical potential.
2. The Fermi energy is constant and far away from the junction is located at the bulk equilibrium distance from conduction and valence band edges [see Eqs. (6.20) and (6.30)].
3. The energy gap E_G is the same throughout except in regions of very heavy doping or large electron (hole) density [see Fig. 6.4 and Eq. (6.14)].
4. At the junction, space charge regions form [see Fig. 12.2 and Eq. (12.16), and both conduction and valence band edges vary considerably.

Consequently the diagrams are plotted best by first drawing a straight horizontal line representing the Fermi energy. Next the junctions are marked and away from the junctions the band edges are drawn. On the sides with donor doping (n sides) the conduction band edge is marked as a line close to and parallel to the Fermi energy. Similarly, on the sides with acceptor doping (p sides) the valence band is marked close to the Fermi energy and (far away from the junction) parallel to it. Knowing E_G, we can then plot parts of the conduction band on the p sides and of the valence band on the n sides. Smooth (except for band edge discontinuities in heterojunctions) connections finish the diagram. The diagram for our abrupt p–n junction is shown in Fig. 13.10.

Figure 13.10 shows clearly that the Fermi energy is distant from both the conduction and valence band edges for the range $-l_p \leqslant z \leqslant l_n$. According to Eqs. (6.20) and (6.26), this range is therefore depleted of electrons and holes, and for some purposes we can approximate $n = p = 0$ in this range. In the same spirit the electric field is put at zero for $z < l_p$ or $z > l_n$. In the depleted range there is fixed space charge, the positively charged donors and negatively charged acceptors as shown in Fig. 13.10. The voltage difference from the n to

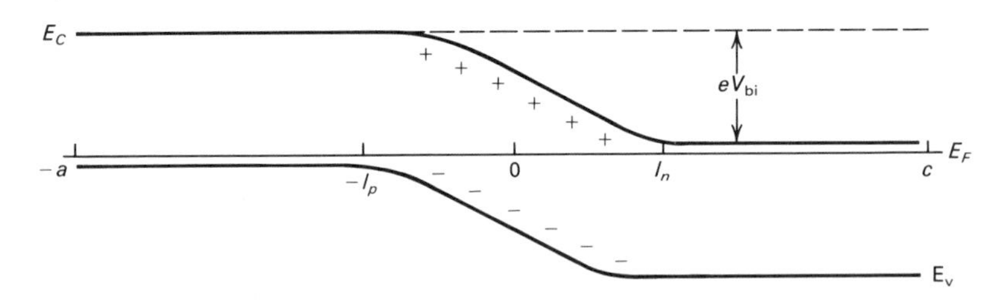

Figure 13.10 Band diagram for an abrupt p–n junction.

the p side is denoted by V_{bi}, which is the built-in voltage totally analogous to the built-in voltage in heterojunction barriers [Eqs. (12.14)–(12.16)]. To obtain $-l_p$ and l_n as a function of the built-in voltage, we proceed as in Eqs. (12.14) and (12.16) and integrate Eq. (12.14) from $-l_p$ to l_n.

In addition we use the fact that the electric field at l_n, $F(l_n)$, or beyond it to the right, is equal to zero. From Eq. (12.11) we have

$$F(z) = -\frac{e}{\epsilon\epsilon_0}\int_{-l_p}^{z}(p - n + N_D^+ - N_A^-)dz \tag{13.15}$$

Then we use the depletion approximation $p = n \approx 0$ to obtain

$$0 = F(l_n) = \frac{e}{\epsilon\epsilon_0}(N_A^- l_p - N_D^+ l_n) \tag{13.16}$$

Using $N_A^- \approx N_A$ and $N_D^+ \approx N_D$ in the depletion region, we further have

$$l_p N_A = l_n N_D \tag{13.17}$$

which we also could have guessed from charge neutrality considerations. This, together with the equation for the built-in voltage [Eq. (12.14)] and $p = n \approx 0$, gives

$$l_p = \left(V_{bi}\frac{2\epsilon\epsilon_0}{e}\frac{N_D}{N_A(N_A + N_D)}\right)^{1/2} \tag{13.18}$$

and

$$l_n = \left(V_{bi}\frac{2\epsilon\epsilon_0}{e}\frac{N_A}{N_D(N_A + N_D)}\right)^{1/2} \tag{13.19}$$

It is easy to verify that for $N_A \gg N_D$ we are led back to Eq. (12.16) by using $W = l_p + l_n$. Our results are summarized in Fig. 13.11. The free carrier concentration is approximately constant outside the depletion region and denoted by p_p, p_n (holes on the p and n sides, respectively) as well as by n_n, n_p (electrons on the n and p sides, respectively). If the doping is homogeneous, the fixed charge is approximately constant between $-l_p$ and 0, as well as between 0 and l_n. The electric field has a maximum, F_{max}, at $z = 0$ and is triangular within the depletion approximation.

From Eq. (13.15), F_{max} is found to be

$$F_{max} = -\frac{e}{\epsilon\epsilon_0}N_D l_n = -\frac{e}{\epsilon\epsilon_0}N_A l_p \tag{13.20}$$

The above equations and figures form the basis for an ideal model of the p–n junction away from equilibrium, that is, with external voltage applied. We will see that the current in a p–n junction can best be understood from the following analogy of high energy physics.

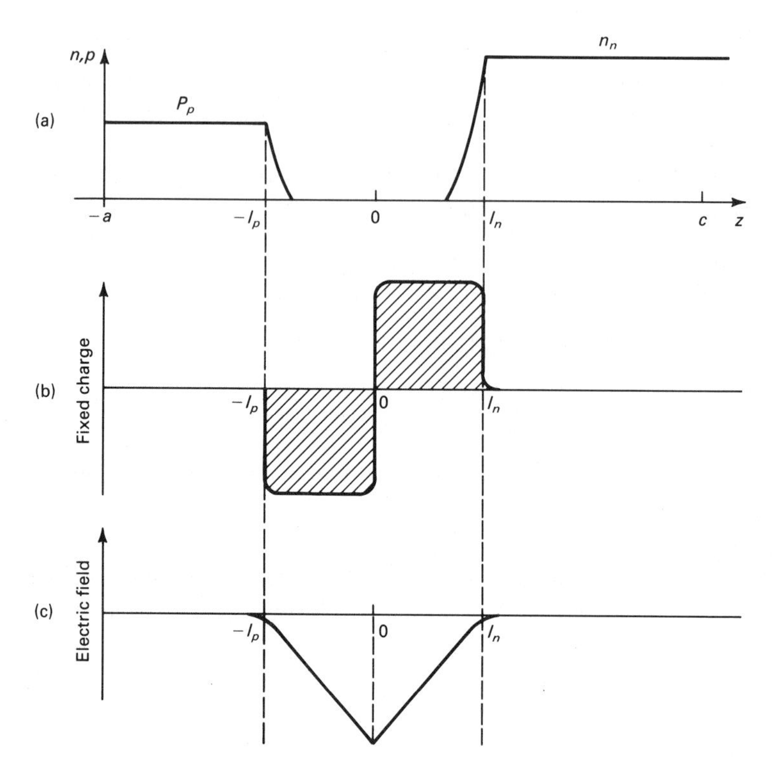

Figure 13.11 (a) Concentration of free carriers (electrons and holes) in a p–n junction at equilibrium; (b) distribution of fixed charge in the depletion layer (notice that charge neutrality requires equal areas); (c) electric field in the depletion region.

Consider the box shown in Fig. 13.12 and a beam of electrons incident from the right side and a beam of positrons (antiparticles) from the left. The electrons and positrons meet in the box, annihilate each other, and emit radiation. Therefore a current is flowing continuously from place a to c, but neither electrons nor positrons make it all the way through the box. It is precisely the same mechanism that explains the current in a p–n junction. (In other words, Dirac discovered a "funny" form of p–n junction.) The fact that

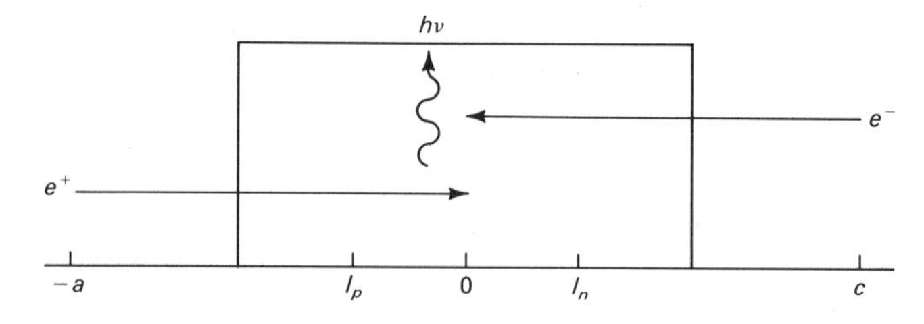

Figure 13.12 High energy physics analogy for p–n junction.

the electrons and holes in a p–n junction need to overcome a potential barrier before recombining is also important and leads to exponential dependencies in the current.

The depletion region is, of course, formed by the recombination of electrons and holes. If external voltages are applied, the equilibrium is perturbed and the current is maintained by continuous additional recombination (or generation).

While the details require more discussion, we can already plot the non-equilibrium band diagram by applying two principles that we have developed previously. The first principle we use is that the difference in the quasi-Fermi levels for electrons and holes is equal to eV_{ext} far away from the junction. Second, we have discussed in Appendix G that the quasi-Fermi levels are also constant in the depletion region provided that certain criteria (for the replenishing of charge carriers) are fulfilled. These criteria (Appendix G) are somewhat different for p–n junctions than for the case of Schottky barriers, since the carriers are not "lost" by emission but by recombination. However, by reasoning as in Appendix G, one easily sees that as long as the recombination rate (time) is weaker (longer) than the rate (time) that is necessary to replenish the carriers, the quasi-Fermi levels will be constant in the depletion region. We will derive these rates below. A detailed examination (which is left to the reader) shows that for small V_{ext} the assumption of constant quasi-Fermi levels is justified. We denote the quasi-Fermi levels of electrons by E_F^n and of holes by E_F^p (the notation involving L far left and R far right is not sufficient now since we will need the extension of electron quasi-Fermi levels for the hole side, and vice versa).

The above discussion leads to the band diagram shown in Fig. 13.13. Beyond the depletion region the quasi-Fermi levels have to merge since far away from the junction electrons *and* holes approach their equilibrium density.

The external voltage appears across the depletion region and, depending on its polarity, adds or subtracts from the built-in voltage. It reduces the barrier if negative voltage is applied to the n side and positive to the p side. This is therefore called forward bias (the opposite is reverse). Note that the

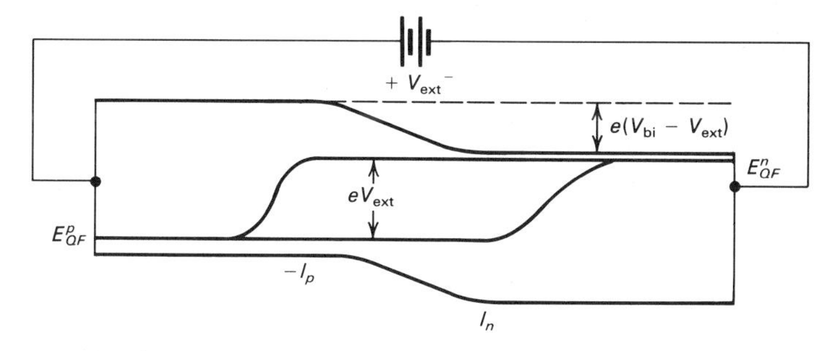

Figure 13.13 Band diagram of p–n junction with external voltage applied.

depletion width $W = l_n + l_p$ changes with external bias. From the derivation [Eq. (12.12)] it can be seen that the external voltage adds (subtracts) to the built-in voltage. In the formulas for l_n and l_p, Eqs. (13.18) and (13.19), we need to replace V_{bi} by $V_{bi} + V_{ext}$. Therefore

$$l_n = \left[(V_{bi} + V_{ext}) \frac{2\epsilon\epsilon_0}{e} \frac{N_A}{N_D(N_A + N_D)} \right] \tag{13.21}$$

Equation (13.21) is of importance for many applications.

Of equal importance is the product of electron and hole concentrations when a voltage is applied. We have learned from Eq. (6.27) that this product is proportional to exp $(-E_G/kT_L)$ (in this equation the lattice temperature must be used since the electrons are excited by lattice vibrations). Following the derivation of Eq. (6.27) with quasi-Fermi levels, we see that

$$n \cdot p = n_i^2 \exp (E_F^n - E_F^p)/kT_L \tag{13.22}$$

which gives us (by use of Fig. 13.13)

$$n \cdot p = n_i^2 \exp (eV_{ext}/kT_L) \tag{13.23}$$

If the bias is in forward direction, eV_{ext} is positive and increases the product.

Equations (13.22) and (13.23) can also immediately be deduced from Fig. 13.13, which shows that the energy gap is effectively reduced (forward bias) by eV_{ext} or increased by this amount in reverse bias.

The current through the junction is twofold in nature. If we have nowhere in the p–n junction the possibility of electron-hole recombination, then the current is given by the thermionic-emission-diffusion theory (now for both electrons and holes), discussed in Chap. 12 and Sec. 13.1. Of course, this case is rather academic. Even if the material does not contain deep traps, and even if recombination by phonon emission is improbable, we still have the possibility of photon emission. If the semiconductor is indirect, however, we can have only combined phonon-photon emission, and recombination is then weak. Still the recombination current will be substantial in a reasonably long p–n junction, and under "normal" conditions the recombination (generation) will be a dominant mechanism to influence the current, as outlined in connection with Fig. 13.12.

To derive the equation for the dc current density j (due to generation recombination) we use the following approach: In steady state j is everywhere the same. We therefore need to calculate only the current at one point, say, at c of Fig. 13.10. There the current is composed of electron j_n and hole j_h current. Since at c the semiconductor is n-type, the hole current will be negligible and we need to calculate the electron current only (this is the trick) (see also Landsberg). Therefore

$$j = j_n(c) + j_h(c) \approx j_n(c) \tag{13.24}$$

From the steady state equation of continuity [Eqs. (11.4) and (10.33)], we have (if there is no additional generation by light, etc.)

$$\frac{\partial j_n(z)}{\partial z} = eU_s \tag{13.25}$$

where U_s is the net electron recombination rate in steady state. This gives

$$j_n(z) = \int_{-a}^{z} eU_s \, dz \tag{13.26}$$

and a similar equation for the hole current. Here we have assumed, consistent with Eq. (13.24), that $j_n(-a) \approx 0$ as the electron current must be very small at the p side. We partition now the integration:

$$j_n(c) = \int_{-a}^{-l_p} eU_s \, dz + \int_{-l_p}^{l_n} eU_s \, dz + \int_{l_n}^{c} eU_s \, dz \tag{13.27}$$

U_s is in general a complicated function of the carrier concentration in the conduction and valence bands. We therefore have to digress at this point and calculate the carrier concentration. It is sufficient to derive the concentration $n(z)$ of electrons. The derivation for holes is entirely analogous.

In the range $-a \leqslant z \leqslant -l_p$, the electric field is negligible (depletion approximation). The electronic current in the conduction band is therefore a diffusion current that has its origin in the concentration gradient caused by the recombination (see also Appendix G). Therefore

$$j_n \approx eD_n \frac{\partial n}{\partial z} \tag{13.28}$$

We now have to use the subscript n for the diffusion constant since the constant for holes is different, and we have both types of carriers present. Using Eq. (13.25), we obtain

$$eD_n \frac{\partial^2 n}{\partial z^2} = eU_s \tag{13.29}$$

This second-order differential equation has no explicit solution for the general expression of U_s. We therefore use the approximation of Eq. (10.36), that is

$$U_s \approx \frac{n - n(-a)}{\tau_n} \tag{13.30}$$

where $n(-a)$ is the electron concentration on the p side. This gives

$$\frac{\partial^2 n}{\partial z^2} = \frac{n - n(-a)}{\tau_n D_n} \equiv \frac{n - n(a)}{L_n^2} \tag{13.31}$$

Because of the constant coefficients in this equation, the solution is

$$n(z) = C_1 e^{z/L_n} + C_2 e^{-z/L_n} + C_3 \tag{13.32}$$

C_1, C_2, and C_3 are constants that can be detrmined by the following boundary conditions: The electron concentration approaches small values at $z = -a$; since a/L_n is large and the exponent becomes very large, C_2 must be 0. Furthermore, since e^{z/L_n} becomes extremely small at $z = -a$, we have $C_3 \approx n(-a)$. Therefore, we simply have to determine C_1. This can be obtained from Eq. (13.23), which reads (at $z = -l_p$)

$$n(-l_p)p(-l_p) = n_i^2 e^{\overline{V}_{\text{ext}}} \tag{13.33}$$

Here $\overline{V}_{\text{ext}} = eV_{\text{ext}}/kT$, and since $p(-l_p) \approx p(-a)$, one obtains

$$n(-l_p) = \frac{n_i}{p(-a)} e^{\overline{V}_{\text{ext}}} \approx n(-a)e^{\overline{V}_{\text{ext}}} \tag{13.34}$$

which completes the boundary conditions and results in

$$C_1 = n(-a)(e^{\overline{V}_{\text{ext}}} - 1)e^{l_p/L_n} \tag{13.35}$$

This finishes the calculation of the carrier concentration and together with Eq. (13.30) gives

$$U_s = \frac{n(-a)}{\tau_n}(e^{\overline{V}_{\text{ext}}} - 1)e^{(z+l_p)/L_n} \tag{13.36}$$

which permits us to calculate the dc current through a p–n junction.

The first contribution to the current is then

$$\int_{-a}^{-l_p} eU_s\, dz = \frac{eL_n n(-a)}{\tau_n}(e^{\overline{V}_{\text{ext}}} - 1)(1 - e^{(-a+l_p)/L_n}) \tag{13.37}$$

Here the last factor is approximately equal to one for long diodes. For short diodes, the boundary conditions need to be changed. This is discussed in the context of the bipolar transistor in Chap. 14.

The second contribution, Eq. (13.27), to the electron current is much easier to calculate since in the space charge region the product $n \cdot p$ is constant and is given by Eq. (13.23). Therefore we can derive this portion of the current using the general expression of Eq. (10.33). The concentrations n and p in the denominator of Eq. (10.33) can be neglected under weak forward or reverse bias conditions. Then U_s is constant over the depletion range, and

$$e\int_{-l_p}^{l_n} U_s\, dz = \frac{ec_n c_p n_i^2 N_{TT} W}{e_n + e_p}(e^{\overline{V}_{\text{ext}}} - 1) \tag{13.38}$$

Notice that W depends on the external voltage according to Eq. (13.21). There is only one term left to calculate. Since the calculation is entirely analogous to the first term, we merely give the result

$$\int_{l_n}^{c} eU_s\, dz = \frac{L_p ep(c)}{\tau_p}(e^{\overline{V}_{\text{ext}}} - 1)(1 - e^{(-c+l_n)/L_p}) \tag{13.39}$$

Note again that the last factor can be approximated by the one for long diodes. The total current is then

$$j = j_{rs}(e^{\overline{V}_{\rm ext}} - 1) \tag{13.40}$$

where j_{rs} is the sum of the constant factors of Eqs. (13.37)–(13.39) and is called the reverse saturation current (since $j \approx j_{rs}$ for reverse bias). This is the famous Schockley equation. Formally it is the same as the equation for the thermionic current, Eq. (13.10). The reason is that although the various constants are very different, the origin of the current is based on the same physics (Boltzmann's law).

Under stronger forward bias the various approximations for U_s do not hold, as n and p in the denominator cannot be neglected (or approximated by a constant). Under symmetric conditions one can deduce from Eq. (13.23) that $n + p \approx 2n_i \exp(\overline{V}_{\rm ext}/2)$, which is the same as Hall's result for the p-i-n diodes. The current then does not depend on only $\exp(\overline{V}_{\rm ext})$. Experimental results show that often $\exp(\overline{V}_{\rm ext})$ must be replaced by $\exp(\overline{V}_{\rm ext}/s)$ where s is a "nonideality" factor, typically of the order of 2. For more ambitious projects on p–n junctions, a numerical integration is recommended using precise equations for U_s and replacing the assumption of constant quasi-Fermi levels in the depletion region by a numerical approach.

Important for device applications is, in addition to the dc current, the ac response, and therefore the differential capacitance per unit area, which is defined as

$$C \equiv \frac{\partial Q}{\partial V_{\rm ext}} \tag{13.41}$$

where Q is the charge per unit area on one side of the junction:

$$Q \equiv -\int_{-l_p}^{0} \rho dz = +\int_{0}^{l_n} \rho dz \tag{13.42a}$$

From Poisson's equation [Eqs. (12.11) and (12.14)] we have

$$V_{\rm bi} + V_{\rm ext} = \frac{1}{\epsilon\epsilon_0}\int_{-l_p}^{l_n} z\rho dz \tag{13.42b}$$

Differentiating $V_{\rm ext}$ by using the chain rule and assuming that only l_n and l_p depend on $V_{\rm ext}$ on the right-hand side of Eq. (13.42b) (this assumption breaks down in forward bias as will be pointed out later), we obtain

$$\epsilon\epsilon_0 = \frac{\partial}{\partial l_n}\left[\int_{-l_p}^{l_n} z\,\rho dz\right]\frac{\partial l_n}{\partial V_{\rm ext}} + \frac{\partial}{\partial l_p}\left[\int_{-l_p}^{l_n} z\,\rho dz\right]\frac{\partial l_p}{\partial V_{\rm ext}} \tag{13.43}$$

which is equivalent to

$$\epsilon\epsilon_0 = l_n\,\rho(l_n)\frac{\partial l_n}{\partial V_{\rm ext}} - l_p\,\rho(-l_p)\frac{\partial l_p}{\partial V_{\rm ext}} \tag{13.44}$$

From the definition of the capacitance we have

$$C = \frac{\partial Q}{\partial V_{\text{ext}}} = \frac{\partial}{\partial V_{\text{ext}}}\left[-\int_{-l_p}^{0} \rho dz\right] = -\rho(-l_p)\frac{\partial l_p}{\partial V_{\text{ext}}} = \rho(l_n)\frac{\partial l_n}{\partial V_{\text{ext}}} \qquad (13.45)$$

Use of Eqs. (13.44) and (13.45) results in

$$\epsilon\epsilon_0 = \left[l_p \frac{\partial Q}{\partial V_{\text{ext}}} + l_n \frac{\partial Q}{\partial V_{\text{ext}}}\right] \qquad (13.46)$$

and

$$C = \frac{\epsilon\epsilon_0}{W} \qquad (13.47)$$

where W is a function of V_{ext} as shown in Eq. (13.21).

This result is, of course, expected and the derivation has only been given to stress its generality, as we have made only the assumption that the charge density $\rho(z)$ does not depend on V_{ext}. This means that the result applies in reverse bias for junctions with arbitrary transitions from n to p doping. However, the general frequency response of a p–n junction needs more detailed treatment because of the existence and dynamics of generation-recombination and because of the dependence of $\rho(z)$ on V_{ext} in forward bias, which have not been included into the above derivation. Before we deal with forward bias we describe the influence of the generation-recombination dynamics in reverse bias and develop, as a side result, the theory of important experimental techniques called transient capacitance methods, that is, deep level transient capacitance spectroscopy (DLTS).

In the case of reverse bias, n and p are extremely small in the depletion region and can be neglected. However, if the p–n junction has been in forward bias previous to the application of reverse bias, there is a memory effect in the reverse bias behavior, which is the basis for DLTS. In the mathematical treatment the earlier forward bias enters only as a boundary condition to n_T (traps are filled by electrons and holes under forward bias conditions); otherwise the electron and hole concentrations can be neglected in the depletion region. Then Eq. (13.20) is valid provided that W is calculated as a time-dependent quantity that includes the dynamics of trapping. These dynamics can be obtained from Eq. (10.28) neglecting the terms proportional to n and p or

$$\frac{\partial n_T}{\partial t} = -(e_n + e_p)n_T + e_p N_{TT} \qquad (13.48)$$

If we assume that all the traps are filled by electrons at $t = 0$ (because of the forward bias), then the solution of Eq. (13.48) is straightforward, and

$$n_T = N_{TT}\left[\frac{1}{1 + e_n/e_p} + \left(1 - \frac{1}{1 + e_n/e_p}\right) e^{-(e_n + e_p)t}\right] \qquad (13.49)$$

Should the traps be filled to any other degree at $t = 0$, similar equations can be obtained with ease. For $e_p \approx 0$ and $n_T = N_{TT}$ at $t = 0$, we have the simple result:

$$n_T = N_{TT} e^{-e_n t} \tag{13.50}$$

which will be used in the following for the purpose of demonstration.

Consider now the switching cycle shown in Fig. 13.14. At $t < 0$ the junction is forward biased as shown in Fig. 13.14a. The donors (acceptors) are

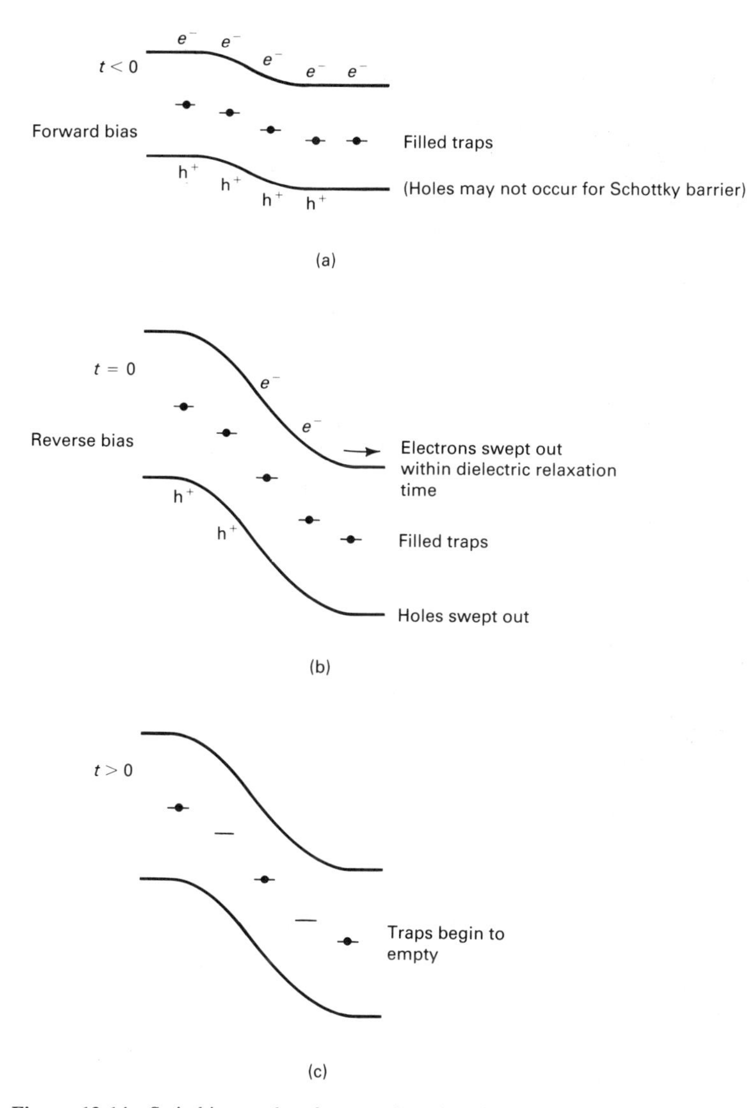

Figure 13.14 Switching cycle of a *p–n* junction (or Schottky barrier) and the corresponding filling and emptying of electron traps: (a) forward bias (filling), (b) reverse bias (electrons are swept out of the conduction band), (c) release of electrons from traps. (Typical time constant for a trap at midgap in silicon at $T = 300$ K is $\approx$1 millisecond.)

partially filled with electrons in accord with the position of the quasi-Fermi levels. The trap, which is shown in the middle of the gap, is almost totally filled with electrons since the conduction band is flooded with electrons that can be captured. At $t = 0$ the junction is switched to reverse bias. The electrons are then removed from the conduction band within the dielectric relaxation time, which for higher electron densities is of the order of 10^{-12}s. Electrons in shallow donors are also emitted with about the same time constant and leave the junction region. To simplify the calculation we assume, as stated above, $e_p = 0$ (electron trap only) and furthermore $N_A \gg N_D$, which means we consider an abrupt p^+–n junction (or Schottky barrier) whose depletion width can be approximated by $W \approx l_n$. We also assume that the trap density is much smaller than the density of shallow donors that form the junction. At $t = 0$ all traps are filled. In the following we will calculate the capacitance change due to the dynamics of electron release.

The depletion width in the absence of these traps is denoted by W_0. Then we have

$$V_{\text{bi}} + V_{\text{ext}} = -\frac{e}{\epsilon\epsilon_0} \int_0^{W_0} z N_D^+ \, dz \tag{13.51}$$

where N_D^+ denotes the density of positively charged donors (the conduction band and donors are emptied immediately after $t = 0$). Our goal is to calculate the depletion width W, including the time-dependent filling of traps, which is given by the equation

$$V_{\text{bi}} + V_{\text{ext}} = -\frac{e}{\epsilon\epsilon_0} \int_0^{W} z\left(N_D^+ + N_{TT}(1 - e^{-e_n t})\right) dz \tag{13.52}$$

We have tacitly assumed that the traps are also donorlike; that is, the positively charged traps are the empty traps whose density is $N_{TT}(1 - e^{-e_n t})$.

Since N_{TT} is small compared to N_D, we have $W = W_0 + \Delta W$ with $\Delta W \ll W_0$ (this can also be seen from our result), and therefore we can use perturbation theory [as for Eq. (1.14)]:

$$-\frac{\epsilon\epsilon_0}{e}(V_{\text{bi}} + V_{\text{ext}}) = -\int_0^{W_0} z N_D^+ \, dz + \int_{W_0}^{W_0 + \Delta W} z N_D^+ \, dz + \int_0^{W_0} z N_{TT}(1 - e^{-e_n t}) \, dz \tag{13.53}$$

where we have neglected the second-order term:

$$\int_{W_0}^{W_0 + \Delta W} z N_{TT}(1 - e^{-e_n t}) \, dz \tag{13.54}$$

Using Eq. (13.53), we obtain

$$\int_{W_0}^{W_0 + \Delta W} z N_D^+ \, dz = -\int_0^{W_0} z N_{TT}(1 - e^{-e_n t}) \, dz \tag{13.55}$$

which gives for small ΔW

$$W_0 \Delta W N_D^+(W_0) = -\int_0^{W_0} z N_{TT}(1 - e^{-e_n t})\, dz \tag{13.56}$$

We can now use the mean value theorem of integral calculus to evaluate the right-hand side of Eq. (13.56) or we can numerically integrate it, or, in case of constant N_{TT}, explicitly integrate. If W_1 is a suitable mean value of z, the result is

$$W_0 \Delta W N_D(W_0) = -W_0 N_{TT}(W_1)(1 - e^{-e_n t}) \tag{13.57}$$

Using the capacitance formula, Eq. (13.47), one gets for the relative capacitance change

$$\frac{\Delta C}{C_0} = \frac{N_{TT}(W_1)\epsilon\epsilon_0}{2eN_D^+(W_0)}(1 - e^{-e_n t}) \tag{13.58}$$

In other words, the time-dependent capacitance gives us information on the density of deep traps as well as the emission rates e_n. This fact is used to determine e_n and N_{TT} by so-called transient capacitance methods which measure capacitance vs. time and temperature. The temperature is varied in order to put e_n into the appropriate range accessible to measurement. Remember that e_n depends exponentially on temperature.

The capacitance under forward bias cannot be accurately calculated from Eq. (13.47) since it has been assumed in its derivation that the charge density ρ does not depend on the external voltage [see Eqs. (13.42) and (13.43)]. This clearly is not the case if we flood the depletion region with electrons and holes. It is therefore necessary to solve the dynamic equations for dc plus small superimposed ac voltage in order to derive circuit elements such as the junction capacitance. We then write

$$\overline{V}_{\text{ext}} = \overline{V}_{dex} + \overline{V}_{1ex} e^{i\omega t} \tag{13.59}$$

It is clear that the carrier concentration $n(z)$ will also have an ac component, and we define

$$n(z,t) = n_d(z) + n_1(z)e^{i\omega t} \tag{13.60}$$

Inserting this concentration into the equation of continuity (now including the term $\partial n/\partial t$) and writing n for $n(z,t)$, one has

$$\frac{\partial n}{\partial t} = \frac{1}{e}\frac{\partial j_n}{\partial z} - \frac{n - n(-a)}{\tau_n} \tag{13.61}$$

Here we have used the steady state formula for the generation-recombination rate, Eq. (10.36), which follows essentially from Eq. (10.33) and assumes that, although n and p change dynamically, the net electron generation-recombination rate is the same as the net hole generation-recombination rate. In other words, the filling of the traps either does not change or does not influence the generation-recombination characteristics.

This assumption is not always justifiable, however; if it is not made, the calculation of the ac current involves more elaborate numerical approaches. The current j_n is the diffusion current of Eq. (13.28) if we neglect the term $\partial j_n/\partial t$ in Eq. (11.5b). Since τ is typically 10^{-13} sec, this approximation is usually justified. We then have from Eq. (13.61)

$$\frac{\partial n}{\partial t} = D_n \frac{\partial^2 n}{\partial z^2} - \frac{n - n(-a)}{\tau_n} \tag{13.62}$$

Inserting Eq. (13.60), one obtains for n_1

$$\frac{\partial^2 n_1}{\partial z^2} = n_1 \left(\frac{1 + i\omega\tau_n}{L_n^2} \right) \tag{13.63}$$

while for the dc component n_d one obtains an equation identical to Eq. (13.31). Actually, Eq. (13.63) is also almost identical to this equation and therefore has the solution

$$n_1(z) = \overline{C}_1 e^{z/\bar{L}_n} + \overline{C}_2 e^{-z/\bar{L}_n} + \overline{C}_3 \tag{13.64}$$

where $\overline{C}_1$, $\overline{C}_2$, and $\overline{C}_3$ are constants of integration and $\overline{L}_n^2 = D_n\tau_n/(1 + i\omega\tau_n)$.

If we now could determine the constants of integration as in Eqs. (13.33) and (13.34), then we could arrive at an equation for the ac current in precisely the same way as in Eq. (13.37). Indeed, we can write

$$[(n_d(-l_p) + n_1(-l_p)][p_d(-l_p) + p_1(-l_p)] = n_i^2 e^{\overline{V}_{\text{ext}}} \tag{13.65}$$

as in Eq. (13.33) under the condition that the quasi-Fermi levels in the depletion region follow instantaneously the ac voltage. The supply of charge carriers in this region is mainly by diffusion through the region of width W. Diffusion over such a distance will typically take the time t_{diff}:

$$t_{\text{diff}} \approx W^2/D_n \tag{13.66}$$

This can be seen from solving the parabolic differential equation [(13.62)] (neglecting the generation-recombination term) and is well known from problems of heat conduction. If the inverse frequency of the ac voltage is longer than t_{diff}, the electrons (holes) follow the ac voltage in the depletion region, and we can indeed use boundary conditions analogous to those used for Eq. (13.32). The coefficient $\overline{C}_1$ can be obtained from Eq. (13.65). We are only interested in the linear response, that is, in the case $V_1/kT \ll 1$, and keep only the dominant terms on the left side of Eq. (13.65). Since $n_d \cdot p_d = e^{\overline{V}_{dex}} n_i^2$, we obtain

$$n_1(-l_p)\, p_d(-l_p) = n_i^2 e^{\overline{V}_{ex}} V_{1ex} e^{i\omega t} \tag{13.67}$$

and

$$n_1(-l_p) = n_d(-a) e^{\overline{V}_{dex}} V_{1ex} e^{i\omega t} \tag{13.68}$$

This gives $\overline{C}_1$, which is the same as C_1 except that the term $(e^{\overline{V}_{\text{ext}}} - 1)$ is replaced by $e^{\overline{V}_{dex}}V_{1ex}e^{i\omega t}$. The constant $\overline{C}_2$ is zero since the perturbing electron concentration $n_1(-a) \approx 0$, and $e^{z/\overline{L}_n}$ will be negligibly small for $z = -a$, which for long junctions gives $C_3 \approx 0$.

We therefore can arrive at the equation for the ac current simply by replacing L_n by $\overline{L}_n$ and $(e^{\overline{V}_{\text{ext}}} - 1)$ by $e^{\overline{V}_{dex}}V_{1ex}e^{i\omega t}$ in the equations for the dc current [Eqs. (13.37)–(13.40)].

The ac current gives us immediately the so-called low frequency diffusion capacitance C_d (see Problems section), which is

$$C_d = \frac{e^2}{2kT}(\overline{L}_p p(c) + \overline{L}_n n(-a))e^{\overline{V}_{dex}} \tag{13.69}$$

Notice that many texts have an error in this equation due to oversimplification in the derivation (the factor ½ is missing).

We assume that the reader is familiar with the application of p–n junctions as rectifiers, voltage variable capacitors, and the like, and for these topics refer to Streetman, where experimental methods to increase the speed of ac response by shortening τ_n are also discussed.

13.3 HIGH FIELD EFFECTS IN SEMICONDUCTOR JUNCTIONS

In the previous sections, we have largely ignored the problem of electron heating, and we have just pointed out on occasions which temperature, the temperature of the charge carriers or the temperature of the crystal lattice, must be used. In this section we treat the high field effects in more detail. On the whole there are four major consequences of electron heating that are of importance in semiconductor devices:

1. The carrier mobility and diffusion constants change because of the higher scattering rates at higher electron temperatures.
2. The electrons (holes) are redistributed in **k** space among the various valleys of the band structure, which leads to changes in the effective mass and therefore to changes in all the transport coefficients.
3. Hot electrons can escape over barriers by thermionic emission easier than cold electrons and transfer to neighboring layers. We call this effect real space transfer.
4. Finally electrons (holes) can, if their energy is very high, impact ionize and even create crystal defects. The most important effect is impact ionization. In very high fields ionization can also occur by band-to-band tunneling (the Zener effect).

All of these effects will be treated in this section by discussing certain examples. We start with the increase of the phonon scattering rate in high fields as a consequence of the higher carrier temperature T_c. This case has already been extensively treated via Eqs. (11.17)–(11.24). However, at junctions there is the additional problem of having a built-in field, as well as an external field, and the question arises whether the built-in field can contribute to electron heating. The answer to this question is not straightforward and will be discussed below.

13.3.1 The Role of Built-in Fields in Electron Heating and *p–n* Junction Currents

There has been some controversy in the literature regarding the effects of built-in fields arising from (mobile or fixed) space charge. The question is: Can a built-in field heat up the energy distribution of the electrons? Additional questions not unrelated to the above have arisen concerning the diffusion current and the modification of the Einstein relation in the presence of high electric fields. While a precise answer to this question will in most cases require Monte Carlo simulations, one can formulate the balance equations in an appropriate way and derive approximate answers on a physical basis. It is the purpose of this section to clarify how these equations should be used in the study of hot electrons in the space charge region of Schottky barriers or *p–n* junctions. The general principles are inherent in Eq. (11.13a). From this equation one can see that it is the term $\mathbf{F} \cdot \mathbf{j}$ that is responsible for the heating of the electrons and not the electric field alone. If $\mathbf{F} \cdot \mathbf{j}$ is zero, then $T_c = T_L$, as can be seen from Eq. (11.13a). The presence of the built-in field alone is therefore not sufficient to heat electrons. As long as the field and diffusion current balance each other and the total current (field plus diffusion) is zero, no average heating effect will be achieved. True, some electrons will slip down the energy band to lower potential energy and get hot in the process. Others, however, will diffuse against the barrier and lose kinetic energy. If the system is not perturbed from the outside, phonons will be emitted and absorbed at the same rate, and a dynamical equilibrium will be achieved. However, as soon as this equilibrium is perturbed by an external voltage (or other factors such as illumination), the situation changes, a current flows and the electrons can be heated. However, Eq. (11.17) cannot be used now, and we have to return to Eq. (11.13a) to calculate the electron temperature. It is important, then, to include the diffusion term in the current. For steady state ($\partial T_c / \partial t = 0$), Eq. (11.13a) reads (in one dimension):

$$\left(en\mu F + e\frac{\partial nD}{\partial z} \right) F = \frac{3}{2}\frac{k}{\tau_E}\, n(T_c - T_L) \tag{13.70}$$

where τ_E is the energy relaxation time, given by Eq. (11.13b), and Eq. (11.6b) has been used for the current density j. Equation (13.70) does not in general have an explicit solution for T_c since the carrier concentration can be a complicated function of z, and the mobility and diffusion constant are functions of the electron temperature also. The point we want to make here is very clear, however: If the diffusion current balances the field current, the left-hand side of Eq. (13.70) vanishes and $T_c = T_L$, as is the case for the unbiased p–n junction.

We have ignored here yet another problem. Since in p–n junctions the electric field is also dependent on the space coordinate (Fig. 13.11), the whole derivation of Eq. (11.13a) does not hold. The drifting carriers transport energy from places of high field to places of low field, and electronic heat conduction plays a role if T_c becomes a function of z. This gives rise to additional terms in the energy balance equation, which are derived in Appendix E. These terms complicate the calculation of T_c greatly, and the numerical solution of the differential equation together with the other device equations is time consuming (in some instances almost as much as the more precise Monte Carlo method). Numerous good approximations are possible and have been discussed in the literature (see Higman and Hess and references therein).

For the calculation of the current in a p–n junction, including hot electron effects, there exists the additional problem that according to Eq. (9.33) the quasi-Fermi level becomes a function of z as the electron temperature does. All of these taken together require a numerical treatment if a precise solution for j in a p–n junction is desired. However, since j depends weakly on T_c, not much has been done with this problem and little appears in the literature.

The reason that j does not depend strongly on T_c is the following. For forward bias the electric fields are small and $T_c \approx T_L$. For reverse bias the fields are quite large and the electrons are heated greatly. However, in this case the speed of the electrons through the depletion layer (which depends on T_c) is not the limiting factor for the current. As can be seen from Eq. (13.38) it only matters how fast carriers are generated. The speed of the carriers does not even enter Eq. (13.38) since the field current is not involved in its derivation within the approximations that lead to Eq. (13.38). As we will see in Chap. 14 the situation is very different for transistors where the increase in T_c influences the current significantly.

For p–n junctions there is, however, a direct influence of high fields on the current via the generation (emission) rates. As can be seen from Eq. (10.26b), the emission rate e_n (and the same is true for holes) depends exponentially on the energy difference E_T between the trap level and the conduction band edge. This difference depends on external fields, as can be seen from Fig. 13.15 for the case of a trap with a square well–like potential. The energy level does not change with external fields in first-order perturbation theory. [M_{mm} in Eq. (1.18) does not change with external fields, as one can see easily by putting the zero of the energy at a reference point in the middle of the well.] However, the distance of the level to the conduction band edge is reduced by the amount $eFd/2$ for wells of width d. This can be seen from

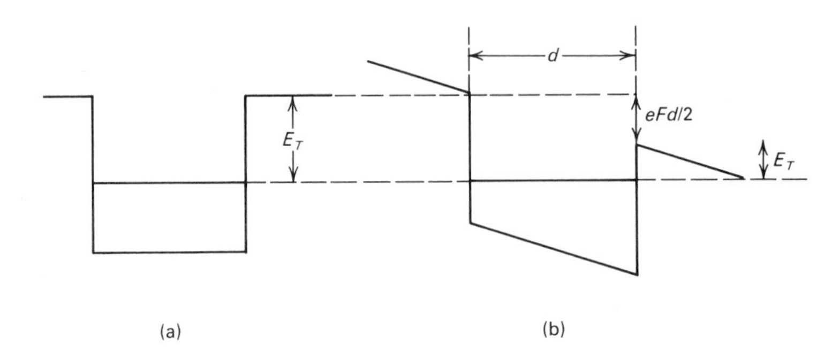

Figure 13.15 Square well trap energy E_T without (a) and with (b) an external electric field.

Fig. 13.15b. The amount of reduction is different for different forms of well potentials but can always be calculated by searching for the maximum of the potential at the right-hand side of the well. In three dimensions the problem is a little more involved. This effect of emission over lowered barriers is sometimes called the Pool-Frenkel effect. The electron can, of course, also tunnel directly out of the well, and this adds to the rate e_n if the tunneling rate becomes high (in high fields). The total rate is then the sum of the Pool-Frenkel emission and tunneling, which can be calculated as shown in Appendix A.

13.3.2 Impact Ionization in *p–n* Junctions

Impact ionization is not an effect specific to p–n junctions. However, since in Ge, Si, and GaAs it occurs only at very high electric fields ($>10^5$ V/cm), it is mostly observed in reverse biased p–n junctions, and we therefore treat it in this chapter.

The impact ionization process is the inverse of the Auger process, that is, an electron (hole) in the conduction (valence) band gains energy by some means and becomes so highly energetic that it can create an electron-hole pair by colliding with an electron in the valence band and exciting it to the conduction band. The process is schematically shown in Fig. 13.16. The process can be viewed as a scattering event; and as long as the Golden Rule can be applied, we expect momentum and energy to be conserved. Energy conservation demands that the energy of the ionizing particle is at least as large as the energy gap E_G. From Fig. 13.16b, however, we can immediately see that if momentum is to be conserved too, the energy of the ionizing particle must be substantially larger than E_G.

For the simplest case of parabolic isotropic bands with equal mass, one would obtain a threshold energy $E_{th} = \frac{3}{2} E_G$ for ionization to be possible. For single-valued (but otherwise arbitrary) band structures (conduction and va-

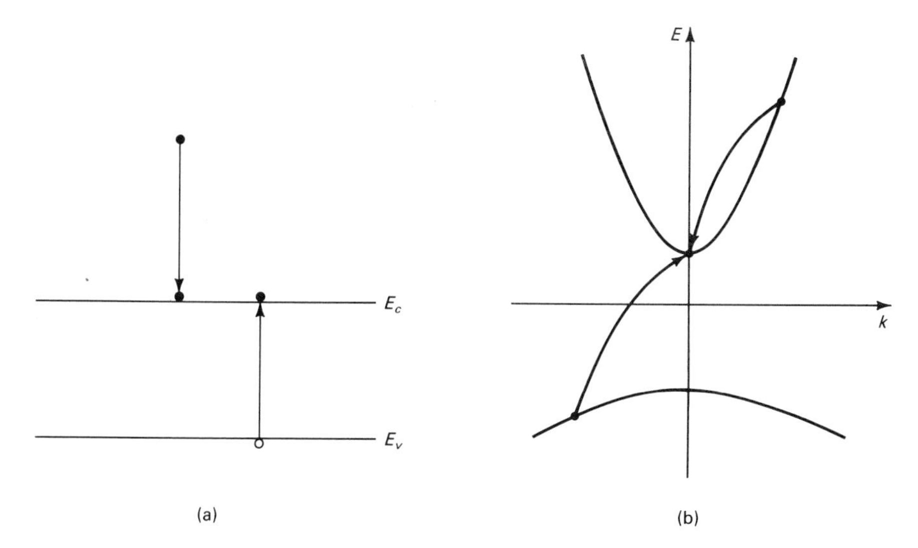

Figure 13.16 The impact ionization process in (a) real space and in (b) $\vec{\mathbf{k}}$ space.

lence bands) it has been shown by Anderson and Crowell that momentum and energy conservation is equivalent to the requirement that the sum of group velocities of the final particles is zero. Using this criterion, threshold energies have been calculated for various materials. These energies become a function of crystallographic direction and assume very complicated shapes if plotted in **k** space in accord with the complicated $E(\mathbf{k})$ relations (Chap. 4). Notice, however, that the $E(\mathbf{k})$ relation is not single valued for most semiconductors in the interesting energy range. Furthermore, the impact ionization rates may be high at high energies, making collision broadening important. We know from the uncertainty principle that as soon as $\hbar/\tau_{\text{tot}}$ reaches values comparable to the energy scale that is interesting in the given problem, the δ function approximation in the Golden Rule becomes questionable [see Eq. (1.21)], and the δ function has to be replaced by a broadened Lorentzian as is well known from optics (see Bethe and Jackiw). The total scattering rate (due mostly to phonons and impact ionization) can be enormous at the energies typical for the ionization process ($\approx$2–3 eV in silicon) as is seen from Fig. 8.8. The broadening effect will then smear out the threshold for impact ionization. One can therefore relax the momentum conservation requirement. Also, Kane has demonstrated that the ionization rate $1/\tau_I$ for electrons in the conduction band of silicon has a rather soft threshold around the energy $E_{th} = E_G$, as is shown in Fig. 13.17.

Even more complicated than the calculation of the threshold is the calculation of the ionization rate per unit time $1/\tau_I$ above threshold. For not too large rates, $1/\tau_I$ can be calculated using the Golden Rule (as we did for impurity scattering) but now including transitions in between different bands.

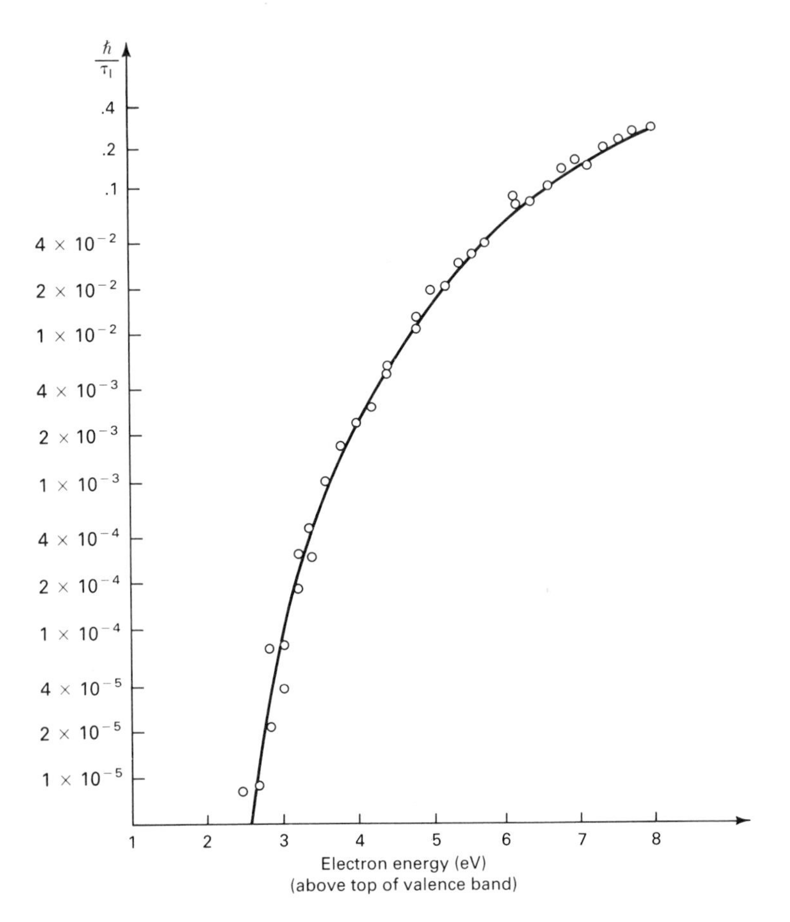

Figure 13.17 Impact ionization rate for electrons in silicon as a function of energy. (After Kane)

A transparent treatment has been given by Ridley, who also discusses a formula previously derived by Keldysh. For real band structures these formulas have only a limited validity and a numerical calculation (as performed by Kane) is necessary to obtain $1/\tau_I$. We use in the following the Keldysh formula for $1/\tau_I$. However, we regard the constants in the formula as freely adjustable parameters, not given by first principle theory.

The Keldysh formula suggests

$$\frac{1}{\tau_I} = \frac{B}{\tau_I(E_{th})}\left(\frac{E - E_{th}}{E_{th}}\right)^p \quad \text{for } E > E_{th} \text{ and zero otherwise} \qquad (13.71)$$

Here B is a dimensionless parameter and $1/\tau_I(E_{th})$ is the scattering rate at threshold. It may be lumped together with B to form an adjustable parameter. The second parameter, the exponent p, is typically between 1 and 2. For more precise purposes this parameterization is insufficiently accurate and the full treatment (as given by Kane) is necessary.

Fortunately, however, the energy dependence of the ionization rate is not the quantity that is needed in detail to understand devices involving impact ionization. What one needs to understand devices is the increase in current I with time (or over distance) due to ionization. This increase is defined by coefficients α_t and α_z as

$$\frac{\partial I}{\partial t} = \alpha_t I \tag{13.72}$$

and

$$\frac{\partial I}{\partial z} = \alpha_z I \tag{13.73}$$

From these definitions it follows that α_t represents an average rate of ionization per unit time or per unit distance (α_z). The average is to be taken over the energy distribution and we therefore have

$$\alpha_t = \int_{-\infty}^{+\infty} d\mathbf{k} \frac{1}{\tau_I} f \Big/ \int_{-\infty}^{+\infty} d\mathbf{k} f \tag{13.74}$$

with $1/\tau_I$ being given by Eq. (13.71) or from Fig. (13.17) and f being the electron (hole) energy distribution function. For the special case of steady state and time and space independence, α_z can be calculated from α_t by dividing α_t by the average drift velocity. This follows from the chain rule of differentiation and the definitions. Therefore, $\alpha_t \propto \alpha_z$.

An inspection of Eq. (13.74) shows that the functional dependence of $1/\tau_I$ on the energy influences α_t only rather weakly as long as $1/\tau_I$ is given by Eq. (13.71), while the energy dependence of the distribution function above E_{th} is exponential and influences α_t greatly. One can put this fact into other words: An electron will go through the history of scattering events and accelerations before the threshold energy is overcome. It is this history that determines the probability of finding an electron above threshold. Then after the electron reaches some energy above threshold, it impact ionizes. It is therefore the history of the electron at the energies below threshold that is of utmost importance for α_t, because it determines the high energy tail of the distribution function. This high energy tail will be discussed now from various points of view.

Wolff assumed that the distribution function in Eq. (13.74) can be approximated by a spherical symmetric distribution (i.e., the drift term does not matter). We derived f_0 in Chap. 11 by using the electron temperature approximation. For simple parabolic bands and phonon scattering, we can therefore use the carrier temperature as given by Eq. (11.18). However, this approach is not self-consistent since we need to include impact ionization itself as a scattering mechanism. In other words, one needs to include an average energy loss due to impact ionization into the energy balance equation. If the electrons lose all their energy due to the ionization (which is only approximately true) the loss is given by

$$\int_{-\infty}^{+\infty} d\mathbf{k} \frac{E}{\tau_I} f_0 \tag{13.75}$$

This term is easy to evaluate, at least for steplike ionization rates $1/\tau_I$, and can then be added to the energy balance equation (see Problems section). Following this approach, one finds

$$\alpha_t \propto \exp\left(-\text{const}/F^2\right) \tag{13.76}$$

where the square of the electric field enters as ordinarily does the carrier temperature. This result compares favorably with experiments only for large F.

For small F a large number of experimental results exhibit a dependence

$$\alpha_t \propto \exp\left(-\text{const}/F\right) \tag{13.77}$$

This dependence was explained by Shockley. Shockley's idea was that only those electrons that do not interact with phonons contribute to impact ionization. This means that impact ionization is not caused by the spherical part of the distribution function, or by high electron temperature, but by those electrons that stream only in the field direction and do not scatter until they are at the ionization threshold where they ionize instantaneously (in Shockley's opinion). The probability P that an electron is not scattered is given by Eq. (8.42). We rewrite Eq. (8.42) by coordinate transformation, and use also (in one dimension):

$$\frac{dE}{dt} = \frac{dE}{dk}\frac{dk}{dt} = \frac{dE}{dk} \cdot \frac{eF}{\hbar} \tag{13.78}$$

This gives, with the definition of the total scattering rate by phonons $1/\tau_{\text{tot}}^{\text{ph}}$ and Eq. (8.35), the probability

$$P = \exp\left\{-\frac{\hbar}{eF}\int_0^{E_{th}} \left(\frac{dE}{dk}\right)^{-1} \frac{1}{\tau_{\text{tot}}^{\text{ph}}(E)}\, dE\right\} \tag{13.79}$$

Taking the lower limit of zero energy was sufficiently accurate for Shockley's purpose. This assumes that the electrons start at $\mathbf{k} = 0$ at $t = 0$. However, the solution of the equation of motion Eq. (4.19) is

$$\mathbf{k} = e\mathbf{F}t/\hbar + \mathbf{k}_0 \tag{13.80}$$

Trivial as the addition of $\mathbf{k}_0$ is, the consequences are very important. The electron does not start from $\mathbf{k} = 0$ but rather from $\mathbf{k}_0$, corresponding to some average energy E of the electrons in the electric field. An energy histogram is shown in Fig. 13.18 to illustrate this point. (Note that the curve lies high above $E = 0$.) The figure shows that the electron energy first increases rapidly within a very short time. At higher energies the electron is scattered frequently.

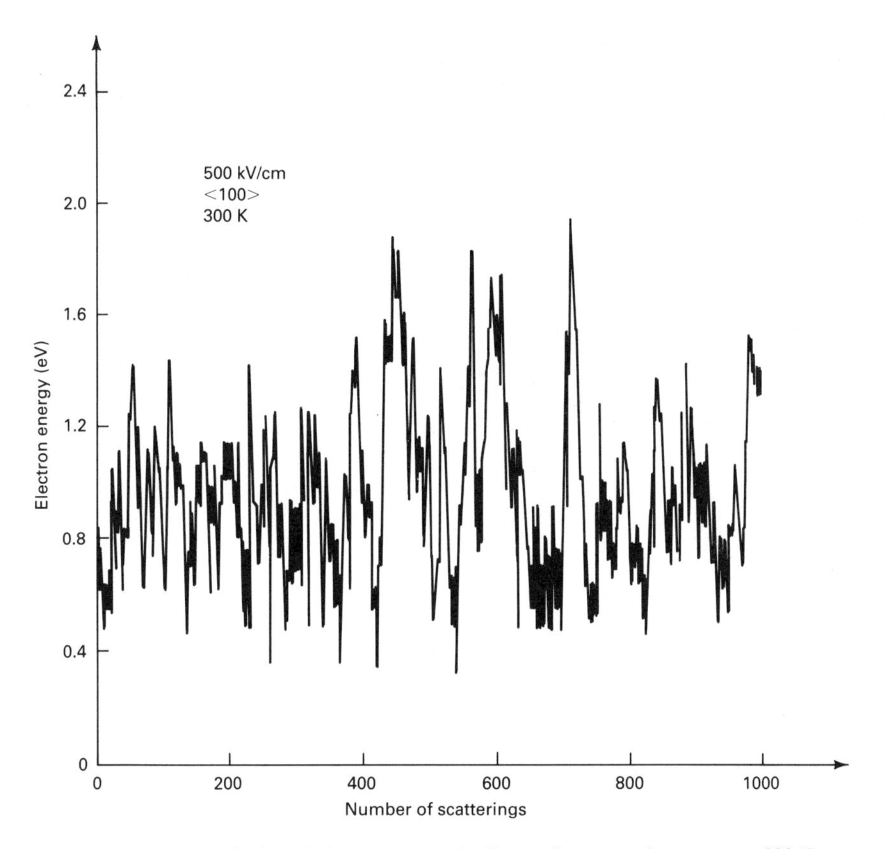

Figure 13.18 Variation of electron energy in GaAs after scattering events at 300 K for $F = 500$ kV/cm. (After Shichijo and Hess)

Whenever it is scattered against the field, it loses much energy and restarts the process. Overall a heated electron gas is established, and at instances electrons escape in field direction toward high energies (Shockley's so-called "lucky" electrons). However, the electrons do not start at $E = 0$ but rather at some average energy that in GaAs and Si is typically 1 eV at $F = 500$ kV/cm. One therefore should replace the lower limit of integration in Eq. (13.79) by an average energy equal to $\frac{3}{2} kT_c$, where kT_c can be calculated from the formalism proposed by Wolff.

There exists a theory of impact ionization (due to Baraff) that accomplishes this inclusion of the starting energy in a more precise way by solving the Boltzmann equation. Baraff's theory improves Shockley's and Wolff's and is very useful since the results are available in terms of polynomial expansions (Sze). It does not do, however, complete justice to the impact ionization effects since the consequences of band structure, which enter into the electron equation of motion and total scattering rate, are difficult to include in Baraff's formalism. The greatest influence of band structure comes from the steep rise in the density of states at higher energies, which is shown in Fig. 6.2. This increase leads in turn to a steep increase of the phonon scattering rate as given in Figs. 8.5 and 8.6. As can be seen from both the Wolff and the Shockley formalisms, the phonon scattering rate influences the electron dynamics and ionization probability very sensitively. It is this fact (together with the material dependence of E_{th}) that leads to the different ionization rates in different materials. Only Monte Carlo simulations (Shichijo and Hess) can include all these factors. However, a reasonable idea of the influence of band structure can also be obtained from Eq. (13.79).

We therefore continue to calculate the ionization coefficient according to Shockley and calculate α_z. The coefficient α_z is, according to Eq. (13.73), the inverse of the mean free distance L_I^* that an electron travels without ionizing. The distance L_I that is necessary to reach threshold without scattering is obtained from

$$E_{th} = e\int_0^{L_I} F(z)\,dz \tag{13.81}$$

To proceed we need to specify the z dependence of the electric field F, which in p–n junctions is given by Eq. (12.11). To avoid the complications of z-dependent F, which also enters Eq. (13.79), we neglect the z dependence here (F = constant) and discuss its consequences briefly below. Then we have

$$L_I = E_{th}/eF \tag{13.82}$$

If we divide L_I by the probability that an electron travels without collision, we obtain L_I^* and α_z:

$$\alpha_z = 1/L_I^* = \frac{eF}{E_{th}}\exp\left\{\frac{\hbar}{eF}\int_{\frac{3}{2}kT_c}^{E_{th}}\left(\frac{dE}{dk}\right)^{-1}\frac{1}{\tau_{\text{tot}}^{\text{ph}}(E)}\,dE\right\} \tag{13.83}$$

The z dependence of F is difficult to include in explicit treatments. Here we discuss two limiting cases.

First, if F increases steeply on a length scale much smaller than the mean free path for ionization, α_z will not immediately follow the field because it takes time to accelerate more electrons to higher energies, and there will be a range of no change in the ionization that can be termed dead space.

Second, if F changes slowly on the scale of L_I^*, then Eq. (13.83) is essentially valid. To illustrate the consequences of the z dependence in this case assume that F is given by

$$F = F_0 + F_1\cos 2\pi z/L_p \tag{13.84}$$

that is, a constant with a periodic component (period L_p) which can be small

($F_1 < F_0$). The average field is F_0 (averaged over distance). However, the averge α_z is not equal to $\alpha_z(F_0)$ since F enters into the exponent and

$$\int_0^{L_p} dz \exp\left(-\frac{C}{F_0}\right) < \int_0^{L_p} dz \exp\left(-\frac{C}{F_0 + F_1 \cos 2\pi z/L_p}\right) \tag{13.85}$$

The proof of this inequality is easily given for $F_1 \ll F_0$, since then

$$\frac{C}{F_0\left(1 + \frac{F_1}{F_0}\cos 2\pi z/L_p\right)} \approx \frac{C}{F_0}\left(1 - \frac{F_1}{F_0}\cos 2\pi z/L_p\right) \tag{13.86}$$

and

$$\frac{1}{L_p}\int_0^{L_p} dz \exp\frac{CF_1}{F_0^2}\cos 2\pi z/L_p = I_0\left(\frac{CF_1}{F_0^2}\right) \tag{13.87}$$

Here I_0 is a modified Bessel function that is always larger than one (Abramovitz and Stegun).

This result is at first sight strange because it is counterintuitive that the average field is the same but the average ionization rate is increased. This is, however, a very general feature of nonlinear transport and applies also for the electron temperature which in our approximations varied with F^2. It is therefore possible to enhance the ionization by superposed fields that actually average to zero. This can be achieved by varying, for example, the alloy composition of a material and changing the band edge, and this can lead to a bandstructure engineering of the ionization coefficient (see review by Capasso).

Let's consider once again the influence of band structure on the ionization coefficient. For this purpose we consider the band structure of GaAs and the influence that a single scattering event can have on the ionization probability. Our sample electron starts at the Γ point and moves along the [111] direction. At point $A\{\mathbf{k} = (\pi/\alpha)(0.3, 0.3, 0.3)\}$, the energy is at the conduction band maximum in this direction, but it is still much less than the threshold energy, and therefore impact ionization is impossible. Now let's assume that the electron is scattered (by a phonon or impurity) to some other point in the Brillouin zone, point B, for example. Following this scattering event, the [111] component of the electron wave vector continues to increase. However, the wave vector points in a direction different from [111], so that the electron can now reach a higher energy because the band is wider in this direction. As shown in Fig. 13.19, the electron can actually exceed the threshold energy for impact ionization ($\approx$2.0 eV) and subsequently can impact ionize. This example shows that, in some crystallographic directions, electrons can impact ionize only if they are scattered, which contradicts Shockley's assumption. It also illustrates the importance of the inclusion of $\mathbf{k}_0$ in the starting conditions of the electrons. This, of course, does not mean that Shockley's ideas are wrong. It only means that they cannot be stretched too far.

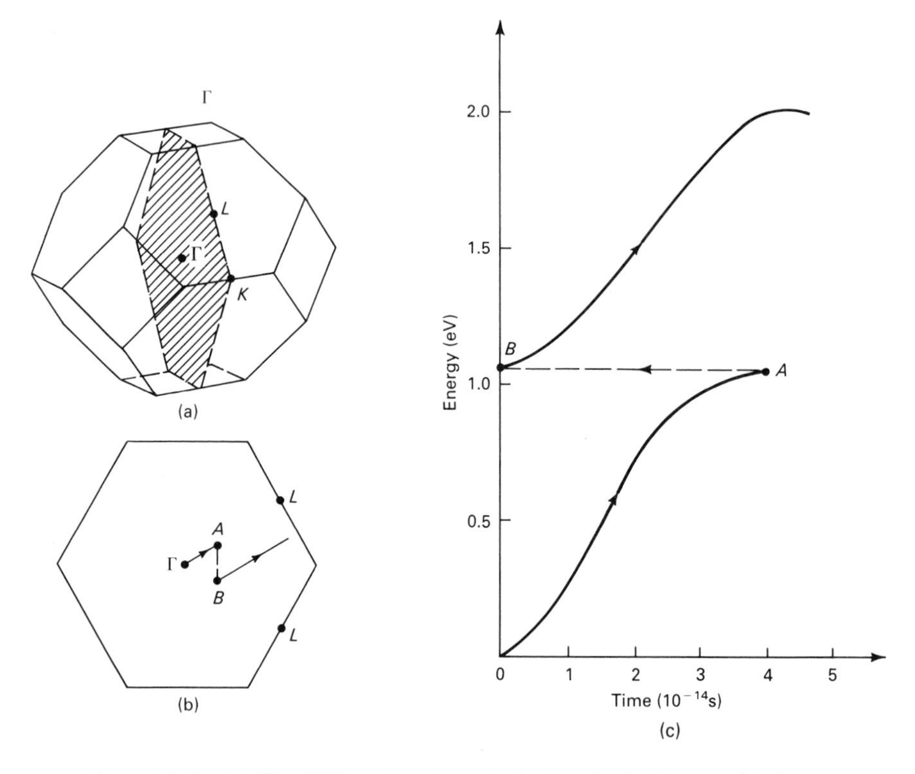

Figure 13.19 (a) The (110) section through the first Brillouin zone. (b) Wave vector trajectory of electron in this plane with F in the [111] direction. The electron is scattered from point A to point B. The energy change in the scattering process has been neglected. (c) Variation of electron energy with time for scattering process illustrated in part b. (After Shichijo and Hess)

Ionization coefficients as obtained experimentally for electrons and holes in GaAs are shown in Fig. 13.20. The measurements have been taken in the [100] crystallographic direction and are virtually independent of the orientation. Monte Carlo simulations fit the experimental data almost perfectly and show little if any dependence on the crystal direction in which the electric field points.

Finally, it should be noted that the two-carrier transport in p–n junctions adds a new feature to the ionization process. If an electron excites another electron to the conduction band a hole is created, which now moves in opposite direction, is then again accelerated, and itself can ionize, thus creating a second hole and an electron in the conduction band. In this way, if the device is long enough, infinite multiplication of one primary carrier can occur, which is termed avalanche breakdown (Streetman).

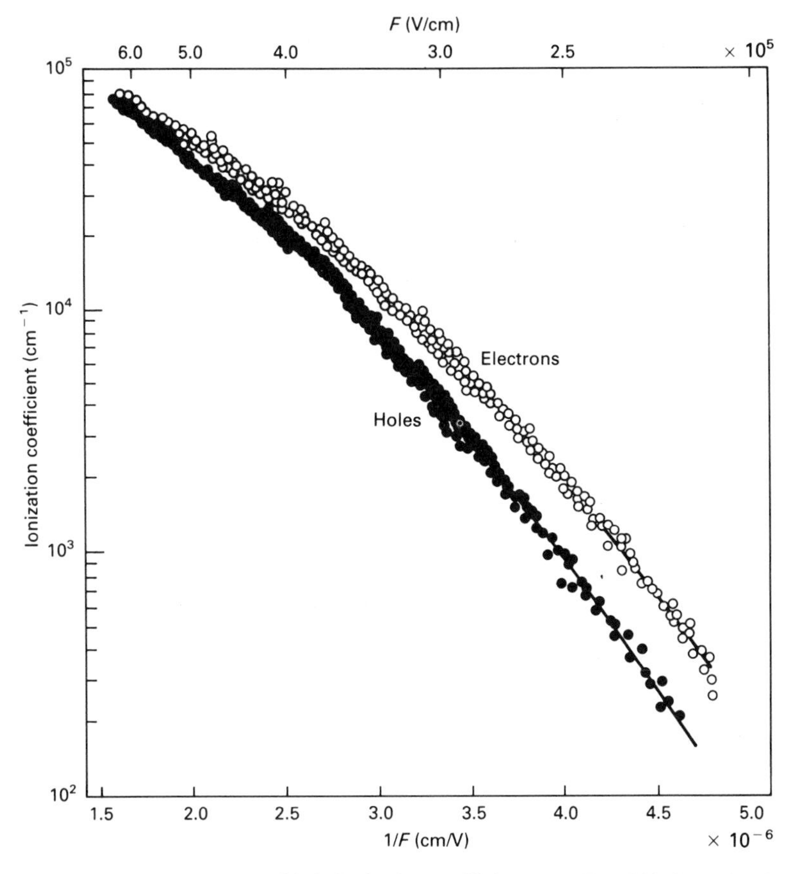

Figure 13.20 Electron and hole ionization coefficient α_z at $T_L = 300$ K vs. electric field. (After Stillman et al.)

13.3.3 Zener Tunneling

Very heavily doped reverse biased p–n junctions show also significant current increase, which, however, is not caused by impact ionization but rather by the tunneling of valence band electrons to the conduction band because of high electric fields (the Zener effect). The principle of the Zener effect is illustrated in Fig. 13.21. The electron tunnels from point z_a to point z_b and changes bands during this process. The effect is, of course, only an extension of the field emission, discussed in Chap. 1 (Fig. 1.2). The calculation of the tunneling current has been outlined in Appendix A and in Eqs. (13.13) and (13.14). Here we evaluate only the exponent (the dominant term) of these equations, which also represents the tunneling probability of a single electron,

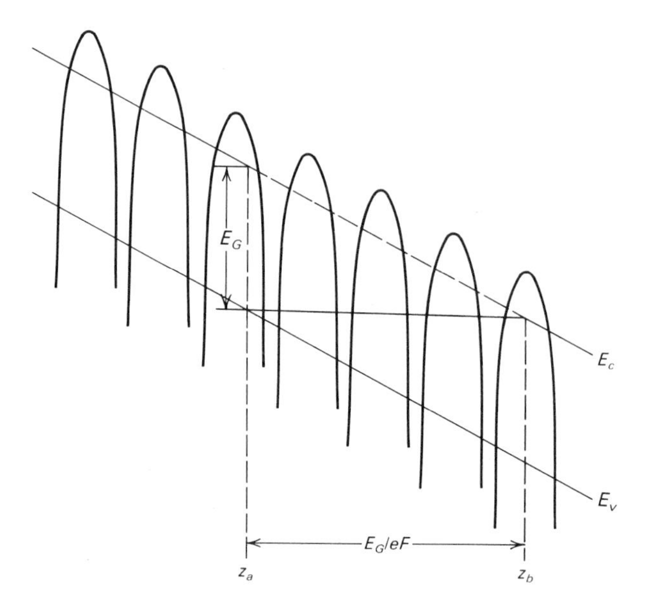

Figure 13.21 Conduction and valence band edges as well as atomic potentials in an electric field. The dashed curve illustrates an idealized potential for the tunneling of an electron from point z_a to point z_b.

and put $z_a = 0$ and $z_b = E_G/eF$ (Fig. 13.21). This exponent becomes (measuring the energy from the top of the valence band at z_a)

$$\exp\left\{-2\int_0^{E_G/eF} \frac{\sqrt{2m}}{\hbar}\sqrt{E_G - eFz}\,dz\right\} \tag{13.88}$$

The integration gives

$$\exp\left\{-\frac{4}{3}\frac{\sqrt{2m}}{eF\hbar}E_G^{3/2}\right\} \tag{13.89}$$

This formula describes the Zener tunneling probability. Its limitations can be seen from the appearance of the mass of the electron, which is generally not the same in the valence and conduction bands. The simple appearance of the free electron mass in Eq. (13.89) arises from our idealization of the potential and the neglect of the atomic potentials and Bragg reflection. Kane has developed an elegant method of including these effects by viewing the $E(\mathbf{k})$ relation as an analytical function of complex variable $\mathbf{k}$ (for details, see Burstein and Lundquist, and Duke).

The tunneling process can be assisted by photons (or phonons, see Holonyak et al.), as shown schematically in Fig. 13.22, and then gives rise to optical absorption below the band edge ($\hbar\omega < E_G$), as seen in the figure. This effect is called the Franz-Keldish effect. Note, however, that the electron-hole interaction (there is also a hole created in the valence band in addition to the

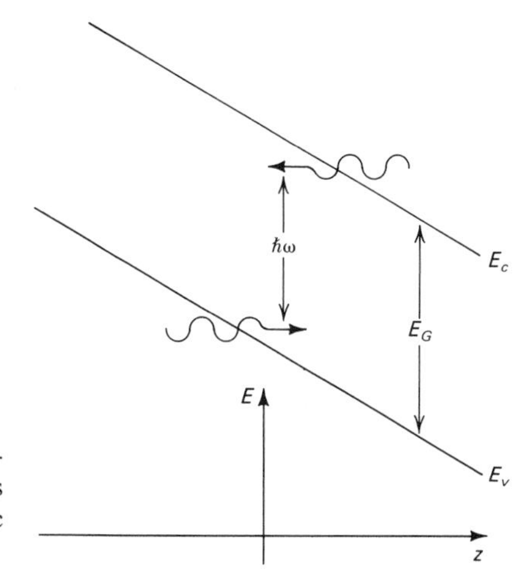

Figure 13.22 Illustration of the possibility of an optical transition for energies $\hbar\omega < E_G$ due to the effect of the electric field and tunneling.

electron in the conduction band) introduces additional subtleties (excitonic effects), which goes beyond the scope of this book.

13.3.4 Real Space Transfer

The transfer of electrons between two different solids was discussed in Sect. 13.1 for the case of electric fields perpendicular to the solid-solid interfaces. This section deals with this effect again, but from a more general viewpoint. It will be shown that transfer of electrons can also be accomplished if the field is parallel to the interface. The electric field heats the electrons, and the increased electron temperature enables them to overcome barriers. This effect is called real space transfer (RST) and is shown schematically in Fig. 13.23.

The emission of hot electrons over barriers (or tunneling through them) is more complicated and more difficult to understand than other effects basic to semiconductor device operation. The reason is that RST can be visualized

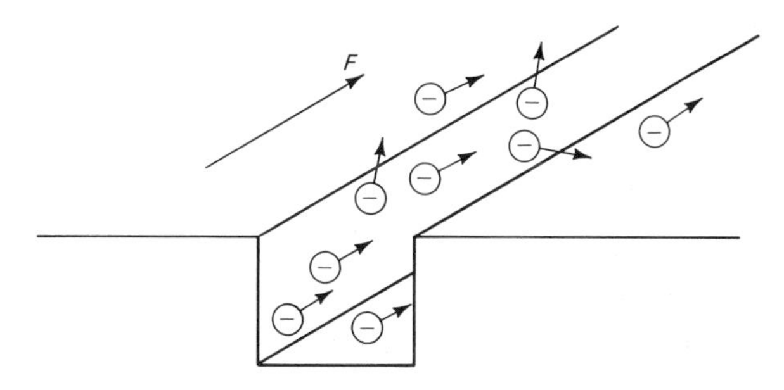

Figure 13.23 Emission of electrons heated by an electric field parallel to a heterolayer interface.

only by the combination of two concepts concerning the energy distribution of electrons. The first concept is the concept of quasi-Fermi levels and the second is the concept of a charge carrier temperature T_c (different from the equilibrium lattice temperature T_L). We have outlined these concepts and their interconnections in Chap. 9 [see Eq. (9.33)]. For RST problems, both concepts matter, and both the carrier temperature and the quasi-Fermi levels are a function of space coordinate and time.

Imagine, for example, electrons residing in a layer of high mobility GaAs neighboring on either side two layers of low mobility AlGaAs. The GaAs equilibrium distribution function f_0 is

$$f_0 \propto \exp(-E/kT_L) \tag{13.90}$$

while in the AlGaAs we have

$$f_0 \propto \exp[-(\Delta E_c + E)/kT_L] \tag{13.91}$$

Here the energy is measured from the GaAs conduction band edge and ΔE_c is the band edge discontinuity between AlGaAs and GaAs. If now the electrons are heated by an external field parallel to the layers, we have to replace T_L in Eqs. (13.90 and 13.91) by a space-dependent carrier temperature T_c. It is clear that for $T_c \rightarrow \infty$ the difference between the AlGaAs and the GaAs population density vanishes. In other words, the electrons will spread out into the AlGaAs layers. This also means that even perpendicular to the layers (z direction) a constant Fermi level cannot exist, and E_F has to be replaced by the quasi-Fermi level $E_{QF}(z)$ as the density of electrons becomes a function of $T_c(z)$. This is unusual since commonly the quasi-Fermi levels differ only in the direction of the applied external voltage V_{ext} (by the amount eV_{ext}). In the present case, a voltage is applied parallel to the layers, the electrons redistribute themselves perpendicular to the layers, and a field (and voltage perpendicular to the layers) develops caused by the carrier redistribution. Basic to the calculation of this process are the thermionic emission currents of hot electrons from one layer to the other j_{RL} and j_{LR}, which have been derived in Eqs. (12.5) and (12.6). Since the external voltage is applied parallel to the layers, we have in steady state:

$$j_{LR} = j_{RL} \tag{13.92}$$

from which we can determine the quasi-Fermi levels if the carrier temperatures are calculated from the power balance equation. If precision is needed, then one needs to use the space-dependent power balance equation derived in Appendix E. For rough estimates, T_c can be obtained from Eq. (11.18) for each layer separately. A further complication is presented by the necessity (in most cases) of having to solve also self-consistently Poisson's equation as charge is transferred.

The time constants for the real space transfer process can be calculated as in Eqs. (12.7)–(12.10). For typical parameters of the GaAs-AlGaAs material system and electric fields of the order of 10^3–10^4 V/cm parallel to the layers, one obtains time constants of the order of picoseconds, which makes the RST effect attractive for device applications (switching and storage in between the layers). The real space transfer effect is also of general importance in all situations when electrons are confined in potential wells and parallel fields are applied (accelerate the carriers), even if the electrons do not get out of the wells but merely redistribute themselves within the well. This is important for the understanding of the influence of transverse fields (such as a "gate field" in a transistor as discussed in Chap. 14) on nonlinear transport parallel to interfaces. The RST effect and the spreading of the electrons are then determined by the transverse field. The quantum analog of this classical picture is the redistribution of hot electrons in the different size quantized subbands. Remember that the carrier concentration in the various subbands depends in principle on the electron temperature [Eq. (12.37b)] and therefore on the parallel electric field, as well as the perpendicular electric field, which determines the energy of the subbands [E_n in Eq. (12.32)].

13.4 OPTICAL RECOMBINATION IN *p–n* JUNCTIONS AND SEMICONDUCTOR LASERS: A GLIMPSE INTO OPTOELECTRONICS

The high energy physics analogy for p–n junctions in Fig. 13.12 makes it clear that p–n junctions can in principle be used for light generation. Of course, a necessary condition will be that the recombination by light emission has a reasonably high probability compared to other mechanisms such as phonon- and trap-mediated recombination. From Eq. (10.8) we deduce therefore that the semiconductor will have to be a direct-gap semiconductor such as GaAs (infrared) or AlGaAs or GaAsP (visible). For optical fiber communications certain wavelengths in the infrared are needed (absorption!), and therefore the InGaAsP system is of special interest. The useful materials are generally ternary and quarternary III–V compounds and lattice-matched combinations of them. These have been reviewed by Casey and Panish (the mole content of the various elements determines also if the alloy is direct or indirect).

The theory of a light-emitting p–n junction could proceed precisely as in Sec. 13.2. However, as a light emitter such a junction would not be very efficient, and it would be difficult to achieve lasing. To achieve lasing, the separation of the quasi-Fermi levels must exceed the photon energy (for the laser emission), which is approximately equal to the band gap. This condition, which will be derived below, is not easily met in an ordinary p–n junction. Extremely high-doping at the n and p sides and extreme forward bias is needed to inject enough electrons and holes to put the quasi-Fermi levels at the band edges (see Fig. 13.13). One, therefore, needs to concentrate the electron and hole densities at some common place in the junction. Such confinement can be achieved by constructing a low-gap region in the junction (i.e., a double heterojunction). The double heterojunction and its band structure in forward bias is shown schematically in Fig. 13.24.

In such a structure most of the recombination will occur in the narrow gap region, since both electron and hole concentrations are highest there and will be confined and not easily overcome the barrier into the n or p region on either side of center.

The narrow gap region offers another advantage: It tends to confine the electomagnetic field (the photons) since its GaAs index of refraction is different (higher) than the index of the surrounding semiconductor layer (AlGaAs). Confinement is crucial for lasers and, therefore, most if not all commercially available semiconductor lasers include heterolayers, in fact, multiple heterolayers. There are currently new and exciting developments based on the use of very thin quantum wells (size quantization!) in laser technology. The interested reader is referred to the review of Holonyak and Hess.

If the narrow gap region is thin enough, the quasi-Fermi levels will be constant in this region, and the theory of the current in such a junction can be developed just by calculating the recombination rates. Before we discuss these rates, however, we will prove the above-mentioned condition for lasing and introduce several important definitions of optoelectronics. The notation will be the same as in Chap. 10 for obvious reasons. We, therefore, denote in the following equations the band edges by E_c^0 and E_v^0 and the energies in the conduction and valence band by E_c and E_v, respectively.

The rate for a transition from an energy E_c in the conduction band to energy E_v in the valence band is [see Eq. (10.11)]:

$$R_{cv} = S(E_c, E_v) f_c(E_c)(1 - f_v(E_v)) \tag{13.93}$$

As we know from our treatment of phonon scattering [e.g., Eq. (8.32)], the rates for emission and absorption carry a factor proportional to the number of scattering agents plus one (stimulated and spontaneous emission) and the number of scattering agents (absorption), respectively. In the case of

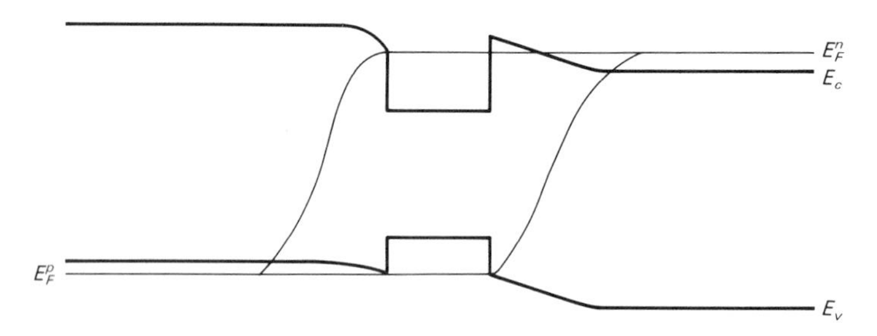

Figure 13.24 Schematic band structure and the quasi-Fermi levels in a forward-biased p–n double heterojunction.

the phonons these factors were $(N_q + 1)$ as well as N_q. The interaction with light follows the same rules as discussed in the context of Eq. (8.21a). In the notation of optics N_q is replaced by the so-called spectral density $P(E)$ having units of number of photons per unit volume and unit energy interval. Again the emission term proportional to this density of photons is called stimulated emission (since the photon density enhances the emission). For stimulated emission we therefore can factorize the transition rate into the product of two coefficients:

$$S(E_c, E_v) = P(E_{cv})B_{cv} \tag{13.94}$$

Here E_{cv} is the photon energy (equal to $|E_c - E_v|$) and B_{cv} is the factor representing the rest of the transition rate that can be found from the Golden Rule and Eq. (10.7). The rate for absorption can be written exactly as Eq. (13.94) with exchanged indices c and v, while the rate for spontaneous emission does not contain the photon density, and we will denote it by A_{cv}. Detailed balance of the emission (stimulated plus spontaneous) and the absorption rates require that $R_{cv} = R_{vc}$ in equilibrium, which leads to the well-known relation

$$B_{cv} = B_{vc} \tag{13.95}$$

and

$$A_{cv} \propto B_{cv} \tag{13.96}$$

Equation (13.95) permits us to deduce the necessary condition for predominant stimulated emission (lasing). If the rate for stimulated emission is larger than absorption, we have

$$B_{vc}f_v(1 - f_c)P(E_{vc}) < B_{cv}f_c(1 - f_v)P(E_{cv}) \tag{13.97a}$$

Using Eq. (13.95), we have

$$f_v(1 - f_c) < f_c(1 - f_v) \tag{13.97b}$$

which is equivalent to

$$E_F^n - E_F^p > E_{cv} = E_c - E_v \tag{13.98}$$

where E_F^n and E_F^p are the quasi-Fermi levels of electrons and holes in the active layer. In other words (and stated in the introduction), for lasing to be possible, the difference between the quasi-Fermi energies must exceed the energy of the emitted photons.

In laser diodes a whole range of states contributes to emission and absorption. This does not alter the reasoning presented above. It does enter, however, into the details of laser characteristics. These characteristics depend sensitively on the functional form of the quantities defined below.

An important quantity is the absorption coefficient $\alpha(E_{vc})$ defined as absorption rate R_{vc} per photon flux. The photon flux can be defined as given by the product of photon density $P(E_{cv} = \hbar\omega)$ and photon group velocity $v_g = d\omega/dq$ (q is the photon wave vector and ω the angular frequency). Thus

$$\alpha(\omega) = \frac{R_{vc}}{P(\hbar\omega)v_g} \tag{13.99}$$

Also of special importance is the so-called total net absorption coefficient, which is defined as the difference between upward and downward transition integrated over all possible states:

$$\alpha_{\text{tot}}^{\text{net}}(\omega) = \int_{-\infty}^{E_v^0} \int_{E_c^0}^{\infty} dE_v dE_c \left\{ \frac{R_{vc} - R_{cv}}{P(\hbar\omega)v_g} \times \delta(E_c - E_v - \hbar\omega) g_v(E_v) g_c(E_c) \right\} \tag{13.100}$$

Negative α means that stimulated emission is predominant and laser action can occur. The gain coefficient $g(\omega)$ is defined as the negative absorption coefficient:

$$g(\omega) = -\alpha_{\text{tot}}^{\text{net}}(\omega) \tag{13.101}$$

A typical curve for $g(\omega)$ is shown in Fig. 13.25.

The value of maximum gain g_m is a function of population inversion (the

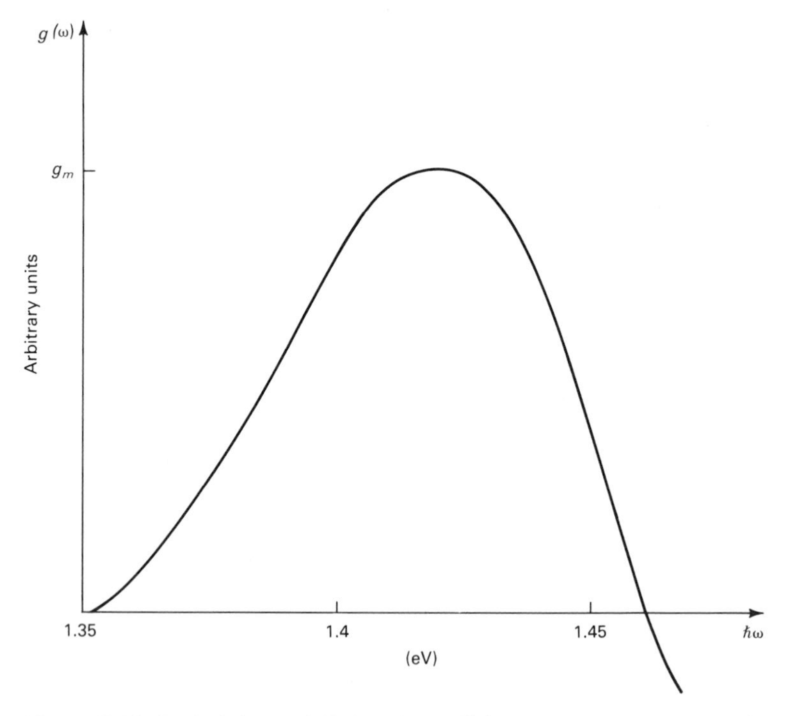

Figure 13.25 Typical shape of GaAs gain coefficient at room temperature. The value of maximum gain is denoted by g_m.

difference in the quasi-Fermi levels) and therefore of the junction current. The energy $\hbar\omega$ corresponding to the maximum has also a current dependence (i.e., so-called carrier band filling).

The calculation of $\alpha_{tot}^{net}(\omega)$ or the gain is, of course, always possible by numerical integration if R_{vc} and R_{cv} are known. These can be obtained from our definitions and the optical matrix element. Since the calculation of all prefactors and so on involves simple but extensive algebra, we refer the reader to Casey and Panish for the details.

The current of the laser junction is, as is generally the case in p–n junction theory, most easily calculated as the total recombination current [Eq. (13.27)]. The net recombination rate must then be calculated for the optical transition. The calculation does not contain any new theoretical aspects except that the recombination rate changes when lasing occurs since stimulated emission becomes strong and the stimulated emission rates are much larger than the spontaneous ones. The current therefore steeply increases beyond the laser threshold.

This section, of course, does not do justice to optoelectronics and the importance of lasers. The field is big enough to devote a separate book to it and a chapter alone to the consequences of heavy doping on the band edges and the energy of the emitted light.

13.5 NEGATIVE DIFFERENTIAL RESISTANCE AND SEMICONDUCTOR DIODES

In Sec. 13.3.3 we discussed Zener tunneling, which becomes important in very high fields and especially for cases of high built-in fields as they occur in heavily doped p–n junctions. This process gives rise to an entirely new feature of the p–n junction current: a negative differential resistance, shown in Fig. 13.26. This negative differential resistance has been explained by Esaki. The Nobel Prize-winning explanation is illustrated in Fig. 13.26.

In reverse bias, electrons can tunnel from the valence band to the conduction band and a relatively large current is flowing. In forward bias, electrons tunnel from conduction states into the hole states of the valence band and the current is large. Then tunneling is prohibited because the electron and hole states are not energetically aligned (to first order only, energy-conserving transitions are possible). Therefore the current decreases and a negative differential resistance (dI/dV is negative) occurs. Finally in extreme forward bias the current increases again.

As is well known, negative differential resistance can be used to convert dc power into ac power and negative differential resistance devices can therefore (and indeed are, if fast enough) be used as microwave generators. The Esaki diode is used for low power applications only since the voltage range in

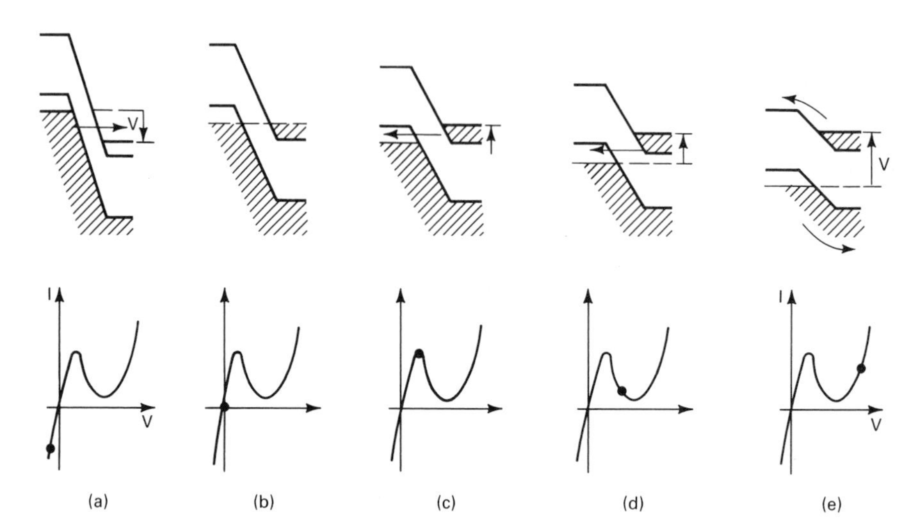

Figure 13.26 Simplified energy band diagrams of tunnel diode at (a) reverse bias; (b) thermal equilibrium, zero bias; (c) forward bias such that peak current is obtained; (d) forward bias such that valley current is approached; and (e) forward bias with thermionic current flowing. (After Hall, © 1960 IRE (now IEEE))

which the negative differential resistance occurs is small (and fixed by the value of E_G/e of the semiconductor). If higher power output is required, two other kinds of negative resistance effects are used, which are discussed below.

We have treated the negative resistance of GaAs already in Chap. 11 (Fig. 11.1). GaAs exhibits this negative differential resistance because of its inherent band-structure properties due to the transfer of electrons in **k** space. No special doping profile or *p–n* junction is necessary in this case. The effect was predicted by Ridley and Watkins and by Hilsum. Devices based on this effect are called transfer devices. When Gunn measured (without knowing the previous theoretical work) the current-voltage characteristic of *n*-GaAs, he found microwave oscillations related to the transit time of electrons through his GaAs samples. This transit time effect is based on the negative differential resistance and can be explained as follows.

Consider a section of semiconductor in the range of negative differential resistance and assume a small increase in resistance in this section, let's say because of fluctuations in the doping. Then, because of its higher resistance, a still higher voltage drops in this section and the electric field is increased. This in turn leads to a lower current, a still higher resistance, and a still larger voltage drop. The process continues until a "domain" of very high field is formed. This domain can "hang" at the positive contact or, if it forms at the negative contact, it will move through the semiconductor with the high field drift velocity ($\approx 10^7$ cm/s). After the domain vanishes at the positive contact, the process starts again and can lead, if the semiconductor is short enough, to microwave oscillations, as observed by Gunn.

Since the effect is dynamic and involves carrier density changes, we add here a few comments on high field domain formation. The fundamentals proceed very much along the line of our deviations of the space charge limited current, which starts from Eq. (11.25). Since now we are also interested in time development, we write the equation of continuity [Eq. (11.4), with generation-recombination neglected] in one dimension:

$$\frac{\partial n}{\partial t} = \frac{1}{e}\frac{\partial j}{\partial z} \tag{13.102}$$

and use Eqs. (11.25) and (11.26), which gives

$$\frac{\partial n}{\partial t} = F\frac{\partial}{\partial z}(n\mu) + \frac{en\mu}{\epsilon\epsilon_0}\,\delta n \tag{11.103}$$

Here $\delta n = n - n_0$ and n_0 is the equilibrium carrier density. The factor $(\epsilon\epsilon_0/en\mu)$ has the dimension of time and

$$\tau_R = \frac{\epsilon\epsilon_0}{en\mu} \tag{13.104}$$

is called the dielectric relaxation time. This time constant appears in all problems of similar type. As can be seen from the form of Eq. (13.103), the carrier concentration n will not change significantly for times much shorter than the dielectric relaxation time. High field domains will therefore only build up on time scales shorter or comparable to τ_R. The transit time τ_{tr} of a domain is approximately given by

$$\tau_{tr} \approx L/v_d \tag{13.105}$$

where v_d is the drift velocity of the domain. If this transit time is shorter than τ_R, obviously domains will not form. Therefore there is a condition for domain formation

$$\tau_R \ll \tau_{tr} \tag{13.106}$$

which is equivalent to

$$L \cdot n \gg \frac{v_d\epsilon\epsilon_0}{e\mu} \tag{13.107}$$

All of these processes are very complicated because of effects occurring simultaneously in **k** space and real space, and a proper description needs involved computation. My preference is again the Monte Carlo method if a quantitative understanding is needed. Note that although many semiconductors have satellite minima, only those with relatively large band gaps are eligible for Gunn devices, since in other materials impact ionization interferes with the transfer effect.

The real space transfer effect mentioned in the previous section can also give rise to negative differential resistance, and real space transfer diodes have also shown current oscillations.

For high power applications, important device types are impact ionization avalanche transit time (IMPATT) devices. These devices are based on negative resistance (not merely negative *differential* resistance), which is achieved by the generation of a carrier avalanche (by impact ionization) and its subsequent transit through a drift region.

The detailed theory of this type of devices is based on the following idea. Consider a structure consisting of an "electron injector" (a range in which electrons are supplied depending on the external voltages, e.g., a p–n junction with high built-in fields in which electrons are generated by impact ionization once an external voltage is applied) and an electron drift region. This latter region is biased toward saturation of the current, and a smaller additional external voltage will not change the drift velocity. Assume, then, that an ac voltage is applied to the structure. This ac voltage will modulate the injection (increase it at a certain phase). Subsequently the injected electron will go through the drift region and increase the current there even if by now the ac voltage lowers the injection and acts against the drift (but does not change the saturated drift velocity). Therefore the device will exhibit truly negative resistance during certain periods of the ac cycle. A comprehensive theory of this process has been developed by Read.

RECOMMENDED READINGS

ABRAMOWITZ, M., and I. A. STEGUN, *Handbook of Mathematical Functions.* New York: Dover, 1965.

ANDERSON, C. L., and C. R. CROWELL, *Physical Review R* (1972), p. 2267.

BEBB, H., and E. WILLIAMS, in *Semiconductors and Semimetals,* vol. 8. ed. R. K. Willardson and A. C. Beer. New York: Academic Press, 1972, pp. 243–53.

BETHE, H. A., and R. JACKIW, *Intermediate Quantum Mechanics.* New York: Benjamin, 1973.

BURSTEIN, E., and F. LUNDQUIST, *Tunneling Phenomena in Solids.* New York: Plenum Press, 1969.

CAPASSO, F., *Physics of Avalanche Photodiodes in Semiconductors and Semimetals,* vol. 22, ed. R. K. Willardson and A. C. Beer. New York: Academic Press, 1985.

CASEY, JR., H. C., and M. B. PANISH, *Heterostructure Lasers Part A: Fundamental Principles.* New York: Academic Press, 1978, (a) pp. 122–27; (b) p. 146; (c) pp. 144–50.

DUKE, C. B., *Tunneling in Solids.* New York: Academic Press, 1969.

HALL, R. N., *I.R.E. Transactions on Electron Devices,* ED-7 (1960), 1.

HESS, K., *Journal de Physique.* C7, Supplement 10 (1981), C7-3–17.

HIGMAN, J., and K. HESS, *Solid-State Electronics* 29 (1986), 915.

HOLONYAK, JR., N., and K. HESS, "Quantum-Well Heterostructure Lasers," in *Synthetic Modulated Structures,* ed. L. L. Chang and B. C. Giessen. New York: Academic Press, 1985.

HOLONYAK, JR., N., I. A. LESK, R. N. HALL, J. J. TIEMANN, and EHRENREICH, H., *Physical Review Letters,* 3 (1959), 167.

KANE, E. O., *Physical Review,* 159 (1967), 624.

LANDSBERG, P. T., *Solid State Theory,* New York: Wiley/Interscience, 1969, pp. 305–6.

READ, W. T., *Bell System Technology J.,* 37, 401 (1958).

RIDLEY, B. K., *Quantum Processes in Semiconductors.* Oxford: Clarendon Press, 1982, pp. 251–63.

SHENAI, K., E. SANGIORGI, R. M. SWANSON, K. SARASWAT, and R. W. DUTTON, *IEEE Transactions on Electron Devices,* ED-32, 793 (1985).

SHICHIJO, H., and K. HESS, *Physical Review B,* 23 (1981), 4197.

STILLMAN, G. E., V. M. ROBBINS, and K. HESS, *Physica,* 134 B-C (1985), 241.

STREETMAN, B. G., *Solid State Electronic Devices,* 2nd ed. Englewood Cliffs, N.J.: Prentice-Hall, 1980.

PROBLEMS

13.1. Derive Eq. (13.5) from Eq. (13.4).

13.2. Calculate the current of Eq. (13.14) for a triangular potential.

13.3. Derive j_{rs} in Eq. (13.40) from Eqs. (13.37)–(13.39). Discuss the limiting cases of short devices ($a/L_n, c/L_p \lesssim 1$) and long devices ($a/L_n, c/L_p > 1$).

13.4. Derive Eq. (13.58) for a graded profile of N_{TT}. Assume linearly decreasing and exponentially decreasing deep trap concentrations N_{TT}.

13.5. Evaluate Eq. (13.75) using Eq. (13.71) for $p = 1$. Discuss the influence of impact ionization on the electron temperature.

13.6. Find the functional form of $R_{vc} - R_{cv}$ from the optical matrix element (see also Casey and Panish). Insert into $\alpha_{\text{tot}}^{\text{net}}(\omega)$ [Eq. (13.100)] and perform at least one energy integration. Discuss the influence of the density of states if band tailing [Eq. (6.14)] is important.

13.7. Calculate the depletion width in a p–n junction that contains an additional "doping spike" $N_D^{ad}\delta(\epsilon)$, where ϵ is a short distance from the junction at the n side.

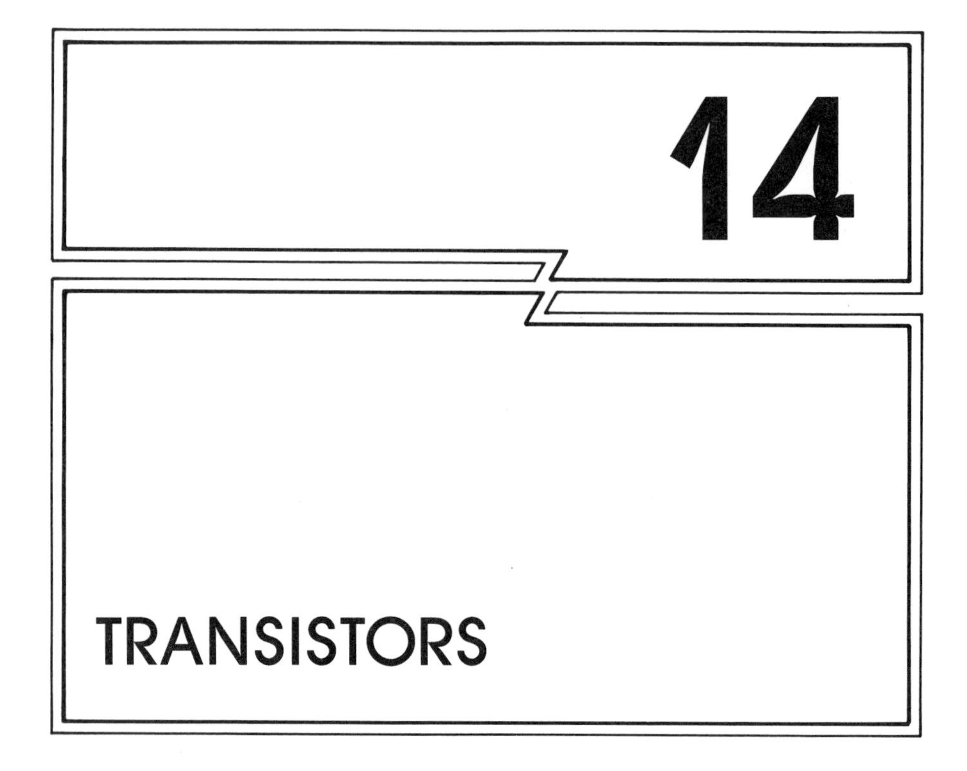

14 TRANSISTORS

Transistors (transfer resistors) are the most important solid state devices and distinguish themselves from the diodes by having a third terminal. In 1947, John Bardeen and Walter Brattain identified minority carrier injection and invented the point contact transistor. Their invention had an unprecedented impact on the electronic industry and on solid state research.

Diodes and their use rely generally on a material having a nonlinearity with a fairly distinct threshold, which will not respond to an input below a certain strength and which turns on above it. Both signal and power supply come into the same port to achieve the desired function. Transistors, on the other hand, have a third terminal and can be compared to a mechanical switch, as shown in Fig. 14.1. (The third terminal is the switching magnet.)

Such a switch allows a small signal to control a large signal input from a power supply. Transistors have, of course, enormous advantages over mechanical switches. They can be very fast (1 picosecond is about the ultimate switching speed limit of transistors), they can be produced in large numbers (many millions on a chip the size of 1 cm^2), and they can be made very small (1 micrometer feature size or smaller). Transistors also are cheap because of the elaborate photolithographic techniques that are available to produce them. As a consequence of all of these advantages, transistors are widely used in analog and digital circuits. The saying goes that any bigger engineering project that involves only diodes must be supervised by fellows of a physical

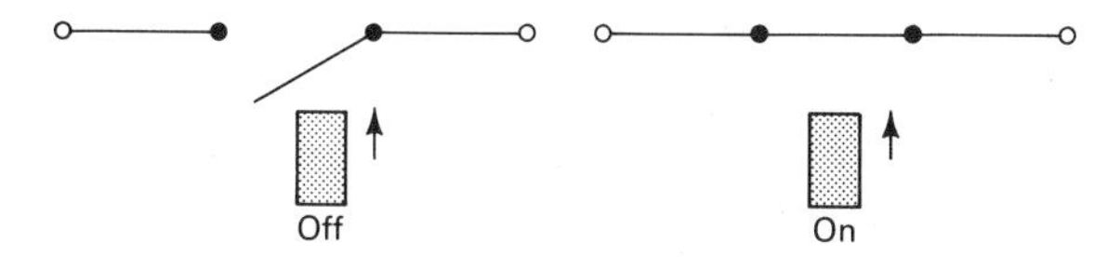

Figure 14.1 Mechanical switch as transistor analogy.

society while fellows of engineering societies prefer three terminal devices (and sometimes this is no joke).

The transistors of today are typically too small or too fast to be described by simple analytical theories and require elaborate models. Nevertheless, one needs to obtain some analytical understanding to appreciate the complications. Therefore, simplified transistor models are presented in the next section. The remainder of this chapter deals with aspects of transistor theory that cannot be analyzed without the help of computational resources. In these sections we develop the main physical picture and point toward the numerical solution by revisiting the basic equations and their solutions in the existing literature.

14.1 SIMPLE MODELS OF TRANSISTORS AND DEVICE MODELING

Although the models of transistors presented here are highly simplified, they work astonishingly well as long as they are not pushed too far. The reason is the following. The equations that are actually used are greatly simplified (assuming, for example, constant field-independent mobility). However, if the various physical constants are used as parameters, the equations often are sufficiently accurate because of the mean value theorem of integral calculus. This theorem is the savior of many device models and works sometimes so well that the scientists who develop the models tend to forget the simplifications. Then, only when some features (size, temperature) are changed drastically does the model fail, and one needs to go back to the drawing board and investigate the neglected physics.

The use of the mean value theorem is so powerful because in the method of moments, Eqs. (11.5a) and (11.5b), mean values are taken. If these mean values are used as parameters, the Boltzmann equation can be highly simplified for narrow (more or less) ranges of variables. An additional help for simple device models is given by "sum rules," such as overall charge neutrality, Eq. (13.17), and the theorem of Gauss, Eq. (12.29). It is no accident that the best models stem from noted scientists who knew the multitude of aspects that are important for semiconductor devices. In retrospect, these models look as if everybody could create them, but this is not the case. On the other hand, the literature contains many device models that everyone can create and there is not really any need for them.

14.1.1 Bipolar Transistors

The idea of a switch using p–n junctions can be understood from Fig. 14.2, which shows a reverse biased p–n junction into which carriers are injected optically. Obviously, the light source can cause a large generation rate and corresponding current and therefore can switch on the reverse current to high levels. The light source can be represented by another p–n junction (light-emitting diode) in forward bias, as shown in Fig. 14.3. (In fact, a device like this has been proposed, called a beam-of-light transistor.)

The charge flow Fig. 14.3 is as follows. Electrons are pushed from the left side of the left-hand p–n junction to its p side, where they recombine and generate light. Then the light creates electrons at the p side of the reverse

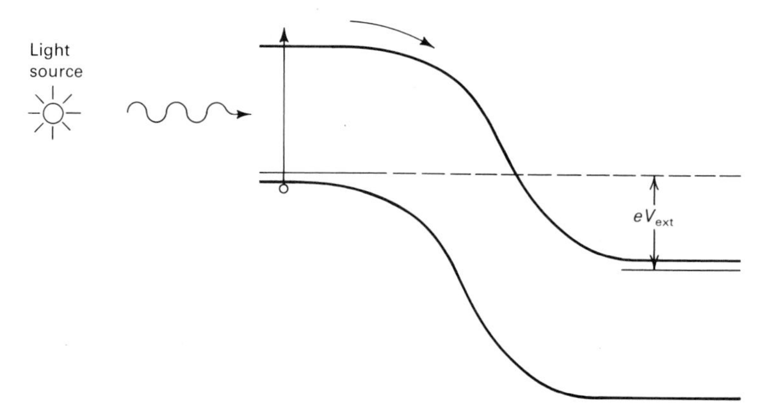

Figure 14.2 p–n junction in extreme reverse bias. The reverse current is caused by thermally generated carriers plus by electrons that are generated optically by a light source and the absorption of light.

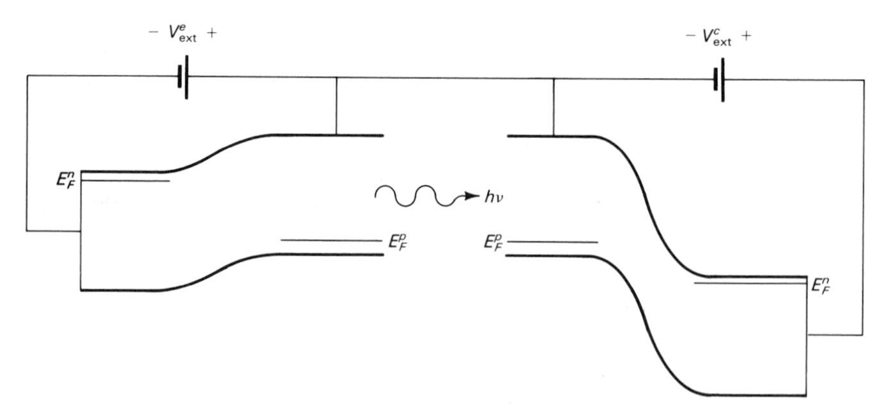

Figure 14.3 Forward biased (light-emitting) p–n junction triggering current in a reverse biased p–n junction through light emission.

biased p–n junction that contribute to the current. The question arises of why we need the light generation and absorption process. Can't we bring the p–n junctions close enough together so that the electrons from the forward biased junction are directly injected and switch on the current in the reverse biased junction? Indeed, one can, and the injection of electrons (minority carriers) into a p–type base is a fundamental transistor principle. The resulting band structure is shown in Fig. 14.4, which represents the band structure of a bipolar transistor with its three parts—emitter (of electrons), base, and collector (of electrons). The voltages V^e_{ext} and V^c_{ext} are called the emitter base and collector base voltage and are sometimes also denoted as V_{EB} and V_{CB}, respectively.

The theory of this device develops essentially in the same way as the p–n junction theory with the significant difference that in the base the boundary conditions change for the injected electrons. Obviously, for switching action to occur, the base must be short enough to let electrons (minority carriers) be injected into the collector junction without recombination with the holes (majority carriers) in the base. Therefore, the integration constants need to be recalculated for Eq. (13.32), as a/L_n cannot be regarded as large (infinity) any longer. Also C_2 is no longer equal to zero. Our new boundary conditions are now determined by the presence of the collector junction. To facilitate the discussion, we repeat Eq. (13.32):

$$n(z) = C_1 e^{z/L_n} + C_2 e^{-z/L_n} + C_3 \tag{14.1}$$

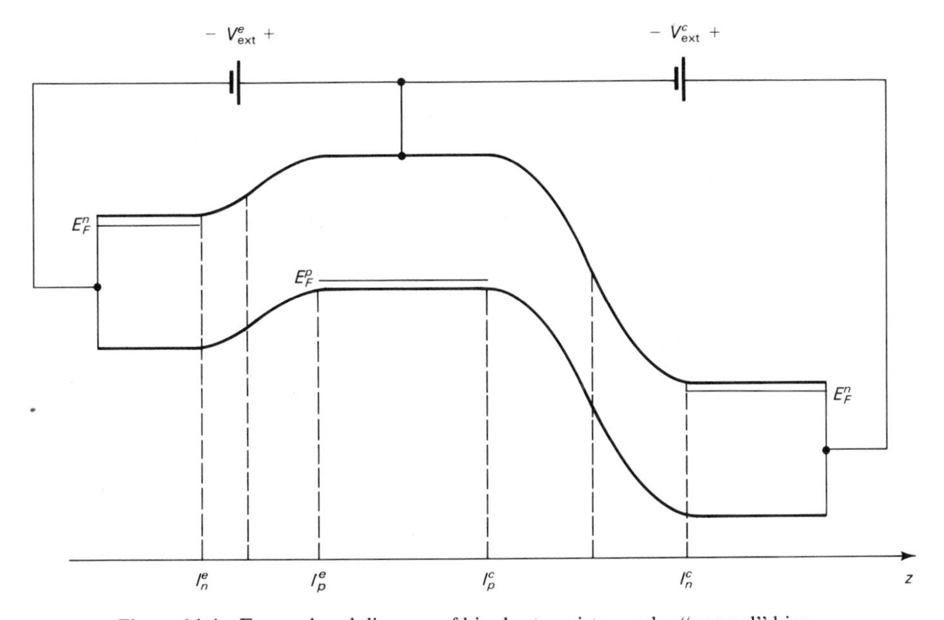

Figure 14.4 Energy band diagram of bipolar transistor under "normal" bias.

At the emitter base borderline l_p^e, we still have [see Eq. (13.34)]

$$n(l_p^e) = n(\text{base})\, e^{\overline{V}^e_{\text{ext}}} \tag{14.2}$$

where we have denoted the equilibrium concentration of electrons in the base by n (base) and the external voltage and depletion distance carry now a label e for emitter. The notation is also explained in Fig. 14.4.

Another boundary condition is obtained from the requirement that C_3 is equal to the equilibrium electron concentration at the p side [$C_3 = n$ (base)], as can be seen from Eq. (13.31) and replacing $n(-a)$ by n (base). The third boundary condition comes from the determination of the electron concentration at the collector base junction l_p^c. This latter concentration is not easy to find since its determination requires the detailed knowledge of electronic transport in the reverse biased collector with the simultaneous injection of electrons. The standard transistor theory inserts for the change Δn from the equilibrium ($\overline{V}^c_{\text{ext}} = 0$) concentration

$$\Delta n(l_p^c) = n\,(\text{base})(e^{\overline{V}^c_{\text{ext}}} - 1) \tag{14.3}$$

without justification. This value represents the true value only if no carriers are injected from the emitter and the collector diode is essentially independent [see Eq. (13.35)]. The reason why this assumption still works is, in my opinion, mainly due to the fact that under normal operating conditions the collector is in considerable reverse bias and the carrier concentration will be very small to the right of the point l_p^c where the electric field becomes large. Its real value then does not matter very much and one can use Eq. (14.3) or any negligibly small value. To avoid unnecessary complications and to highlight the essence of transistor theory, we put $n(l_p^c) = 0$. To simplify the notation, we chose as the zero of the coordinate system l_p^e (i.e., $l_p^e = 0$) and denote l_p^c as the (voltage-dependent) base width W_b. Then

$$\Delta n(0) = n\,(\text{base})(e^{\overline{V}^e_{\text{ext}}} - 1) = C_1 + C_2 \tag{14.4a}$$

and

$$C_1 e^{W_b/L_n} = -C_2 e^{-W_b/L_n} \tag{14.4b}$$

which gives

$$C_1 = -\frac{n\,(\text{base})(e^{\overline{V}^e_{\text{ext}}} - 1)}{e^{2W_b/L_n} - 1} \tag{14.5a}$$

and

$$C_2 = \frac{n\,(\text{base})(e^{\overline{V}^e_{\text{ext}}} - 1)}{1 - e^{-2W_b/L_n}} \tag{14.5b}$$

Insertion of Eqs. (14.5a) and (14.5b) into Eq. (14.1) or Eq. (13.32) and integrating over the generation-recombination rate U_s as in Eq. (13.37) leads to an emitter current contribution that is quite similar to Eq. (13.37) and for $W_b \ll L_n$ practically identical. Notice that the condition $W_b \ll L_m$ is typical for transistors since the electrons need to be injected into the collector without much recombination in the base. This means that the emitter diode functions basically independent of the collector (it does not care much where its electrons are dumped). The collector current density, on the other hand, can be significantly increased. The increase due to emitter injection can be calculated by assuming that the additional collector current density j_c is a diffusion current at $z = W_b$ and, therefore

$$j_c = -eD_n \, dn(z)/dz|_{W_b} \tag{14.6}$$

where the subscript W_b means that the derivative is taken at W_b. $n(z)$ can be obtained from Eqs. (14.1) and (14.5). From Eq. (14.5) it is immediately evident that

$$j_c \propto (e^{\overline{V}^e_{\text{ext}}} - 1) \tag{14.7}$$

which shows that the collector current density is controlled by the emitter voltage (or emitter current).

One can, therefore, write for the collector current I_C (I denotes current in contrast to j, which denotes current density) of an arbitrary bipolar transistor

$$I_C = -I_{CO}\,(e^{\overline{V}^c_{\text{ext}}} - 1) + \alpha_N I_E \tag{14.8}$$

where α_N is a transfer coefficient which in principle can be obtained from the theory outlined above and I_{CO} is the reverse saturation current of the collector (with zero emitter base).

In this discussion one can interchange emitter and collector and then obtain

$$I_E = I_{EO}\,(e^{\overline{V}^e_{\text{ext}}} - 1) + \alpha_I I_C \tag{14.9}$$

where we have included an influence term of the collector on the emitter. This time a coefficient α_I is used where the I stands for inverted (since in the normal operation electrons are not transferred from collector to emitter). Equations (14.8) and (14.9) together with Kirchhoff's law)

$$I_B = I_E - I_C \tag{14.10}$$

(where I_B is the base current) are the so-called Ebers-Moll equations. The four constants α_N, α_I, I_{EO}, I_{CO} are usually regarded as parameters and determined by experiments. The Ebers-Moll equations represent an example for an excellent device model that is greatly simplified but retains much of the device physics.

The ac characteristic of the bipolar transistor is more difficult to understand and involves the (not entirely independent) emitter and collector capacitance (Sec. 13.2.). The interested reader is referred to the charge control model of Gummel and Poon (Streetman).

We finish this section by discussing some principles that are important for the operation of bipolar transistors. Device performance necessitates high doping in the base since the resistance-capacitance products need to be kept small if high speed is to be achieved. High doping leads to a narrowing of the band gap for several reasons. In Chap. 6 we discussed the effect of band tailing, Eq. (6.14), and its consequences for the effective narrowing of the band gap. The high density of holes or electrons in the base leads to an additional band gap shrinkage due to many body interactions. This many-body shrinkage must be added to the tailing of Chap. 6. We have not placed much emphasis on many body effects in this book because the normally relatively low densities of conduction electrons (holes) in semiconductors justify one electron approximations. The exception has been the treatment of screening effects. These many-electron effects are always important for scattering problems, as we saw from the treatment of impurity scattering. The screening of impurities gives us a natural idea of the many body effects that have bearing on the total energy of an electron gas and therefore also to the value of the band gap. Charged impurities are not any different from the electrons except that they are localized. The coulombic repulsion will therefore also lead to screening of the electron potential itself. In other words, each electron will attempt to correlate the other electrons (pushing them away and thus create a "correlation hole" around itself). This of course will lead to differences in the total energy of the system. The Pauli principle adds to this classical coulombic effect as an electron repulses another electron with equal spin. The total effect, the "exchange correlation effect," has been subject to many studies and is by now well understood (Inkson).

Exchange correlation effects can be approximated by adding an exchange correlation potential as in Eq. (12.33). The simultaneous presence of crystal imperfections, however, gives additional complications to the theory and consequently the experimental results (shown in Fig. 14.5) are not well understood. The puzzling fact of the experiments is that the change in band gap ΔE_g increases much more rapidly than with the third root of the doping density N_D while all "simple" theories give an approximate $N_D^{1/3}$ dependence since the average distance between dopants is $\approx N_D^{-1/3}$.

Band gap differences in the bipolar transistor, especially a smaller band gap in the base relative to the emitter, have numerous consequences on transistor operation. The effects are especially pronounced in heterolayer transistors which feature large band gap differences. If, for example, the emitter is $n - Al_xGa_{1-x}As$ and the base p-GaAs with a smaller band gap, then holes are blocked to flow from the base to the emitter (holes are minority carriers for the emitter), which gives rise to high "emitter efficiencies." By the same token the base can be doped heavier (lower base resistance) without sacrificing emitter efficiency. As a consequence, large current gain factors

$$\beta = \frac{I_c}{I_B} \tag{14.11}$$

can be obtained.

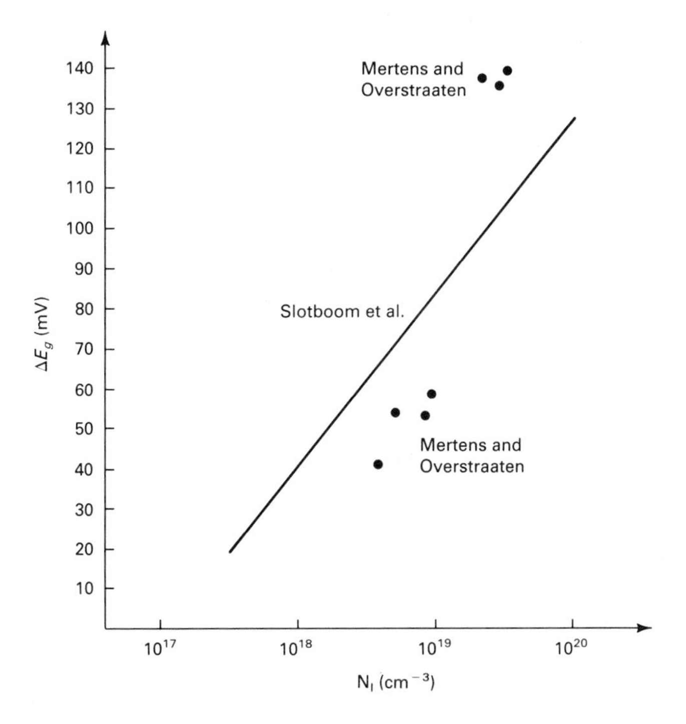

Figure 14.5 Measured band gap narrowing as a function of impurity concentration. (After Mertens and v. Overstraaten, © 1978 IEEE.)

Hot electron effects can also be of some importance to bipolar transistor operation as the velocity at the base boundaries (immediately to the right of l_p^c in Fig. 14.3) can exhibit saturation and even overshoot effects due to the high electric fields. The effects, however, are not as large and significant as hot electron effects are in the field effect transistors. Of course, in extreme reverse bias impact ionization can play some role and hot electrons can even be emitted in neighboring oxides. This latter effect is treated in Sec. 14.2 (real space transfer). If such effects must be included in the transistor theory, elaborate numerical calculations are necessary and have been performed, for example, by Cook. Let us note finally that for high doping levels Zener tunneling can also interfere with normal transistor operation.

14.1.2 Field Effect Transistors (FETs)

We have treated the field effect in Secs. 12.3 and 12.4 and have shown in Eq. (12.44) that the application of a voltage to a metal gate will lead to charge accumulation in a semiconductor that is separated from the metal by another semiconductor (insulator) with larger energy gap. This effect can be used as another transistor principle, as illustrated in Fig. 14.6.

Two heavily doped n^+ regions (contacts) are diffused (or implanted) into

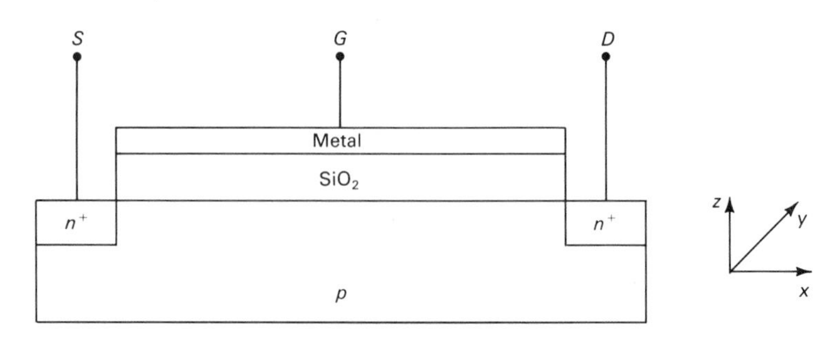

Figure 14.6 Schematic plot of a metal-insulator semiconductor (MIS) field effect transistor.

a p-type semiconductor. These contacts are usually denoted by S (for source) and D (for drain). Whatever bias is applied to these contacts, one n^+–p junction will always be in reverse bias and only a small reverse saturation current will flow. However, as soon as the gate electrode G is given a positive bias, a sheet of inversion electrons (Fig. 12.7) will form at the interface and source and drain will be connected by this sheet, which now forms a n^+nn^+ diode (Chap. 11) which for long and homogeneous n regions represents a conducting semiconductor element. In other words, we can switch from a reverse bias operation to a conducting channel. This is the field effect switching principle. In this discussion n and p are, of course, interchangeable.

The transistor type shown in Fig. 14.6 is called an n-MIS, or if the insulator is silicon dioxide, an n-MOS (NMOS) transistor. It is the insulating quality of SiO_2 (quartz) that makes the MOS system so valuable. Also, the system can be fabricated with few interface states (see Fig. 5.6). No semiconductor other than silicon has a comparable oxide. With n and p interchanged, it would be a p-MOS (PMOS) transistor. Of special importance for logic applications are circuits that contain both n-MOS and p-MOS devices. One then talks about complementary MOS devices or c-MOS (CMOS).

Before we proceed with the theory of n-MOS transistors, we ask ourselves the question: Where do the inversion electrons (the electrons that switch the transistor on) come from? In the good old days, people talked about "induced" electrons at the surface. To understand what "induced" means, we discuss again the band structure (see Chaps. 12 and 13) for the various regions separately (Figs. 12.1, 12.4, and 13.1). An ideal combination for all regions (metal-insulator semiconductor) is shown in Fig. 14.7. The insulator can, of course, also be a semiconductor with larger energy gap, although in insulators the electron mobility is typically much lower than in semiconductors.

Application of a positive voltage eV_{ext} to the metal gate will separate the Fermi energies (now quasi-Fermi energies) by eV_{ext} and will also lead to a redistribution of charge in the semiconductor, as shown in Fig. 14.8 and discussed in Chap. 12.

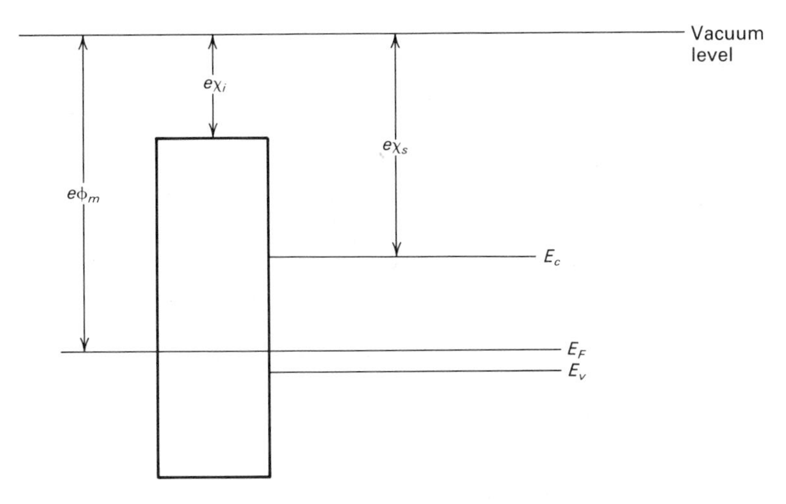

Figure 14.7 Band structure of an ideal metal-insulator semiconductor structure. Ideal means that the Fermi energy is lined up (straight line) automatically without charge transfer and band distortions (such as in Fig. 12.3).

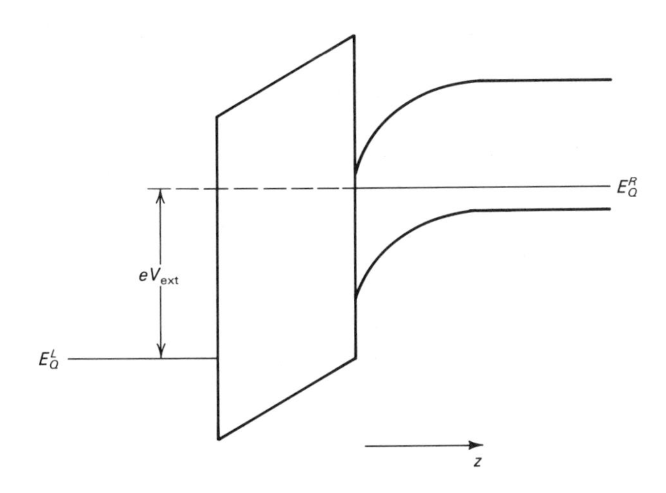

Figure 14.8 Metal-insulator semiconductor structure with applied voltage.

The quasi-Fermi level in the semiconductor is a straight line because in the ideal case no current flows through the insulator and therefore no external voltage drops in the semiconductor. As already known from Chap. 12, electrons accumulate at the interface and an electron channel forms at the interface. If the structure consists only of metal-insulator semiconductor, the electrons at the interface have to be generated thermally (or optically), as discussed in Chap. 10 on generation-recombination. This process is, of course, slow. Especially at low temperatures it would take years to establish equilibrium. Such a switch would therefore be useless. Fortunately, there are electrons available in a transistor structure in the n^+ contact regions (Fig. 14.6) and these can be pulled into the channel so that the switch on time in the ideal case is equal to channel length divided by an average electron velocity. This time is for channel length of 10^{-4} cm (μm) of the order of picoseconds. The total charge "induced" by the gate voltage has been derived in Eq. (12.44).

We subdivide this charge now into two contributions. One (Q_i) is the charge of the (mobile) inversion layer electrons. It is this charge that contributes to the current. The second part is fixed charge of acceptors Q_A in the p–n junction depletion layer, which is formed in the region where E_Q^R is close to the middle of the band gap in Fig. 14.8. Since we need to apply a voltage between source and drain in addition to the gate voltage, the surface potential becomes a function of x (the direction of current parallel to the interface). Eq. (12.44) reads then

$$-Q_i - Q_A = C_{ins}\,(V_G - \phi_s^R(x)) \tag{14.12}$$

It is important that high voltages V_G can be applied to the SiO_2 without breakdown (the field when impact ionization becomes strong is 2×10^7 V/cm) and therefore high densities of inversion charge can be achieved. This is unique for the MOS system.

We now assume that the source electrode is grounded and the gate voltage is such that we have achieved inversion. The interface potential is then equal to the saturated value ϕ_s^R as given by Eq. (12.30) plus an external x-dependent potential equal to 0 at the source, equal to the drain voltage at the drain and x-dependent $V(x)$ in between such that

$$\phi_s^R(x) = 2kT_c \ln\,(N_A/n_i)/e + V(x) \tag{14.13}$$

and the acceptor charge is

$$Q_A = -eN_A W = -\sqrt{2\epsilon_0\epsilon e N_A (V(x) + 2kT_c \ln\,(N_A/n_i)/e)} \tag{14.14}$$

where use has been made of Eq. (12.16) for the depletion width W with the built in potential being now equal to $\phi_s^R(x)$.

The channel (inversion layer) conductance Δg_c of a small section of length Δx then becomes

$$\Delta g_c = \mu |Q_i| \frac{W_y}{\Delta x} \tag{14.15}$$

where W_y is the width of the transistor in y direction. We now assume that the mobility μ is constant (independent of the electric field) and neglect all other hot electron effects and the diffusion current. Then we have from Ohm's laws

$$\Delta g_c\, \Delta V(x) = I_D \tag{14.16}$$

where I_D is the drain current, which is in steady state constant and the same everywhere. Collecting all terms in differential form ($\Delta \rightarrow d$), we get

$$W_y \mu[C_{\text{ins}}(V_G - \phi_s^R - V(x)) + Q_A]\, dV(x) = I_D\, dx \tag{14.17}$$

Since I_D is constant the integration of this equation is straightforward (see Problems section) and gives for small drain voltages V_D

$$I_D \approx \frac{W_y}{W_x} \mu C_{\text{ins}} (V_G - V_T) V_D \tag{14.18a}$$

with

$$V_T = 2kT_c \ln (N_A/n_i)/e + \sqrt{2\epsilon_0 \epsilon N_A (2kT_c \ln (N_A/n_i)}/C_{\text{ins}} \tag{14.18b}$$

V_T is called the threshold voltage, which represents the voltage below which the channel is essentially nonconducting.

There exists a subthreshold current, however, which is not included in Eq. (14.18a) because it is essentially a diffusion current and diffusion has been neglected so far (see next section). For large V_D values, I_D approaches a saturated value at the moment the inversion charge goes to zero at the drain. It is clear that as V_D increases to more positive values, the potential difference between channel and gate approaches zero and the carrier concentration at the drain side decreases. Therefore the resistance increases steeply, giving rise to I_D values below the Ohmic straight line. It is easy to show from the integration of Eq. (14.17) (see Problems section) that the saturation current $I_{D\text{sat}}$ is proportional to

$$I_{D\text{sat}} \propto (V_G - V_T)^2 \tag{14.19}$$

For small devices the saturation is strongly influenced by hot electron phenomena, which is described in the next section, and Eq. (14.19) has little validity. However, the reason for the saturation of Eq. (14.19) without hot electron effect deserves discussion. We have mentioned already that the density of inverted electrons decreases as the drain voltage approaches the gate voltage. At $V_D = V_G$, the concentration of mobile carriers at the drain side approaches zero. One calls this phenomenon the "pinch-off" effect. Beyond the point at which the mobile carrier concentration at the drain approaches zero, the pinch-off point, the current I_D saturates. In this regime, the channel can be essentially subdivided into two parts: the part that still contains conduction electrons and is described by the above equations and the pinched-off part, which, according to the above calculation, does not contain much mobile charge, if any. If we view the existing electron channel (inversion layer) as an extension of the source n^+ contact, then the channel represents essentially a n^+nn^+ diode with the n region being eventually depleted. We treated the n^+nn^+ diode in Chap. 11 and found the space charge limited current density with [Eq. (11.30)] and without [Eq. (11.32)] hot electron effects included.

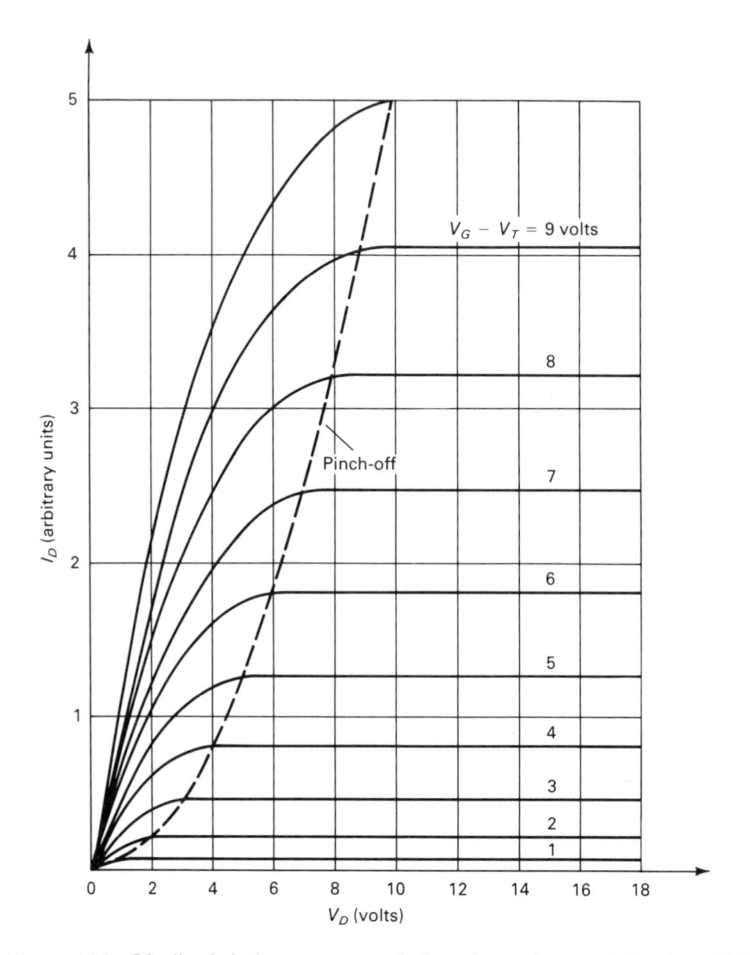

Figure 14.9 Idealized drain current vs. drain voltage characteristic of a MOS transistor for various values of the gate voltage V_G.

The voltage V_{sp}, which controls the space charge limited current, is given by the drain voltage minus the voltage $V(x)$ at the pinch-off point, which is denoted by V_{pi}:

$$V_{sp} = V_D - V_{pi} \tag{14.20}$$

We now perform a "gedanken" experiment. If the space charge limited current were smaller than the saturation drain current, we could have the case of negative differential resistance, which means that as we increase the drain voltage, the drain current decreases. In such a situation, however, the voltage drop would continue to increase over the region of negative resistance and decrease in the rest of the channel region, leading to an increase in space charge limited current to the value of the saturation current (an increase beyond I_{Dsat} is limited by the channel current). In other words, the pinched-off range will adjust itself so that the current through the device stays saturated and we arrive at a transistor current-voltage characteristic as shown in Fig. 14.9.

14.2 NONLINEAR TRANSPORT EFFECTS AND THE TRANSISTOR CHARACTERISTICS (SMALL DEVICES, HIGH ELECTRIC FIELDS)

The reduction in device size as required by attempts to integrate large circuits and systems on a chip, and the desire of high performance, lead to complications on the device level. For example, if the device should have a size of 1 μm and the various "voltage swings" are of the order of 1–5 V, electric fields of up to 5×10^4 V/cm and more will be present in the device. The idealized theory of the previous sections will therefore be invalidated as the mobility is not constant and other high field effects become important. In the following we discuss some of the effects encountered because of the (partly conflicting) demands of large systems and high speed.

14.2.1 Scaling Down Devices

How many devices fit on the tip of a pin? We can give a rough estimate as follows. Most semiconductor devices contain at least one p–n junction. Any p–n junction has at least a length of its depletion layer width W, which is for a p^+–n junction given by Eq. (12.16). The integration of devices uses mostly planar technology, that is, the devices are essentially in a plane and the third dimension is not as critical and can only help (by stacking up devices) for size reductions. If we assume, then, a minimum device area of W^2, we can fit on a square centimeter $N_{max} = 1\ \text{cm}^2/W^2$ devices, and N_{max} is given by

$$N_{max} = \frac{e^2 N_D}{2\epsilon\epsilon_0 E_G} \tag{14.21}$$

We have replaced here the built-in voltage by its approximate value of E_G/e. Equation (14.21) demonstrates the dependence of "integrability" on materials constant (E_G, ϵ, and the donor concentration). In particular, the maximum number of devices on a chip is directly proportional to the doping concentration. Therefore, if we know an ideal chip that makes optimal use of the area and if we want to increase the number of devices by a factor of K, then we have to scale (increase) the doping by the same scaling factor as long as Eq. (14.21) determines the maximum number of devices.

Scaling requirements may be derived from other considerations as well. If, for example, our devices were all designed to follow the ideal device equations with constant mobility and if we have a working chip, then we might postulate that it is scaled by keeping the electric field constant since the mobility will then stay field independent. This means that if we scale down the device area by a factor of K^2, the voltages have to be reduced by a factor of K. The doping needs to increase by K [Eq. (14.21)] and the transit time through the channel will decrease by K automatically (higher speed). Also, the power dissipation per device will be reduced by K^2. These latter facts are, of course, very desirable. However, the scheme will work only down to certain voltages. As long as p–n junctions are involved, the supply voltages need to be larger than E_G/e (or the built-in voltage) to achieve strong forward bias. Also, the gate voltage needs to be higher than the threshold voltage. Practical consideration may give even higher limits to the supply voltages. Therefore, when these limits are reached, different scaling laws must be used.

There is no universal scaling approach satisfying all demands, and each case must be considered separately. To optimize the scaling procedure, good two- and three-dimensional numerical models are necessary that can be used as computer-aided engineering design tools. The basic equations and making such models are discussed in Sec. 14.3. Here we proceed with some analytical considerations concerning effects that lead to deviations from the ideal transistor behavior and need to be taken into account when the sizes are reduced.

14.2.2 The Subthreshold Current

In the subthreshold region $V_G < V_T$, the channel is essentially a n^+pn^+ structure with p being close to the intrinsic concentration. Therefore, the conducting part of the MOS transistor is actually a bipolar transistor with the bulk of the channel forming the base. The current in the bipolar structure is a diffusion current and can be calculated from Eq. (14.6). Notice, however, that the carrier concentration depends on the surface potential.

It is clear that for a small device the calculation has to be performed numerically since the results will depend sensitively on the charge distribution around the source and drain contact and not only at the interface. The appropriate device model has to be, at least, two dimensional. We refer the interested reader to the discussion of numerical device models in Sec. 14.3.

14.2.3 Hot Electron Effects

As discussed, the mobility and diffusion constant are functionals of the energy distribution of electrons and holes. In a homogeneous semiconductor and under steady state conditions (time independence) the mobility μ and diffusion constant D become functions of the local electric field because the average electron energy is a single valued function of the local electric field only [Eqs. (11.6) and (11.8)]. This ceases to be true if the semiconductor is inhomogeneous and the electric field varies with distance or time. Velocity overshoot, a nonlocal effect, can then occur, as shown in Fig. 11.2, as a function of time when the electric field is switched on at $t = 0$. Even for time-independent fields overshoot is possible provided that the field varies spatially fast enough to "trick" the electron ensemble into acceleration without giving it time to establish a random motion by scattering and thus having an elevated mobility-degrading temperature. Figure 14.10 shows the typical functional form of electron temperature and drift velocity in such a situation.

Note that to the right of the maximum field, the electron temperature is higher than for the same value of a homogeneous field, while to the left it is lower. This arises from the fact that energy (temperature) is transported with the drifting electrons from left to right and the temperature follows delayed due to the energy relaxation time. The complicated process is described well by the moment equation derived for energy in Appendix E.

The drift velocity follows a corresponding yet different pattern. As long as T_c is lower than it would be in the homogeneous case, the drift velocity will exhibit overshoot effects since μ is a function of T_c and is, for phonon scattering, higher at lower values of T_c [see Eqs. (9.46) and (9.47)]. Following the maximum of the electric field, T_c is higher than it would be for the homogeneous case. Therefore the mobility becomes smaller and the velocity "undershoots." The undershoot, however, is usually not as pronounced as the overshoot.

In the following, the effect of the nonlocal field dependence of μ and D on transistor performance is described for various device sizes. We do not give a precise size classification because the given results depend on many factors.

The large device is identical to the ideal device discussed in the previous

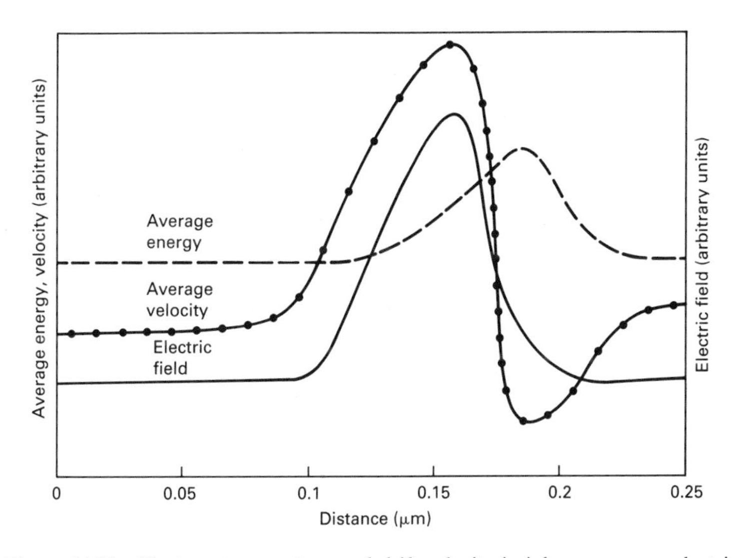

Figure 14.10 Electron temperature and drift velocity in inhomogeneous electric fields.

sections, and μ and D are constant. At a size of about 10 μm channel length, the electric fields at the drain (when pinch-off occurs) become large enough in MOS transistors to lead to a velocity saturation as described by Eq. (11.19) and the current close to the pinched-off region must be calculated by assuming a saturated velocity.

As the device sizes approaches 1 μm, saturation can occur everywhere in the channel. Overshoot and undershoot in silicon are still not very important. This is not so for GaAs devices, since GaAs exhibits a much higher overshoot than silicon due to its specific band-structure properties. The current saturates, then, typically before pinch-off occurs while the carrier concentration is still approximately homogeneous in the channel and given by Eq. (14.12), with $\phi_s^R(x)$ being approximately equal to $\phi_s^R(0)$ (i.e., its value at the source side). Using Q_i from Eq. (14.12) and Eq. (11.19) one obtains the current by multiplying Eq. (11.19) by the channel cross section:

$$I_D \approx eC_{\text{ins}}(V_G - \phi_s^R(0))\left[\frac{1}{1 + e\mu_0 F^2\left(\dfrac{300\text{ K}}{T_L C}\right)}\right]^{1/2} \mu_0 F \tag{14.22}$$

where $F \approx V_D/W_x$ is the electric field.

For still smaller device sizes overshoot effects are increasingly important. Since these effects are nonlocal and μ ceases to be a function of the local electric field, explicit calculations are difficult. As already mentioned, overshoot is much more important in GaAs than in silicon because of the delayed scattering of the electrons to the X and L minima and therefore the delayed increase in mass in addition to the delayed increase of T_c, which is the only reason for the overshoot in silicon. Therefore overshoot must be taken into account in GaAs devices and, in fact, can be used to increase the device speed, whereas in silicon these effects are small. The best way to account for the overshoot is explained in the next section, which briefly describes numerical device models.

Two more high field effects are of importance in MOS transistor operation. A very important effect is the emission of hot electrons from silicon into silicon dioxide. The attentive reader has, of course, noticed by now that the electron temperature in small transistors is 1000 K or higher under typical operating conditions. This temperature is totally unnoticeable to the user of the device or chip, who in principle can touch them without feeling heat. The reason is, of course, the large electron affinity $e\chi$ (Fig. 13.1a) which the electron needs to overcome to reach the vacuum level (or the fingers of the observer). The barrier that carriers need to overcome to propagate from silicon into silicon dioxide is lower than $e\chi$, and is about 3.6 eV for electrons and ~3.1 eV for holes. Therefore very hot electrons or holes can overcome this barrier. The number of such electrons can be estimated as in our treatment of impact ionization Eq. (13.83) by replacing the ionization threshold by the SiO_2 barrier height. Instead of L_I^*, one obtains, then, the mean free path for emission from silicon into silicon dioxide. A more precise treatment has been given by Tang and Hess. Note that this effect of emissions is, in principle, a form of real space transfer.

The treatment of impact ionization in MOS transistors is described in principle by our treatment in Sec. 13.3. However, if the electric fields vary very rapidly on the scale of L_I^* [Eq. (13.83)] transient effects are important. These cannot be estimated with any precision without taking recourse to sophisticated Monte Carlo simulations (Kim et al.).

14.2.4 High Frequency Response

Under high frequency operating conditions, there are two limiting factors for the FET response: the transit time and the resistance capacitance (RC) time constants. The average transit time from source to drain is responsible for switch on and is given by

$$\tau = W_x \Big/ \left[\frac{1}{W_x} \Big/ \int_0^{W_x} \langle v \rangle dx \right] \tag{14.23}$$

where $\langle v \rangle$ is the average drift velocity, which can include overshoot. For the simple case of constant mobility μ_0, we have

$$\int_0^{W_x} \langle v \rangle dx = \mu_0 \int_0^{W_x} F dx = \mu_0 V_D \tag{14.24}$$

In some instances $\langle v \rangle$ can be replaced by the constant saturation velocity and the result in Eq. (14.24) is then $v_s W_x$. If overshoot is important, more elaborate numerical computations are necessary.

The important and relevant capacitances are the source gate capacitance C_{SG}, the drain gate capacitance C_{DG}, the drain capacitance C_{DC}, and the source drain capacitance C_{SD}. In long devices these capacitances can be calculated by standard methods, as have been described in Sec. 13.2. For small devices these capacitances are not independent and the detailed charge distribution in the device is important. For accurate results one needs to employ numerical methods such as described in the next section.

14.3 NUMERICAL SIMULATIONS OF TRANSISTORS

We have seen that although essential parts of the theory of semiconductor devices can be understood analytically, the multitude of nonlinear transport effects and the complicated geometries make numerical calculations necessary if quantitative understanding is to be achieved. Large-scale computer simulations have become a cost-effective and predictive tool of device design and related computer-aided engineering questions. Our treatment of principles was aimed toward highlighting the most precise treatments and the various conditions for excellent approximations, and we will give a short summary in this section. Table 14.1 shows the "ultimate" system of equations, which, of course, already contains numerous approximations. For example, the Boltzmann equation is semiclassical and does not account for the uncertainty principle.

TABLE 14.1 Optimum Set of Device Equations

Maxwell equations	Including time- and space-dependent dielectric constant (matrix)
Boltzmann equation	Including all scattering mechanisms and the Pauli principle. Can be replaced by Monte Carlo simulations (which are totally equivalent)
Schrödinger equation	Used indirectly to determine properties of electronic states, band structure, effects of size quantization (quantum wells) and of wave functions for scattering matrix elements

Another set of device equations is shown in Table 14.2. This set represents intermediate accuracy. The Maxwell equations are replaced by Poisson's equation, which assumes a time- (space-) independent dielectric constant. It is still important to include the displacement current $\partial \epsilon \epsilon_0 F / \partial t$. The Boltzmann equation (Monte Carlo) is replaced by a set of moment equations, including energy balance, and the Schrödinger equation is used only conceptually. For example, one could model the electrons at the interface or in a quantum well as a two-dimensional electron gas. This set of equations is not only of intermediate accuracy but also of intermediate status with respect to computer time consumption.

It should be noted that the numerical methods that can be applied to this set of equations converge only if the electron energy is limited. It is "physical" to set as an upper bound of this energy the energy of steep increase in the

TABLE 14.2 Intermediate Set of Device Equations

Poisson's equation	(Displacement current included for short time processes)
Equation of continuity	Eq. (11.4)
Equation for current	Eqs. (11.6) and (11.6), plus term $\tau \partial j / \partial t$, as described in context with Eq. (11.6)
Energy balance equation	Eqs. (11.8a) and (11.13a) and Appendix E
Choice of constants (μ, D, etc) in accordance with scattering theory (perturbational solution of the Schrödinger equation). One band approximation [Eqs. (4.18) and (4.19)]. Size quantization effects are included phenomenologically, for example, by declaring the electrons in a quantum well two-dimensional.	

density of states (steep increase in scattering). For example, in GaAs, this could be (for practical transistor applications) the energy of the X valleys.

The intermediate set, although still very accurate (it includes, for example, overshoot phenomena), is in a way "neither fish nor flesh" and in the long run may be used even less than it currently is.

Table 14.3 shows device equations that contain a minimum of physics. It is these equations that are most frequently used and whose numeral solutions are most developed. These equations have been solved using finite difference and finite element methods (Selberherr). The resulting algebraic equations can be solved, for example, by successive overrelaxation, full Newton or special methods (Gummel) that have been outlined in detail by Selberherr. The corresponding numerical programs are used in various computer-aided engineering tools for device design. Examples are the MINIMOS and the PISCES codes.

These codes are very effective in modeling silicon devices with significant accuracy. They can deal with almost arbitrary geometries (two dimensional, three dimensional), and are reasonably accurate down to micrometer sizes. For very small transistors (submicrometer) a more precise evaluation of μ and D and the corresponding rearrangement of the carrier distribution is desirable. This is especially true for materials other than silicon. GaAs calls for more precise treatment even in the micrometer range because of the negative differential resistance and its pronounced overshoot effects.

TABLE 14.3 Minimum Set of Device Equations

Poisson's equation	(Displacement current included)
Equation of continuity	Eq. (11.4)
Equation for current and μ and D chosen as functions of the electric field, as in Eq. (11.19)	Eq. (11.6)
Effective mass approximation	Eq. (4.22)

One way of dealing with the complications is to use the intermediate set of equations. Another way that I feel is promising is the following. One can use a very precise set of equations, including Monte Carlo methods. The numerical solution is, of course, computer time consuming. However, as always has been done in such cases, one can create "look-up tables" for μ and D, and so on, and use these tables in a "minimum set equation" program. These look-up tables will not be general enough to be used for all temperatures, geometries, and times, and if parameters change too drastically, new look-up tables need to be created and checked by experiments. The system is illustrated in Table 14.4. The precise set in Table 14.4 is not identical

TABLE 14.4 Combination of Precise and Minimum Set of Equations

Precise Set (to be executed only to create look-up tables)
Poisson equation and displacement current
Monte Carlo
One band approximation [Eq. (4.18) and (4.19)]
Size quantization effects by including many quasi two-dimensional subbands, as described in Chap. 12
Create look-up tables for μ, D, generation-recombination rates, average effective masses as functions of F, $\partial F/\partial x$, and, if necessary, t
Minimum Set
Poisson equation + displacement current
Equation of continuity [Eq. (11.4)]
Equation for current [Eq. (11.6) plus term $\tau \partial j/\partial t$, if necessary]
Effective mass approximation

to the optimum set. The slight deviations from the optimum set are proposed to make the problem tractable.

The look-up tables need to be created in an appropriate way. If overshoot is to be included, μ and D stop to be a function of F only and need to be included as appropriate functions of several variables such as F, $\partial F/\partial x$, and even t. It is easy to convince oneself (by inspection) that an equation such as

$$j = en(\mu F + W(F)\partial F/\partial x) \tag{14.25}$$

is capable of describing overshoot effects. The function $W(F)$ can be approximated, for example, from Monte Carlo simulations. A more complex equation has been given by Thornber in terms of five tabulated functions.

Another way of combining precise and less precise methods to achieve accurate and fast simulations tools has been suggested by Dutton, who uses Monte Carlo methods for certain geometric ranges within the PISCES code.

This concludes the detailed treatment of transistors. Let me emphasize that I have omitted some details of computation which are usually treated in device-theory books. This concerns the effects of degeneracy which means that the energy distribution is of Fermi-type. Mobility and diffusion constant are then related by a modified Einstein relation and also the calculation of the electron concentration involves Fermi-integrals. All that means is that some of the presented integrations over the Maxwellian exponential function need to be replaced by integrations over Fermi-functions and the integration cannot be done explicitly. However, it can easily be performed numerically.

RECOMMENDED READINGS

ARNOLD, D., K. KIM, and K. HESS. *Journal of Applied Physics,* to be published.

COOK, R. K., *IEEE Transaction on Electron Devices,* ED-30 (1983), 1103.

INKSON, JOHN C., *Many-Body Theory of Solids.* New York: Plenum Press, 1984.

KIM, K., and K. HESS, *Journal of Applied Physics,* 60 (1986), 2626.

MERTENS, R., and VAN OVERSTRAATEN, R., *IEEE Technical Digest, International Electron Devices Meeting,* 1978, p. 322.

SELBERHERR, S., *Analysis and Simulation of Semiconductor Devices.* New York: Springer-Verlag, 1984.

STREETMAN, B. G., *Solid State Electronic Devices.* Englewood Cliffs, N.J.: Prentice-Hall, 1980.

TANG, J. Y., and K. HESS, *Journal of Applied Physics,* 54 (1983), 5145.

THORNBER, K. K., *IEEE Electron Device Letters,* EDL-3 (1982), 69.

PROBLEMS

14.1. Derive the emitter current by inserting Eqs. (14.5a) and (14.5b) into Eq. (14.1). Simplify the result for the case of $W_b \ll L_n$.

14.2. Integrate Eq. (14.17) and derive the drain current I_D. Simplify for the limiting case $V_D \ll V_G - V_T$. Derive the saturation current and show that it is proportional to $(V_G - V_T)^2$.

15

NEW TYPES OF DEVICES—AND CHAPTERS THAT HAVE NOT (YET) BEEN WRITTEN

A newer type of transistor is the metal semiconductor field effect transistor (MESFET) proposed originally by C. Mead. The principle is shown in Fig. 15.1.

A semiconductor layer (in our example, n-GaAs) is grown on an insulating substrate. The transistor has a diffused n^+ source and drain contact and a Schottky barrier gate. Depending on design, the depletion layer width W of the gate is, without gate bias, larger than the semiconductor layer width W_z (normally off transistor) or smaller than the layer width (normally on transistor). Switching is achieved by applying a gate bias and varying the depletion layer width W to switch the conduction on or off. The theory is developed in analogy to the MOS transistor [especially Eq. (14.17)] and leads to very similar expressions and behavior of the drain current (Streetman). One obvious advantage of the device is its structural simplicity and the presence of one carrier type only.

The high electron mobility and high field velocity of GaAs make it attractive for microwave applications as well as for high speed, low power logic. However, the doping level in the n-GaAs layer, which has to be increased when scaling down devices, gives rise to impurity scattering and a correspondingly lowered mobility. This problem is circumvented by the design of the high electron mobility transistor, shown in Fig. 15.2. The effect of modulation doping is discussed in Sec. 12.6. The electrons in the AlGaAs

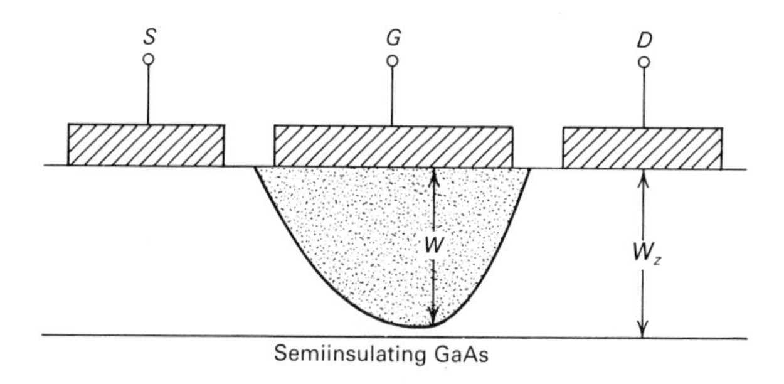

Figure 15.1 Metal semiconductor transistor.

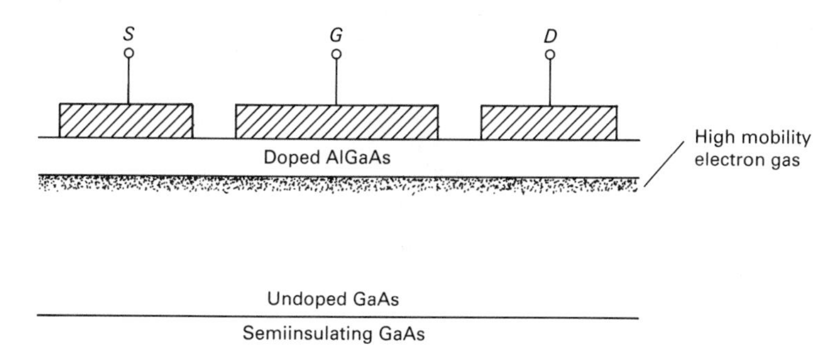

Figure 15.2 The high electron mobility transistor. The AlGaAs is heavily doped, whereas the GaAs is not intentionally doped.

leave their "parent" donors and transfer to the GaAs where they form a high mobility channel. The AlGaAs is then mostly depleted and has the properties of a (not too excellent) insulator. This type of transistor is therefore between MOSFET and MESFET.

The high mobilities in the channel make the device very fast. It should be noted, however, that hot electron effects play a major role and the high low field mobility accounts only for transport very close to the source. Toward the drain the mobility is degraded, while close to the drain the mobility degradation is partly undone by velocity overshoot effects. A typical situation is shown in Figs. 15.3a and 15.3b. The figures show the electron velocity in the GaAs in between the source and drain contact. The top of each figure represents the GaAs-AlGaAs interface.

The values of the electron velocities are indicated by the shading. Bright areas denote very high velocities (up to 6×10^7 cm/s), while dark areas indicate low velocities (below 10^7 cm/s). Such plots (or color-coded movies of this kind if animation is important) are very helpful to see at a glance the major ongoing physical processes. Figures 15.4a and 15.4b show contourplots of the carrier concentration for the switch on-off cycle of the transistor. Again the bright

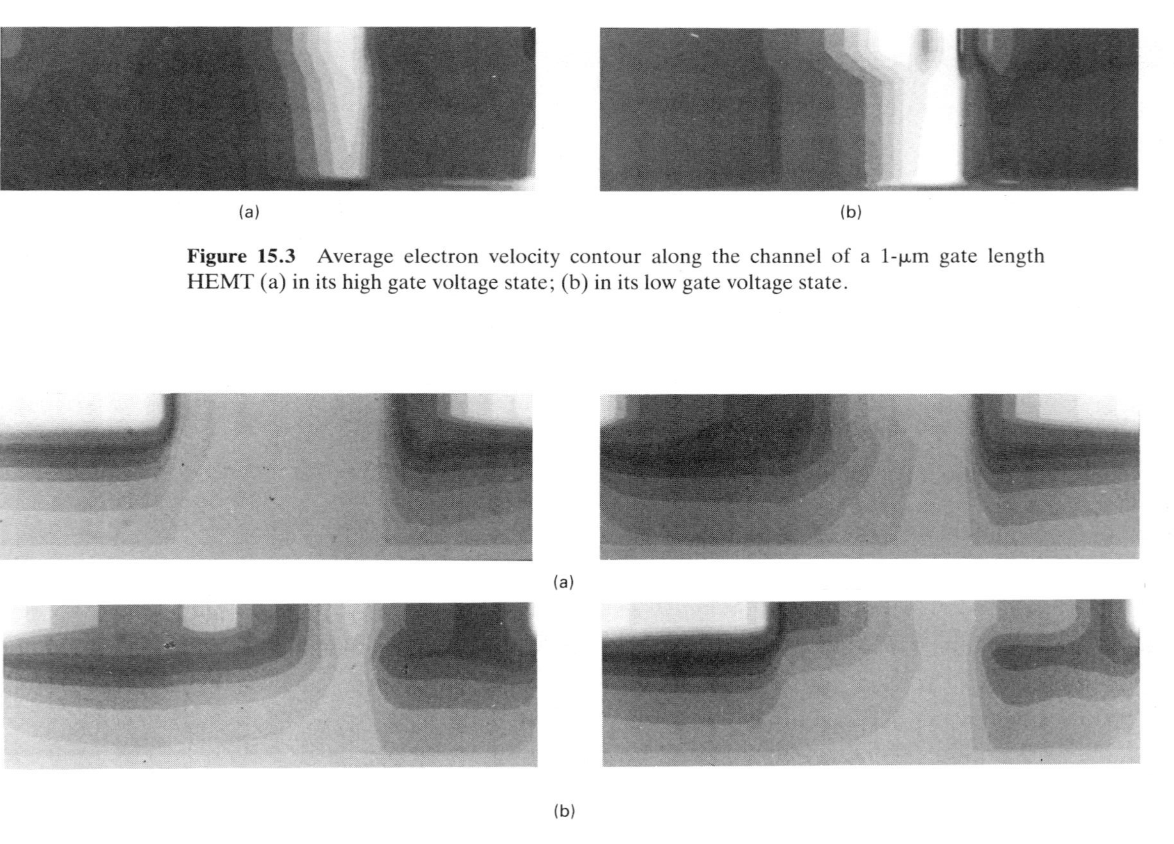

Figure 15.3 Average electron velocity contour along the channel of a 1-μm gate length HEMT (a) in its high gate voltage state; (b) in its low gate voltage state.

Figure 15.4 Electron concentration contours (a) during switch-on of a 1-μm gate length HEMT; (b) during switch-off of a 1-μm gate length HEMT for two different times.

areas denote high values and the dark areas low values, this time of the carrier concentration. Note that switching the transistor off is faster than switching it on. As evident from the figures, the electrons can distribute themselves to the two electrodes (source and drain) in the switch off cycle, while they have to move from source to drain when switching on. Also, as shown, the overshoot is more pronounced when switching off since the electric fields are typically higher during this cycle of operation. Large computational resources are usually necessary to produce the large amount of information that is stored in these figures.

The high electron mobility transistor is between MOSFET (the depleted AlGaAs replaces the insulating SiO_2) and MESFET (the Schottky barrier gate). The theory in its simplest form is identical to that of MOSFET. However, modulation doping and size quantization give rise to very pronounced hot electron effects, as shown by Widiger et al.

MESFET and HEMT are devices of proven usefulness. Below a few devices are described that show new and promising qualities. Their usefulness on a broader scale for applications is not entirely clear. However, these devices are useful from the viewpoint of investigating interesting physical phenomena.

The real space transfer effect can be used to produce a new type of transistor and in fact, represents a new transistor principle (in addition to bipolar injection and the field effect). The principle, proposed by the author and realized by Kastalsky and Luryi is shown in Fig. 15.5.

AlGaAs is grown on top of heavily doped GaAs and lightly doped n-GaAs on top of the AlGaAs. This GaAs layer is again topped by doped AlGaAs, which supplies the electrons. Contacts are alloyed to the sandwiched

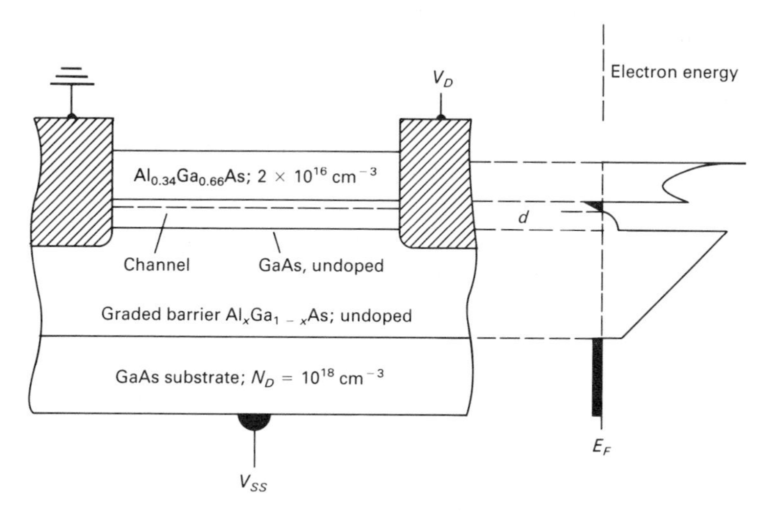

Figure 15.5 Principle of a real space transfer transistor. (After Kastalsky and Luryi, © 1983 IEEE.)

n-GaAs layer, which takes now the role of a glow cathode in a vacuum tube. Of course, the crystal lattice stays cold and only the electrons are heated. The drain voltage V_D heats the electrons in the n-GaAs, which can then transfer (via the real space transfer mechanism) over the AlGaAs barrier to the conducting substrate (the analogy of the anode) and a current, due to the source substrate voltage V_{SS}, is switched on.

The effect is very fast since these electrons can be heated within the energy relaxation time, which in GaAs approaches ~1 picosecond, whereas the heating of the glow cathode in a vacuum tube is slow because the lattice of the metal is heated. Transistor structures with astonishing high quality and performance have been made by Kastalsky.

The production of these new transistor types is possible as a consequence of the new crystal-growth techniques: molecular beam epitaxy introduced by A. Y. Cho as well as metal organic chemical vapor deposition introduced by Manasievit and modified by Dupuis. These crystal-growth techniques and variations of them permit the growth of essentially monatomic layers of semiconductors on top of each other. Therefore, one can vary the crystal boundaries and the electronic boundary conditions on an angstrom level. It is precisely this variability of the boundary conditions that enables the construction of new devices.

The transistor structures discussed above involve still fairly thick layers. Most recently, interest has focused on structures involving resonance tunneling barriers, as shown in Fig. 15.6.

Such a structure can achieve negative differential resistance since the tunneling current exhibits a maximum if the incident electron has the energy of the size quantized state between the two barriers (transmission resonance analogous of the Fabry-Perot effect). The precise theory of such a simple diode is not simple at all due to the fact that the self-consistent field of the electrons must be included. This makes the effect different from the Fabry-Perot effect. The electrons also undergo transitions from a three-dimensional system (contact) to a two-dimensional system (in between barriers) back to a three-dimensional electron gas system. Numerical simulations are necessary to understand the details of this effect. It is, however, known from experiments that the effect is very fast and Terraherz performance has been proven.

Numerous other devices involving thin layers are currently under consid-

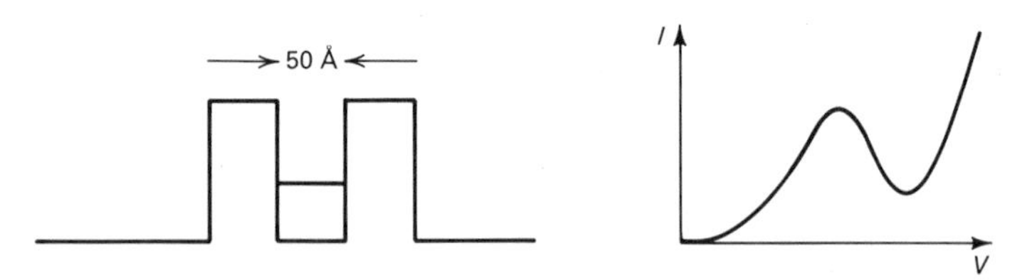

Figure 15.6 Double barrier structure, the electronic analogy to a Fabry-Perot interferometer, and corresponding I-V characteristic of a diode combining such barriers.

eration. We mentioned only the quantum well laser, a heterojunction laser with very thin layer width. As shown in Fig. 10.3, the size quantization effect gives rise to an increased photon energy. There are many other advantages of quantum well layers in addition to the variable photon energy such as the improved laser threshold. A particularly creative discussion of these effects has been given by Holonyak.

These new opportunities should be a lesson to those who have declared the semiconductor field mature some twenty years ago and have announced that new technologies will never work (e.g., the semiconductor laser) and who try to convince us that new inventions will not be made in the semiconductor field and, if made, will not sell. The field of semiconductor electronics and optoelectronics is too rich with respect to theoretical possibilities (band structure, heterolayers, multitude of electron-phonon-photon, etc., interactions), too rich in its technological possibilities (molecular beam epitaxy and other crystal-growth techniques), too rich in useful materials, and too needed in its applications to have a fate like the vacuum tube or the steel industry, for instance. It takes young talent to form the symbiosis of science and engineering which is necessary for milestone progress and sometimes it takes new impetus from other areas such as computer architecture or chemistry. It also takes the trust and belief of management that research in new areas is important to "produce real things well." In this respect, thin heterolayers offer a new world of research. Several other areas will have a major impact on semiconductor research as well.

Optical communications (fiberoptics) currently have a major influence on "semiconductor thinking" and this influence may even broaden to include, for example, the optically interconnected computer or an optoelectronic chip. The first models of neural networks and certain areas of brain research may find a helpful associate in semiconductor models. These more exotic areas may develop as we proceed to use the maturing silicon technology and to develop III–V compound technology.

The author hopes that the principles described in this book will be helpful to reseachers in all of these endeavors of present and future semiconductor devices from their basic physics to their final function.

ADDITIONAL READINGS

HESS, K., "Aspects of High-Field Transport in Semiconductor Heterolayers and Semiconductor Devices," in *Advances in Electronics and Electron Physics,* vol. 59, ed. P. W. Hawkes. New York: Academic Press, 1982.

KASTALSKY, A., and S. LURYI, *IEEE Electron Device Letters,* EDL-4 (1983), 334.

KIZILYALLI, I. C., K. HESS, J. L. LARSON, and D. WIDIGER, *IEEE Transactions on Electron Devices,* ED-33 (1986), 1427.

WIDIGER, D., K. HESS, and J. J. COLEMAN, *IEEE Electron Device Letters,* EDL-5 (1984), 266.

TUNNELING AND THE GOLDEN RULE

Consider the following model Hamiltonian:

$$H_M = \begin{matrix} H_L & \text{for } r \epsilon R_L \\ H_R & \text{for } r \epsilon R_R \end{matrix} \tag{A.1}$$

where R_L denotes a left side region and R_R a right side region. In our example of the tunneling problem in Fig. 1.2, R_R would correspond to the region out of the well, that is

$$H_R = -\frac{\hbar^2 \nabla^2}{2m} - eFz \tag{A.2}$$

while H_L would be given by the Hamiltonian in the well:

$$H_L = -\frac{\hbar^2 \nabla^2}{2m} + WPE \tag{A.3}$$

where WPE is the well potential energy.

The total Hamiltonian is

$$H = -\frac{\hbar^2 \nabla^2}{2m} - eFz + WPE \tag{A.4}$$

which is apparently not entirely equal to the sum $H_L + H_R$. Assume that we know solutions of the partial Hamiltonians H_L and H_R (in ordinary perturbation theory only one partial solution needs to be known). We denote these solutions by

$$H_L \psi_0 = E_0 \psi_0 \tag{A.5}$$

and

$$H_R \psi_\nu = E_\nu \psi_\nu \tag{A.6}$$

where ν is an integer. In other words, we assume we know the solution for the electron in a given state at the left side and in a set of states on the right side. We now attempt to obtain an approximate solution of H by expanding the wave function that satisfies H in terms of the known wave functions:

$$\psi(r,t) = \psi_0 e^{-iE_0 t/\hbar} + \sum_{\nu=1}^{\infty} a_\nu(t)\psi_\nu(r)e^{-iE_\nu t/\hbar} \tag{A.7}$$

The expansion coefficients $a_\nu(t)$ have to vanish at $t = 0$ since we assume that at $t = 0$ the electron wave function is ψ_0.

The Hamiltonian needs to be brought into a form that permits perturbational solutions. In the example of the well emitting electrons, the perturbation is given by the electric field and therefore we can write the perturbing part H' of the Hamiltonian as

$$H' = H - H_L = -eFz \tag{A.8}$$

As we know from Eq. (1.21), or from insertion of Eq. (A.7) into the Schrödinger equation, all that is needed to solve quantum problems to first order is the matrix element, which we can write in shorthand notation as

$$\langle \nu | H - H_L | 0 \rangle \equiv \int_{V_{\mathrm{ol}}} \psi_\nu^* (H - H_L)\psi_0 \, d\mathbf{r} \tag{A.9}$$

The matrix element can, of course, be easily calculated if ψ_ν and ψ_0 are known, which they are for the case of the perturbed quantum well. There is, however, a possibility of rewriting Eq. (A.9) in a more general and useful form. We will outline this "rewriting" below. It is not easy to understand why things should be done just that way, but the derivation is clear and the final equations are very useful and the fact that the method is due to Bardeen says, I believe, everything.

We replace H in Eq. (A.9) by our model Hamiltonian of Eq. (A.1) and write

$$\langle \nu | H - H_L | 0 \rangle \approx \int_{R_R} \psi_\nu^* (H_M - H_L)\psi_0 \, d\mathbf{r} \tag{A.10}$$

The integration now goes only over the right half space since $H_M - H_L$ vanishes to the left. $H_M - H_R$ is identical to zero in R_R and we therefore can write

$$\langle \nu | H - H_L | 0 \rangle \approx \int_{R_R} [\psi_\nu^* (H_M - H_L)\psi_0 - \psi_0 (H_M - H_R)\psi_\nu^*] d\mathbf{r} \tag{A.11}$$

Using Eqs. (A.5) and (A.6), we have

$$\langle v|H - H_L|0\rangle \approx \int_{R_R} [\psi_v^* H_M \psi_0 - \psi_0 H_M \psi_v^* - \psi_v^* E_0 \psi_0 + \psi_0 E_v \psi_v^*] d\mathbf{r} \tag{A.12}$$

We are later only interested in elastic tunneling processes and therefore $\psi_0 E_v \psi_v^* = \psi_v^* E_0 \psi_0$, as demanded by the energy-conserving δ function in Eq. (1.22). Under most circumstances of interest to us, the model Hamiltonian H_M can be formally separated into two factors:

$$H_M = -\frac{\hbar^2 \nabla^2}{2m} + FPE \tag{A.13}$$

where FPE is a function of $\mathbf{r}$ representing the potential energy. Therefore $\psi_v^* FPE \psi_0 - \psi_0 FPE \psi_v^*$ also vanishes and we have

$$\langle v|H - H_L|0\rangle \approx \int_{R_R} \left[\psi_v^* \left(\frac{-\hbar^2 \nabla^2}{2m}\right) \psi_0 - \psi_0 \left(\frac{-\hbar^2 \nabla^2}{2m}\right) \psi_v^*\right] d\mathbf{r} \tag{A.14}$$

The tunneling problem is therefore solved as soon as we know ψ_0 and ψ_v.

The integral in Eq. (A.14) can be evaluated using Green's theorem. Since we discuss later only one-dimensional applications, we reproduce the equations in one dimension only:

$$\begin{aligned} \langle v|H - H_L|0\rangle &\approx \frac{-\hbar^2}{2m} \int_0^\infty \left[\psi_v^* \frac{\partial^2}{\partial z^2} \psi_0 - \psi_0 \frac{\partial^2}{\partial z^2} \partial\psi_v^*\right] dz \\ &= \frac{-\hbar^2}{2m} \int_0^\infty \frac{\partial}{\partial z} \left[\psi_v^* \frac{\partial}{\partial z} \psi_0 - \psi_0 \frac{\partial}{\partial z} \psi_v^*\right] dz \end{aligned} \tag{A.15}$$

where we have assumed the right side to be in the interval $0 < z < \infty$. This gives

$$\langle v|H - H_L|0\rangle \approx -\frac{\hbar^2}{2m} \left[\psi_v^*(0) \left.\frac{\partial \psi_0}{\partial z}\right|_0 - \psi_0(0) \left.\frac{\partial \psi_v^*}{\partial z}\right|_0\right] \tag{A.16}$$

$\left.\frac{\partial \psi_0}{\partial z}\right|_0$ and $\left.\frac{\partial \psi_v^*}{\partial z}\right|_0$ denote the derivatives at $z = 0$, that is, at the point (or surface) separating the regions R_L and R_R. Bardeen's model of tunneling[1] differs from the one described above by a slight generalization. He lets R_R and R_L overlap in the tunneling region (the rest of the derivation is similar). The functions in Eq. (A.16) can then be evaluated at any point of the overlapping region with the same result.

Below we illustrate the use of Eq. (A.16) for the case of a simple rectangular tunneling barrier, as shown in Fig. A.1.

We need to find ψ_0 and ψ_v to obtain the matrix element. Since the problem corresponding to Fig. A.1 is symmetric, the wave functions will be identical except for the sign of the space coordinate z and it suffices therefore

[1] J. Bardeen, *Physical Review Letters,* 6 (1961), 57.

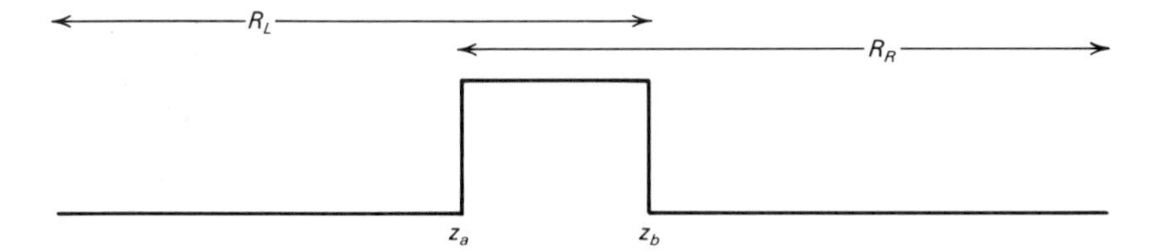

Figure A.1 Tunneling barrier with left and right space division indicated.

to calculate ψ_0. Writing formally for $-E_0 + FPE$ the term $\hbar^2\bar{k}_z/2m$ (this is the definition of $\bar{k}_z$), the Schrödinger equation reads

$$\frac{\partial^2\psi_0}{\partial z^2} + \bar{k}_z^2\psi_0 = 0 \tag{A.17}$$

One has to find solutions for $z < z_a$ and for $z_0 \leqslant z \leqslant z_b$ and then connect the solutions. This connection is a difficult problem and is discussed in most quantum mechanics courses. Thus we give only a simplified result here:

$$\psi_0 = \sqrt{\frac{2}{L_a}}\cos\left(\int_z^{z_a} \bar{k}_z\, dz - \frac{\pi}{4}\right) \quad \text{for } z < z_a \tag{A.18a}$$

and

$$\psi_0 = \frac{1}{\sqrt{2L_a}}\exp\left(-\int_{z_a}^{z} |\bar{k}_z| dz\right) \quad \text{for } z_0 \leqslant z \leqslant z_b \tag{A.18b}$$

where L_a is the length of the crystal left to z_a. Below we will use L_a also for the crystal length to the right of z_b. Also we have $\bar{k}_{za} = \bar{k}_{zb}$ because of the symmetry of the problem. Using Eqs. (A.18a) and (A.18b), one obtains the square of the matrix element from Eq. (A.16), which is

$$|\langle v|H - H_L|0\rangle|^2 \approx \left(\frac{\hbar^2}{2m}\right)^2 \frac{|\bar{k}_{z_a}|^2}{L_a{}^2}\exp\left(-2\int_{z_a}^{z_b} |\bar{k}_z| dz\right) \tag{A.19}$$

A generalization of the approach for z-dependent crystal structure (allloys!) has been given by Harrison.[2]

The calculation for a three-dimensional barrier proceeds along the same lines. For specular flat parallel barriers, one obtains the same results as in one dimension. In addition, one notices that the wave vector component parallel to the barrier is conserved in the tunneling process [compare with Eq. (1.28)]. We note also that tunneling treated in this form can easily be included in Monte Carlo simulations as an additional scattering mechanism.

[2] W. Harrison, *Physical Review*, 123 (1961), 85.

THE ONE BAND APPROXIMATION

The proof of Eq. (4.17a) rests on the following identity:

$$E(-i\nabla)\psi(\mathbf{k}, \mathbf{r}) = E(\mathbf{k})\psi(\mathbf{k}, \mathbf{r}) \tag{B.1}$$

which can be shown by Fourier expanding E in terms of lattice vectors, which, according to Eq. (3.11), is always possible:

$$E(\mathbf{k}) = \sum_l E_l e^{i\mathbf{R}_l \cdot \mathbf{k}} \tag{B.2}$$

Therefore

$$\begin{aligned} E(-i\nabla)\psi(\mathbf{k}, \mathbf{r}) &= \sum_l E_l e^{\mathbf{R}_l \cdot \nabla}\psi(\mathbf{k}, \mathbf{r}) \\ &= \sum_l E_l(1 + \mathbf{R}_l \cdot \nabla + \tfrac{1}{2}(\mathbf{R}_l \cdot \nabla) + {}^2 \ldots)\psi(\mathbf{k}, \mathbf{r}) \end{aligned} \tag{B.3}$$

We notice now that the term $(1 + \mathbf{R}_l \cdot \nabla + \frac{1}{2}(\mathbf{R}_l \cdot \nabla)^2 + \ldots)\psi(\mathbf{k}, \mathbf{r})$ is just the Taylor expansion of the function $\psi(\mathbf{k}, \mathbf{r} + \mathbf{R}_l)$, and therefore

$$E(-i\nabla)\psi(\mathbf{k}, \mathbf{r}) = \sum_l E_l \psi(\mathbf{k}, \mathbf{r} + \mathbf{R}_l) \tag{B.4}$$

which, with the help of Bloch's theorem, becomes

$$E(-i\nabla)\psi(\mathbf{k},\mathbf{r}) = \sum_l E_l e^{i\mathbf{R}_l\cdot\mathbf{k}}\psi(\mathbf{k},\mathbf{r}) = E(\mathbf{k})\psi(\mathbf{k},\mathbf{r}) \tag{B.5}$$

and the proof is complete.

As evident, then, from Eq. (B.1), the operator $E(-i\nabla)$ replaces the operator $-\hbar^2/2m\ \nabla^2 + V(\mathbf{r})$ in the Schrödinger equation, which can now be written without the appearance of the crystal potential $V(\mathbf{r})$.

The addition of an external potential presents no problem, as can be seen from the following arguments:

Let the solution of the Schrödinger equation, which includes V_{ext}, be a combination of all Bloch functions of all bands with index n:

$$\psi(\mathbf{k},\mathbf{r}) = \sum_n a_n\,\psi_n(\mathbf{k},\mathbf{r}) \tag{B.6}$$

We then have

$$\left(-\frac{\hbar^2}{2m}\nabla^2 + V(\mathbf{r}) + V_{ext}\right)\sum_n a_n\,\psi_n(\mathbf{k},\mathbf{r}) = \sum_n a_n\left(-\frac{\hbar^2}{2m}\nabla^2 + V(\mathbf{r}) + V_{ext}\right)\psi_n(\mathbf{k},\mathbf{r})$$

and using the above identity

$$= \sum_n a_n\,(E_n(-i\nabla) + V_{ext})\psi_n(\mathbf{k},\mathbf{r}) \tag{B.7}$$

If, as assumed, only the energy of one band is important and interband transition can be neglected, then $E_n(-i\nabla)$ can be replaced by a single energy $E(-i\nabla)$ and (B.7) becomes

$$(E(-i\nabla) + V_{ext})\sum_n a_n\psi_n(\mathbf{k},\mathbf{r}) = (E(-i\nabla) + V_{ext})\psi(\mathbf{k},\mathbf{r})$$

which completes the proof. The condition of one band being important only means in the semiconductors GaAs, Si, and Ge that the external electric fields are smaller than about 10^6 V/cm. The rigorous proof of this fact is involved.

TEMPERATURE DEPENDENCE OF THE BAND STRUCTURE

It is convenient to separate the temperature dependence of the energy band into contributions arising from the volume change and contributions related to the lattice vibrations.

The volume change is easily accounted for. One only has to insert the temperature-dependent lattice constant into the pseudopotential calculation, Eq. (4.15). The change in lattice constant can be calculated from the volume expansion coefficient $\partial V_{ol}/\partial T$. This coefficient and the pressure coefficient $\partial P/\partial V_{ol}$ (P is the pressure) can be found from data in the *Handbook of Chemistry and Physics*. Since the change of the energy gap with pressure is known from experiments, the change of the energy gap with temperature due to the volume increase can be calculated from

$$\Delta E_G^{V_{ol}} = (\partial E_G/\partial P)(\partial P/\partial V_{ol})(\partial V_{ol}/\partial T)\Delta T \tag{C.1}$$

There is also a change of the value of the energy gap due to the lattice vibrations that is more difficult to calculate. Fan used second-order perturbation theory, including the potential of the phonons [as given, for example, by Eq. (8.23)]. The energy change due to the phonons ΔE_G^{phl} is then given by

$$\Delta E_G^{phl} = \sum_{n,q} \frac{|M_q|^2}{E_m(\mathbf{k}) - E_n(\mathbf{k}+\mathbf{q}) \pm \hbar\omega_q} \tag{C.2}$$

Equation (C.2) follows immediately from Eq. (1.18) with the phonon

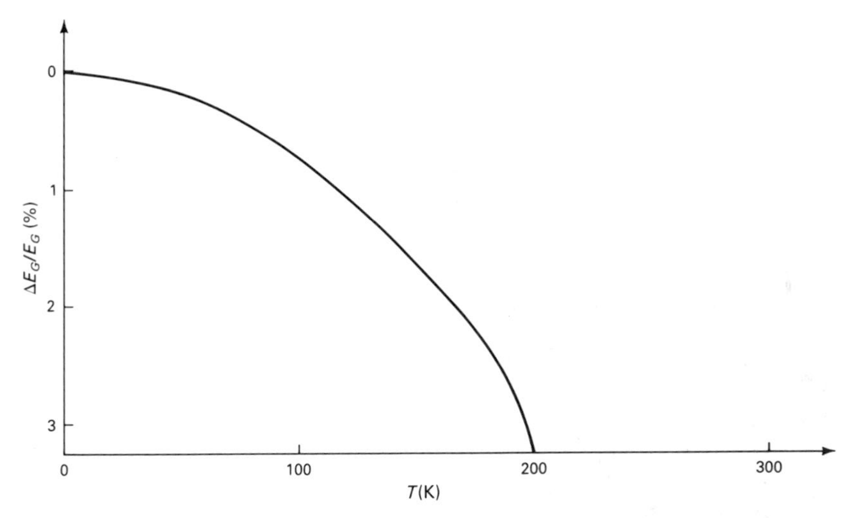

Figure C.1 Typical form of the relative change of energy gap with temperature for III–V compound and group IV (Si, Ge) semiconductors. The energy gap decreases with temperature.

energy included in the denominator. The plus sign is appropriate for phonon absorption and the minus sign for phonon emission. In practice, only a single band needs to be considered—that is, $n = m$ [=conduction (valence) band]—and the summation over q is converted into an integration according to Eq. (6.8). Values for M_q for the various phonon coupling mechanisms are given in Chap. 8.

Another contribution to the change of the energy gap as a consequence of the lattice vibrations arises from the influence of vibrations on the structure factor $e^{-iK\tau}$ in Eq. (4.15). Because of the vibrations this structure factor is "washed out." Antonchik (and later Brooks and Yu) proposed, therefore, to multiply the factor $e^{-iK\tau}$ by $\exp(-\frac{1}{6}|K|^2\langle u^2\rangle)$, where $\langle u^2\rangle$ is the mean square displacement of the atom in question [see Eq. (2.14) for the definition of u]. To obtain the change in energy gap ΔE_G^{phII} due to this effect, one has to calculate $E(\mathbf{k})$ with the temperature-dependent structure factor according to the method described in Chap. 4.

Experimental results of changes of the energy states with temperature can be found in the *Handbook on Semiconductors* (1980).

In comparing the experimental results with the theory outlined above, one finds that ΔE_G^{phII} gives the most important contribution for III–V compound and group IV semiconductors (which are the most important materials from a device point of view). ΔE_G^{phI} and $\Delta E_G^{V_{ol}}$ contribute only about 20 percent to the total change in the energy gap. A typical result for the total change

$$\Delta E_G = \Delta E_G^{V_{ol}} + \Delta t_G^{phI} + \Delta E_G^{phII} \tag{C.3}$$

is plotted in Fig. C.1.

HALL EFFECT AND MAGNETO RESISTANCE FOR SMALL MAGNETIC FIELDS

The precise theory of the classical Hall effect (the quantum Hall effect will not be treated here) involves solution of the Boltzmann equation using the method of moments as described in Eqs. (11.1)–(11.5). The force term is now given by

$$-\frac{e}{\hbar}(\mathbf{F}+\mathbf{v}\times\mathbf{B})\nabla_{\mathbf{k}}f \tag{D.1}$$

where $e(\mathbf{F}+\mathbf{v}\times\mathbf{B})$ replaces the force $\mathbf{F}_0$ in Eq. (9.12). This term complicates the method of moments significantly and we have to proceed slightly different than in Chap. 11. To facilitate the notation we perform the calculation in two dimensions. This treatment applies, for example, to the two-dimensional electron gas in MOS transistors.

The treatment in three dimensions can be done in totally analogous fashion using polar coordinates instead of the cylindrical coordinates which are used below. For the $E(\mathbf{k})$ relation we choose a parabolic band. However, for the sake of generality, we let the surface of constant energy be an ellipse with masses m_x and m_y along the main axes [see Eq. (4.24)]. Our coordinate system for the wave vector $\mathbf{k}$ is chosen along these main axes. In the method of moments one multiplies the Boltzmann equation by $\mathbf{k}$ (to obtain the current) and integrates over the cylindrical coordinates $d\phi k dk$. We integrate first over angle ϕ and define

$$dj_i \equiv \frac{e}{\sqrt{m_i^*}} \int_0^{2\pi} k_i f d\phi \qquad i = x, y \tag{D.2}$$

and

$$dn \equiv \int_0^{2\pi} f d\phi \tag{D.3}$$

The differential notation dj and dn means that integration over kdk (the energy) is still necessary to obtain macroscopic current and carrier concentration. Assuming now the magnetic field in z direction, one obtains for the differential current component in x direction:

$$\frac{dj_x}{\tau_{\text{tot}}} = E \frac{e}{m_x^*} \frac{F_x}{kT_c} dn + \frac{e}{m_x} B_z dj_y \tag{D.4}$$

and a similar equation for dj_y. τ_{tot} is the total scattering time, as given, for example, in Eq. (8.19).

In vector and matrix notation we obtain, then, the following differential current:

$$d\mathbf{j} = \begin{bmatrix} \bar{\mu}_x - B_z^2 \bar{\mu}_x^2 \bar{\mu}_y & \bar{\mu}_y \bar{\mu}_x B_z \\ -B_z \bar{\mu}_y \bar{\mu}_x & \bar{\mu}_y - B_z^2 \bar{\mu}_x \bar{\mu}_y^2 \end{bmatrix} \mathbf{F}\, Eedn/kT_c \tag{D.5}$$

where $\bar{\mu}_i = e\tau_{\text{tot}}/m_i^*$ and T_c is the electron temperature (which in principle also depends on the electric and magnetic fields).

The terms proportioned to B_z^2 are responsible for the magneto resistance, that is, the increase in sample resistance due to the magnetic field. In the explanation of the classical Hall effect, these terms are usually ignored, as B_z is small and B_z^2 therefore negligible. If the sample is elongated in x direction and no net current flows in y (and z) directions, Eq. (D.5) gives the two equations

$$dj_x = \int_0^\infty dkk\, \bar{\mu}_x F_x Eedn/kT_c \tag{D.6a}$$

and

$$\int_0^\infty dkk\, B_z \bar{\mu}_y \bar{\mu}_x F_x Eedn/kT_c = \int_0^\infty dkk\, \bar{\mu}_y F_y Eedn/kT_c \tag{D.6b}$$

which gives the Hall field F_y:

$$F_y = F_x B_z \frac{\int_0^\infty dkk\, \bar{\mu}_y \bar{\mu}_x E\, dn}{\int_0^\infty dkk\, \bar{\mu}_y E\, dn} \tag{D.7}$$

With the definitions of averages in two dimensions, this is equivalent to

$$F_y = F_x B_z \mu_x \frac{\langle \tau_{tot}^2 \rangle}{\langle \tau_{tot} \rangle^2} \tag{D.8a}$$

and

$$\mu_x = \frac{e}{m_x^*} \langle \tau_{tot} \rangle \equiv \langle \bar{\mu}_x \rangle \tag{D.8b}$$

We therefore obtain Eq. (8.12) modified by a "statistics factor" $\langle \tau_{tot}^2 \rangle / \langle \tau_{tot} \rangle^2$, which is equal to one if τ_{tot} is independent of energy (or k). This result is general and holds also in three dimensions with the averages being three-dimensional averages, as in Eq. (9.42). Notice that for large magnetic fields ($(\mu B)^2 \gtrsim 1$), a quantum treatment is necessary.

THE POWER BALANCE EQUATION FROM THE METHOD OF MOMENTS

In order to derive the power balance equation, we need to take the third moment of the Boltzmann transport equation. Beginning with Eq. (11.5b), we let $Q(\mathbf{k}) = E(\mathbf{k}) = \hbar k^2/2m^*$. We will disregard for the moment the time derivative. Equation (11.5b) then becomes

$$\frac{\hbar}{m^*}\nabla\{n\langle E\mathbf{k}\rangle\} + \frac{e}{\hbar}\mathbf{F}\cdot n\langle\nabla_{\mathbf{k}}E(\mathbf{k})\rangle$$
$$= \frac{1}{4\pi^3}\int d\mathbf{k}E(\mathbf{k})\left.\frac{\partial f_0}{\partial t}\right|_{\text{coll}} - \frac{1}{4\pi^3}\int d\mathbf{k}\frac{f_1}{\tau_{\text{tot}}(\mathbf{k})}E(\mathbf{k}) \tag{E.1}$$

We now perform the integrals in Eq. (E.1): The first term on the left is given by

$$\langle E(\mathbf{k})\mathbf{k}\rangle = \int \frac{\hbar^2k^2}{2m^*}\mathbf{k}f\,d\mathbf{k} = \frac{1}{n}\int\frac{\hbar^2k^2}{2m^*}\mathbf{k}(f_0+f_1)d\mathbf{k} \tag{E.2}$$

The term $k^2\mathbf{k}f_0$ is odd and does not contribute to the integral, therefore

$$\langle E(\mathbf{k})\mathbf{k}\rangle = \frac{1}{n}\frac{\hbar^2}{2m^*}\int k^2\mathbf{k}f_1\,d\mathbf{k} = \frac{1}{n}\int E(\mathbf{k})\mathbf{k}f_1\,d\mathbf{k} \tag{E.3}$$

The second term on the right-hand side of Eq. (E.1) represents the average velocity, while the first term, $\frac{1}{4\pi^3}\int d\mathbf{k}\, E(\mathbf{k})\left.\frac{\partial f_0}{\partial t}\right|_{\text{coll}} \equiv -nB(T_c)$, represents the energy loss to the lattice for a carrier temperature T_c. The last remaining integral in Eq. (E.1) vanishes if we assume that τ_{tot} is a function of energy only, since then the integrand is an odd function. Substituting these results back into (E.1), one obtains[1]

$$\frac{\hbar}{m^*}\nabla\left\{\int E(\mathbf{k})\mathbf{k}f_1\, d\mathbf{k}\right\} + e\mathbf{F}\cdot(n\langle\mathbf{v}\rangle) = -nB(T_c) \tag{E.4}$$

which is equivalent to

$$\mathbf{j}\cdot\mathbf{F} = nB(T_c) + \nabla\cdot\mathbf{S}(T_c) \tag{E.5}$$

where we have used the definition of the current density and the energy flux: $\mathbf{S}(T_c) = \frac{\hbar}{m^*}\int d\mathbf{k}\, E(\mathbf{k})\mathbf{k}f_1$.

To evaluate this flux term we have to assume $f_1 \ll f_0$. In order to simplify calculations, we proceed with a one-dimensional analysis. We can write the linearized Boltzmann equation as

$$f_1 = \tau(E)\left\{\frac{eF\hbar k}{m^*}\frac{\partial f_0}{\partial E} - \frac{\hbar k}{m^*}\frac{\partial f_0}{\partial x}\right\} \tag{E.6}$$

Substiututing f_1 into the equation for $\mathbf{S}(T_c)$, one has

$$S(T_c) = \frac{\hbar}{3m^*}\int d\mathbf{k}E(k)k\tau(E)\left\{\frac{eF\hbar k}{m^*}\frac{\partial f_0}{\partial E} - \frac{\hbar k}{m^*}\frac{\partial f_0}{\partial x}\right\} \tag{E.7}$$

The integrations over $\mathbf{k}$ are still performed in three dimensions since the one-dimensional density of states diverges as $E \to 0$ and would therefore introduce nonphysical results. We further use Eqs. (11.6) and (11.7), as well as Eq. (9.40). After some algebra, one obtains

$$S(T_c) = \frac{-2j}{3m^*n\mu kT_c}\int_{①}\tau(E)E^2 f_0 d\mathbf{k} + \frac{2e}{3m^*n\mu kT_c}\int_{②}\tau(E)E^2 f_0\frac{\partial}{\partial x}(Dn)d\mathbf{k} - \frac{2}{3m^*}\int_{③}\tau(E)E^2\frac{\partial f_0}{\partial x}d\mathbf{k} \tag{E.8}$$

The three integrals ①, ②, ③ are evaluated to be

① $$\frac{-2j}{3m^*n\mu kT_c}\int \tau E^2 f_0 d\mathbf{k} = \frac{-2j}{3m^*\mu kT_c}\langle\tau E^2\rangle = \frac{-j\langle\tau E^2\rangle}{e\langle\tau E\rangle} \tag{E.9a}$$

[1] R. Stratton, *Physical Review*, 126(1962), 2002.

$$\textcircled{2}\ \frac{2e}{3m^*n\mu kT}\int \tau E^2 f_0 \frac{\partial}{\partial x}(Dn)\,d\mathbf{k} = \frac{1}{n\langle\tau E\rangle}\int \tau E^2 f_0 \frac{\partial}{\partial x}\left(\frac{2n}{3m^*}\langle\tau E\rangle\right)d^3\mathbf{k}$$

$$= \frac{\langle\tau E^2\rangle}{\langle\tau E\rangle}\frac{2}{3m^*}\frac{\partial}{\partial x}(n\langle\tau E\rangle)$$

$$= \frac{\langle\tau E^2\rangle}{\langle\tau E\rangle}\frac{2}{3m^*}\frac{\partial}{\partial T_c}(n\langle\tau E\rangle)\frac{\partial T_c}{\partial x} \qquad \text{(E.9b)}$$

$$= \frac{\langle\tau E^2\rangle}{\langle\tau E\rangle}\frac{2}{3m^*}\left[\frac{\langle\tau E^2\rangle}{kT_c} - \frac{3}{2}\langle\tau E\rangle\right]\frac{1}{T_c}\frac{\partial T_c}{\partial x}$$

$$= \frac{2n}{3m^*}\left[\frac{\langle\tau E^2\rangle^2}{\langle\tau E\rangle kT_c} - \frac{3}{2}\langle\tau E^2\rangle\right]\frac{1}{T_c}\frac{\partial T}{\partial x}$$

$$\textcircled{3}\qquad \frac{2}{3m^*}\int \tau E^2 \frac{\partial f_0}{\partial x}\,d\mathbf{k} = \frac{2}{3m^*}\int \tau E^2 \frac{\partial f_0}{\partial T}\frac{\partial T_c}{\partial x}\,d\mathbf{k}$$

$$= \frac{2}{3m^*}\int \tau E^2 f_0\left[\frac{E}{kT_c^2} - \frac{3}{2T_c}\right]\frac{\partial T_c}{\partial x}\,d\mathbf{k} \qquad \text{(E.9c)}$$

$$= \frac{2n}{3m^*}\left[\frac{\langle\tau E^3\rangle}{kT_c} - \frac{3}{2}\langle\tau E^2\rangle\right]\frac{1}{T_c}\frac{\partial T_c}{\partial x}$$

Putting all these terms together, we have

$$S(T_c) = \frac{-j}{e}\frac{\langle\tau E^2\rangle}{\langle\tau E\rangle} + \frac{2n}{3m^*}\left[\frac{\langle\tau E^2\rangle^2}{\langle\tau E\rangle kT_c} - \frac{3}{2}\langle\tau E^2\rangle\right]\frac{1}{T_c}\frac{\partial T_c}{\partial x}$$

$$- \frac{2n}{3m^*}\left[\frac{\langle\tau E^3\rangle}{kT} - \frac{3}{2}\langle\tau E^2\rangle\right]\frac{1}{T_c}\frac{\partial T_c}{\partial x} \quad \text{and} \qquad \text{(E.10)}$$

$$S(T_c) = \frac{-j}{e}\frac{\langle\tau E^2\rangle}{\langle\tau E\rangle} + \frac{2n}{3m^*kT_c}\left[\frac{\langle\tau E^2\rangle^2}{\langle\tau E\rangle} - \langle\tau E^3\rangle\right]\frac{1}{T_c}\frac{\partial T_c}{\partial x}$$

F

THE SELF-CONSISTENT POTENTIAL AT A HETEROJUNCTION (QUANTUM CASE)

This appendix describes a block diagram of the self-consistent calculations that are necessary to include size quantization effects at a heterolayer interface. References to important papers are also given, especially for the complications that arise in connection with many valley problems (X valleys in silicon).

To compute the energy eigenvalues of Eq. (12.32) one needs to go through an iterative procedure and therefore starts with an initial guess of the solution. A good initial guess can be obtained by using a triangular approximation for the potential $\phi(z)$. The solution of the Schrödinger equation for triangular potentials is well known and has been described, for example, by Ando, Fowler, and Stern.[1] From this treatment one obtains an initial guess of Fermi energy, subband energies, electron concentration $n(z)$, and wave functions (in the effective mass approximation). Using this initial guess, the Poisson equation is then solved in the three regions: GaAs, undoped AlGaAs (undoped space layer of 50–150 Å at the interface), and doped AlGaAs. The finite difference method can be used, which gives a tridiagonal matrix equation that is easily solved (even for very large matrices). The resulting potential (the contributions of exchange and correlation added) is then used to solve the Schrödinger equation, which is accomplished by the Numerov process, for

[1] T. Ando, A. B. Fowler, and F. Stern, "Electronic Properties of Two-Dimensional Systems," *Review of Modern Physics,* 54 (1982), 466.

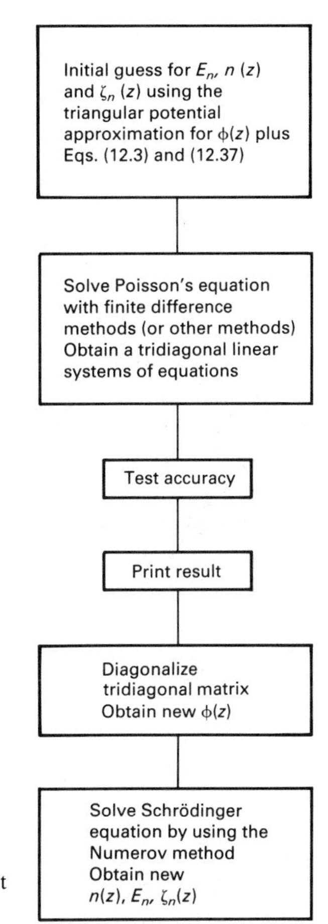

Figure F.1 Flowchart of self-consistent solution for heterolayer potentials.

example. A good review on this process and on computer solutions of the Schrödinger equation has been given by Chow.[2] This combined gives the block diagram in Fig. F.1.

While the formalism described above works for GaAs, silicon needs further considerations because of the six equivalent X valleys (instead of the one Γ valley in the case of GaAs).

Stern and Howard[3] have shown that the subbands have ellipses (circles) as lines of constant energy in the interface plane which are obtained by the intersection of the interface plane with the ellipsoids. If the effective mass perpendicular to the interface is the same, all ellipses are equivalent and belong to the same energy eigenvalue. If the perpendicular masses differ, the ellipses with higher mass belong to lower energies than the one with smaller mass [the reason for this is evident from Eq. (1.12)].

[2] P. C. Chow, *American Journal of Physics,* 40 (1972), 780–784.

[3] F. Stern and W. E. Howard, *Physical Review,* 163 (1967), 816–35.

DIFFUSIVE TRANSPORT AND THERMIONIC EMISSION IN SCHOTTKY BARRIER TRANSPORT

In this appendix we examine again the assumptions of our calculation of transport over barriers and especially the assumption of constant quasi-Fermi levels through the barrier region. We have developed the following physical picture. Below the energy of the barrier no current flows, as electrons cannot propagate into neighboring layers and the distribution function is spherically symmetric. Above the barrier a thermionic current can flow without collisions like a jet stream of air. The density of this jet stream is constant everywhere because of the assumption of constant quasi-Fermi levels. In equilibrium, this assumption is indeed exact, and the jet stream is balanced by an opposing stream from the other material. If we apply an external forward voltage, however, we raise the quasi-Fermi level on one side, therefore increasing the jet stream and generating a net current. [In the reverse direction we soon reach current saturation, as can be seen from Eq. (13.10).]

These assumptions oversimplify the physics of the process, as can be seen from Fig. G.1, where we have divided space into a finite mesh of side length L_m, where L_m is the mean free path.

The electron transport between the rectangles is dominated by certain rates. A vertical transition of electrons requires electron-electron interaction or phonon emission and absorption, which is characterized by the time constants $\tau_{tot}^{e\text{-}e}$ or $\tau_{tot}^{e\text{-}ph}$, respectively.

A horizontal transition at the right side is characterized by the velocity of

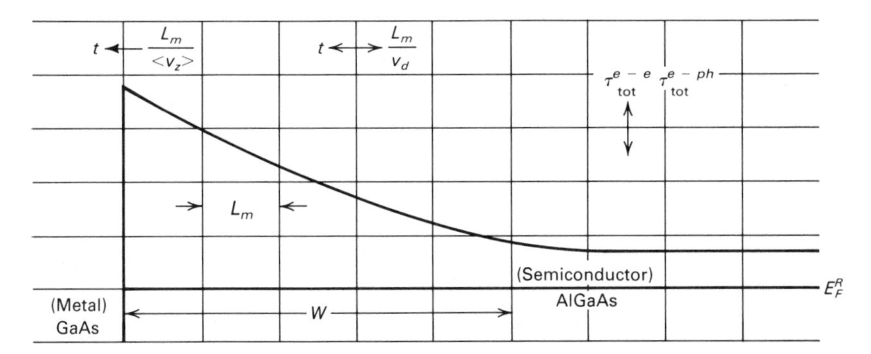

Figure G.1 Schematic of electron supply rates in heterojunctions (Schottky barriers).

electrons and the corresponding time constant L_m/v_d, where v_d is the drift velocity. The thermionic emission over the last partition of the barrier does not involve diffusion and the corresponding time constant is determined by the z component of the electron velocity v_z, which means by the time of collisionless flight of the electrons toward the left side. There will be few reflections of electrons from the left since electrons find themselves at extremely high energy in the GaAs (metal) and will spontaneously emit a phonon after $\approx 10^{-13}$ sec. If the electron energy is above the L valleys in GaAs, the phonon emission is much enhanced because of intervalley deformation potential scattering and the high density of final states, which can reduce the time constant to less than 10^{-14} sec.

In the space elements away from the junction, electrons can be reflected, for example, by phonon scattering or at low energies by the barrier itself. Therefore, the time constant away from the junction will be given rather by the distance divided by the drift velocity, that is, by L_m/v_d, where v_d is the position-dependent average electron velocity (average over positive and negative values), which is much smaller than the average of the absolute value of the electron velocity or the average taken in one direction only as is done for calculating the thermionic emission current. We can see now that our thermionic jet stream theory can work only if the depletion width is very short and comparable to the mean free distance L_m of collisionless travel.

If W is longer, then electrons are lost at a rate $L_m/\langle v_z \rangle$ at the barrier, where $\langle v_z \rangle$ is the average velocity of electrons going in negative z direction (if none returns), while away from the barrier they are replenished at a rate of L_m/v_d only (electrons are scattered and reflected back), which is usually much smaller (see Fig. G.1). Since these two rates are very different, the assumption of a constant electron density above the barrier height breaks down and so does the assumption of a constant quasi-Fermi level. In fact, the whole concept of a quasi-Fermi levels may not be applicable because the distribution function may not even have a Fermi shape if the rates shown in Fig. G.1 do not permit it. Only a very strong electron-electron interaction can help to establish at all times a Maxwellian high energy tail. However, close to the junction the electron density is low and therefore electron-electron interactions are rare.

Let us summarize the main point: The electrons need to be replenished from the bulk of the semiconductor. If the junction is rather broad, the replenishing current is of "diffusive nature." This diffusion current may be smaller than the thermionic current and therefore the limiting factor. Then if W is very large, a "diffusion solution" will be more appropriate than a thermionic current. The thermionic current then only presents a boundary condition, as it represents the current close to the interface (the last rectangle of Fig. G.1).

Below we treat this case following Schottky as well as Crowell and Sze. Assuming a constant mobility and diffusion constant ($T_c = T_L = T$), the current can be written as Eq. (11.6):

$$j = en\mu F + eD\,\partial n/\partial z \tag{G.1}$$

Outside the depletion region we have $F = 0$. Inside the region $F \neq 0$ and we will denote it by $-\partial\phi/\partial z$, where ϕ is the potential. Since in steady state the current is constant, we can integrate Eq. (G.1) using $-e\phi(z)/kT$ as integrating factor. For brevity, we denote $e\phi(z)kT$ by $\overline{\phi}$; then using the Einstein relation for the diffusion constant [Eq. (11.7)], we have

$$j = eD\left(-n\frac{\partial\overline{\phi}}{\partial z} + \frac{\partial n}{\partial z}\right) \tag{G.2}$$

Multiplying Eq. (G.2) by $e^{-\phi}$, we can write

$$je^{-\overline{\phi}} = eD\frac{\partial ne^{-\overline{\phi}}}{\partial z} \tag{G.3}$$

In steady state the current density j is constant and has the same value everywhere. Therefore

$$j\int_0^W e^{-\overline{\phi}}\,dz = eD\,[n(W)e^{\overline{\phi}(W)} - n(0)e^{\overline{\phi}(0)}] \tag{G.4}$$

$\phi(z)$ is known from the solution of the Poisson equation in the depletion region to be

$$\phi(z) = \frac{e}{2\epsilon\epsilon_0}N_D^+z^2 + C_1z + C_2 \tag{G.5}$$

where C_1 and C_2 are constants which can be determined from the boundary conditions $\overline{\phi}(W) - \overline{\phi}(0) = e(V_{bi} - V_{ext})/kT$ and $\left.\frac{\partial\phi(z)}{\partial z}\right|_W = 0$. Here we have made use of the fact that all of the external potential drops in the depletion region. Denoting the function $e^{\overline{\phi}(0)}\int_0^W e^{\overline{\phi}}\,dz$ by $F(N_D^+)$, we have from Eq. (G.4)

$$jF(N_D^+) = eD\,[n(W)\exp(\overline{V}_{bi} - \overline{V}_{ext}) - n(0)] \tag{G.6}$$

where $F(N_D^+)$ can easily be calculated from (G.5) and the boundary conditions. $n(W)$ is equal (in the spirit of the depletion approximation) to the equilibrium value of the carrier concentration at W, which we denote by $n_c(W)$. The equilibrium concentration $n_c(0)$ at the junction ($z = 0$) is then

$$n_c(W)\exp(-\overline{V}_{\text{bi}}) = n_c(0)$$

and therefore

$$jF(N_D^+) = eD(n_c(0)e^{\overline{V}_{\text{ext}}} - n(0)) \tag{G.7}$$

We have yet to determine $n(0)$, the nonequlibrium concentration, at the junction. We know that the current flowing to the left at the junction is given by the thermionic emission current, which corresponds to the carrier density $n(0)$, and zero barrier height (because we are considering now the current at the junction on the top of the barrier).

From the definition of the quasi-Fermi level, Eqs. (10.10 and 6.7) and Eq. (12.18) this current is given by

$$j_{th}^0 = A^*T^2\left(\frac{n(0)}{N_c} - \frac{n_c(0)}{N_c}\right) \tag{G.8}$$

Here N_c is the so-called effective density of states $N_c = 2(2\pi m^*kT/h^2)^{3/2}$ as can be derived from Eq. (6.22) and the definition $n = N_c\exp(E_F - E_c)/kT$. [In Eq. (6.22) E_F is measured from E_c.]

The second term in Eq. (G.8) represents the current coming back from the left side (GaAs, metal), which is equal to the equilibrium current since no significant external voltage drops at the left side.

The term $A^*T^2/N_c e$ is sometimes called the interface recombination velocity v_R (it has the correct dimension). In other words, the left side is viewed as a trap where the electrons are captured and cannot return. Now the fact is used that in steady state the current is the same everywhere and $j = j_{th}^0$. Combining Eqs. (G.7) and (G.8), we have then

$$j = \frac{eDn_c(0)(e^{\overline{V}_{\text{ext}}} - 1)}{F(N_D^+) + \dfrac{D}{v_R}} \tag{G.9}$$

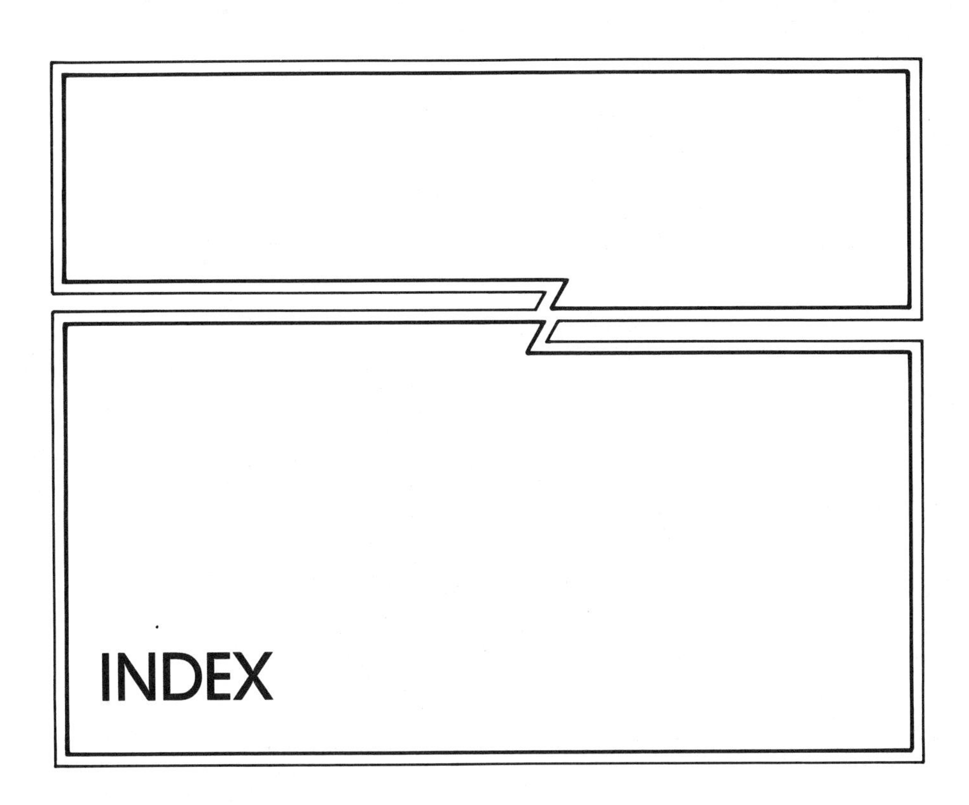

INDEX

A

B

D

E

L

M

N

O

P

Q

R

S

T